NMR Spectroscopy in the Undergraduate Curriculum

ACS SYMPOSIUM SERIES **1128**

NMR Spectroscopy in the Undergraduate Curriculum

David Soulsby, Editor
University of Redlands
Redlands, California

Laura J. Anna, Editor
Montgomery College
Rockville, Maryland

Anton S. Wallner, Editor
Barry University
Miami Shores, Florida

Sponsored by the
ACS Division of Chemical Education

American Chemical Society, Washington, DC

Distributed in print by Oxford University Press

Library of Congress Cataloging-in-Publication Data

NMR spectroscopy in the undergraduate curriculum / David Soulsby, editor, University of Redlands, Redlands, California, Laura J. Anna, editor, Montgomery College Rockville, Maryland, Anton S. Wallner, editor, Barry University, Miami Shores, Florida ; sponsored by the ACS Division of Chemical Education.

pages cm. -- (ACS symposium series ; 1128)

Includes bibliographical references and index.

ISBN 978-0-8412-2794-1 (alk. paper)

1. Nuclear magnetic resonance spectroscopy. 2. Chemistry, Physical and theoretical--Study and teaching. I. Soulsby, David, 1974- editor of compilation. II. Anna, Laura J., editor of compilation. III. Wallner, Anton S., editor of compilation.

QD96.N8N588 2013

543′.66--dc23

2013003382

The paper used in this publication meets the minimum requirements of American National Standard for Information Sciences—Permanence of Paper for Printed Library Materials, ANSI Z39.48n1984.

Distributed in print by Oxford University Press

Foreword

The ACS Symposium Series was first published in 1974 to provide a mechanism for publishing symposia quickly in book form. The purpose of the series is to publish timely, comprehensive books developed from the ACS sponsored symposia based on current scientific research. Occasionally, books are developed from symposia sponsored by other organizations when the topic is of keen interest to the chemistry audience.

Before agreeing to publish a book, the proposed table of contents is reviewed for appropriate and comprehensive coverage and for interest to the audience. Some papers may be excluded to better focus the book; others may be added to provide comprehensiveness. When appropriate, overview or introductory chapters are added. Drafts of chapters are peer-reviewed prior to final acceptance or rejection, and manuscripts are prepared in camera-ready format.

As a rule, only original research papers and original review papers are included in the volumes. Verbatim reproductions of previous published papers are not accepted.

ACS Books Department

Contents

Physical Chemistry and Biochemistry

NMR Spectroscopy Across the Undergraduate Curriculum

NMR Spectroscopy: Collaborative Strategies, Resources, and Grant Writing

Indexes

Preface

In 1955, the Journal of Chemical Education reprinted an article from the Industrial Bulletin of Arthur D. Little Inc. on the subject of nuclear magnetic resonance spectroscopy, which had been discovered only eight years earlier by Bloch and Purcell (*1*). The article described the principles underlying NMR spectroscopy, how this new technique can be used as a nuclei detector, and concluded with this remarkably prescient paragraph:

> "When the sensitivity of the n-m-r spectrograph is pushed as far as possible, one can observe minute differences caused by the chemical environment of the atom...It is then possible to determine the position of the protons within a complex molecule and arrive at a better understanding of its structure. This particular aspect of nuclear magnetic resonance spectroscopy may provide an entirely new research tool for structural and organic chemistry."

Even the most cursory survey of the chemical literature reveals that modern NMR spectroscopy has indeed fulfilled its potential as a powerful and indispensable tool for probing molecular structure, providing detail that is comparable to, and sometimes surpasses that, of X-ray crystallography. As NMR spectroscopy's 70th anniversary approaches, the diversity of chemical problems to which this technique can be applied continues to grow across many scientific fields. Beyond the laboratory setting, the technology underlying NMR is now a widely used and critical medical diagnostic technique, Magnetic Resonance Imaging (MRI). Unfortunately, the number of applications of NMR spectroscopy across so many STEM-related fields presents significant challenges in how best to introduce this powerful technique in meaningful ways at the undergraduate level.

In the past two decades, NMR spectrometers have become increasingly standard instrumentation at most undergraduate institutions, in part through the leadership and direction of the American Chemical Society (*2*). This has not only resulted in increasing numbers of students gaining hands-on experience with NMR spectroscopy, but has also resulted in increased usage by faculty from disciplines that have not traditionally used this technique. As familiarity and understanding have grown, so have the development, implementation, and dissemination of curricular strategies that use NMR spectroscopy to investigate relevant chemical problems in meaningful ways.

Inspired by the development of the field, and building upon the work of previous symposia and an ACS symposium series book on this topic (*3*), we (DS, ASW, LJA) developed a symposium entitled "NMR Spectroscopy in the

Undergraduate Curriculum," for the 239th American Chemical Society National Meeting in San Francisco. We were delighted by the strong interest in this first symposium, and in the breadth, quality, and sheer passion for the subject that our presenters brought to the meeting. The numerous high-quality presentations we witnessed demonstrated to us the continuing innovations that drive NMR spectroscopy pedagogy innovation. As a result of the success of the first symposium, ACS was gracious enough to host us again in successive years as we continued to provide opportunities for faculty to broadly share their knowledge and experiences. When the opportunity for this book arose we were delighted to be able to bring together all of our presenters who have been successful in developing and successfully integrating NMR spectroscopy pedagogy across their undergraduate curriculums. We hope that their knowledge and experiences will aid all of our readers who are interested in expanding and invigorating their own curriculum. For those without access to an NMR spectrometer, or who are looking to purchase a replacement spectrometer, we also hope that the ideas contained within will be helpful in crafting successful proposals.

In any project of this size, success is always the result of the tireless work of a large team of dedicated individuals. As editors we are of course extremely grateful to all of our contributing authors whose passion for the discipline, innovative ideas, and clarity of prose will inspire others to adopt and innovate in this important field. We are also thankful to the numerous peer-reviewers, who by the sheer nature of their duty must remain anonymous, but whose thoughtful comments helped improve and strengthen each and every contribution. We are especially grateful to Bob Hauserman, Tim Marney, Arlene Furman, Kat Wilson, Mary Calvert, Pamela Kame, Cynthia Porath and the entire staff at ACS Books whose technical expertise, kind spirit, and prompt responses to our queries made this entire endeavor possible. Finally, we acknowledge the financial assistance of Bruker and Anasazi Instruments who were inaugural co-sponsors of our NMR Spectroscopy in the Undergraduate Curriculum symposium at the 243rd American Chemical Society National Meeting.

References

1. *J. Chem. Educ.* **1955**, *32*, 79.
2. *Undergraduate Professional Education in Chemistry: ACS Guidelines and Evaluation Procedures for Bachelor's Degree Programs*; American Chemical Society: Washington, DC, 2008
3. *Modern NMR Spectroscopy in Education*; Rovnyak, D., Stockland, R., Jr., Eds.; ACS Symposium Series 969; American Chemical Society: Washington, DC, 2007.

Dr. David Soulsby
Chemistry Department, University of Redlands
1200 E. Colton Ave., Redlands, CA 92373
909-748-8546 (telephone)
David_Soulsby@redlands.edu (e-mail)

Dr. Laura J. Anna
Chemistry Department, Montgomery College
51 Mannakee Street, Rockville, MD 20850
240-567-5489 (telephone)
Laura.Anna@montgomerycollege.edu (e-mail)

Dr. Anton S. Wallner
Department of Physical Sciences, Barry University
11300 NE 2nd Ave., Miami Shores, FL 33161
305-899-3433 (telephone)
twallner@mail.barry.edu (e-mail)

Chapter 1

Introduction to NMR Spectroscopy in the Undergraduate Curriculum

Anton S. Wallner,[*,1] Laura J. Anna,[2] and David Soulsby[3]

[1]Department of Physical Sciences, Barry University, Miami Shores, Florida 33161
[2]Department of Chemistry, Montgomery College, Rockville, Maryland 20850
[3]Department of Chemistry, University of Redlands, Redlands, California 92374
[*]E-mail: twallner@mail.barry.edu

Since its discovery in 1946, NMR spectroscopy has developed into a powerful tool that is used across many disciplines. In the past twenty years NMR spectroscopy has become increasingly important at the undergraduate level, as evidenced by an increasing number of publications in the field. In this book we bring together experts in the field to discuss the latest strategies and techniques for effectively integrating NMR spectroscopy in the undergraduate curriculum so that students can use this powerful spectroscopic tool to investigate an array of relevant problems. Finally, in a area that has developed as fast as NMR spectroscopy, we conclude by considering the benefits that advances in areas such as hardware, software, and increased use of the Internet could mean for the future of the field.

Introduction

When nuclei with spin are placed in a strong magnetic field and perturbed with radio waves, the resulting emission spectrum provides detailed information about the environment of the nuclei. In 1936, Gorter was the first to describe such experiments in his attempts to observe a sudden rise in the temperature of the sample upon slowly varying a transverse magnetic field for lithium fluoride and potassium alum. Gorter searched the radio frequency regions where he predicted that nuclear magnetic resonance signals of ^{7}Li and ^{1}H nuclei would be expected, but failed to observe any signals. He described how the expected increase in occupation of the higher energy levels had been negated by the increase in spin temperature (*1*). Though his experiments were ultimately unsuccessful in observing nuclear magnetic resonance signals, his efforts encouraged further development in the field. In 1946 these efforts culminated in the simultaneous reporting of nuclear magnetic resonance (NMR) signals in water by Bloch, Hanson, and Packard (*2*), and in solid paraffin by Purcell, Torrey, and Pound (*3*). Bloch and Purcell's efforts were awarded with the 1952 Nobel Prize in physics (*4*).

The development of NMR spectroscopy into a mature technique with applicability across numerous disciplines has been described in great detail by others (*5–12*). Any historical summary would no doubt include the development and introduction of superconducting magnets in the 1960's, the introduction of FT-NMR (Fourier Transform-NMR) in the 1970's, and shielded magnets, improved electronics and probe design in later years. Indeed, improvements in instrumentation, field strength, and acquisition capabilities mean that NMR spectrometers have now become a routine method of analysis for both industrial and academic chemists, allowing them to analyze smaller sample sizes, less sensitive nuclei, and complicated structures (e.g., proteins, nucleic acids, carbohydrates, etc.).

The impact of NMR spectroscopy can be seen by simply reviewing the literature. An assessment of SciFinder using "NMR" as the search topic shows a steady increase in articles that use that term, in going from 7,700 articles in 1980 to over 32,300 articles in 2011 (*13*). Yet, clearly this most recent number undercounts the overall impact of NMR spectroscopy since it is now such a routine measurement that it is rarely mentioned in the abstract any more. NMR spectroscopy has also become a significant component in undergraduate education, with an NMR spectrometer being listed as a required component for an American Chemical Society (ACS) certified program by the Committee on Professional Training (CPT) (*14*). Furthermore, a review of the *Journal of Chemical Education* using the search term "NMR", found that for the decade 1980-89, 364 research articles used that term, this grew to 499 for the next decade, and to 571 for the most recent decade. Indeed, the number of NMR spectrometers being used in academia has significantly increased over the past 25 years (*15*).

Overview of NMR Spectroscopy in the Undergraduate Curriculum

The organization of this book guides both new and experienced NMR users towards innovative methods that incorporate NMR spectroscopy in the undergraduate chemistry curriculum. The introductory chapter "Modern NMR Experiments: Applications in the Undergraduate Curriculum" provides a comprehensive overview of routine ^{1}H and ^{13}C NMR spectroscopy and advanced one-dimensional and two-dimensional experiments that can be performed to provide structural details to answer questions of molecular structure, to quantitate the amounts of individual compounds in mixtures, to measure molecular relaxation times, and to probe reaction rates, particularly in biological reactions. While this review does not cover the theoretical aspects behind these NMR experiments, it does provide users with relevant examples from the undergraduate literature to act as a guide in selecting the most useful type of experiment to select for a particular problem.

The book is then organized into sections based on more traditional curricular areas. These sections illustrate how NMR instruction has been effectively integrated into the undergraduate curriculum. The first section is devoted to the application of NMR spectroscopy in organic chemistry lecture and laboratory courses. Contributing authors share pedagogical ideas to show how NMR data can be used in an organic chemistry lecture to support and provide evidence of fundamental concepts of organic structures such as resonance, conformational analysis and stereochemistry. Numerous examples of laboratory experiments are provided, giving readers tested methods for introducing new or supplementing existing laboratory experiments with the analysis of organic products by NMR methods. Specific strategies that promote active learning in the laboratory, that advocate the use of non-deuterated solvents, and that offer practical solutions to streamline NMR sample acquisition and data processing are shared. Two-dimensional NMR methods, such as Correlation Spectroscopy (COSY) and Heteronuclear Single Quantum Coherence (HSQC) experiments, are also gaining a foothold in the organic chemistry laboratory curriculum, and methods are presented that use these experiments to explore and unequivocally solve structural problems.

Recognizing that NMR technology is not limited solely to the ^{1}H and ^{13}C NMR analysis of organic compounds, the next section of the book focuses on hetereonuclear applications and how NMR spectroscopy is finding a place in the inorganic chemistry curriculum. Contributing authors report a variety of experiments conducted by undergraduate students in both research and teaching laboratories using ^{31}P, ^{195}Pt and other heteronuclear NMR analysis to characterize inorganic compounds. With the introduction of NMR spectroscopy to the inorganic laboratory, students can receive broader training in NMR instrumentation and applications that complement traditional ^{1}H and ^{13}C NMR spectroscopy experience in the organic chemistry curriculum.

Progressing through the undergraduate curriculum to upper-level courses, the next section of the book is grouped by the involvement of NMR spectroscopy in physical chemistry and biochemistry courses. Contributing authors show how students couple NMR data with quantum mechanical calculations to examine electronic effects in aromatic compounds and the chemical shift effects of methanol solutions. Students in these upper-level courses also build upon their qualitative NMR analysis skills through the investigation of kinetic rate experiments, the quantitative examination of substituent effects on keto-enol equilibria and gain the unique experience of characterization of phase behavior in lipids using ^{1}H NMR magic-angle spinning.

Other author contributions showcase how the acquisition of a new NMR spectrometer has transformed their overall undergraduate curriculum from non-majors courses through senior-level research experiences and relate the challenges and successes regarding the impact of NMR technology on their teaching and student learning.

The concluding chapters in the book provide strategies and information on resources that can assist faculty and NMR users in a variety of situations. The National Science Foundation (NSF) encourages collaboration between institutions and one author describes the implementation of an NMR consortium, and the challenges and outcomes resulting from its formation. Faculty and students are increasingly using online sources to augment their NMR spectroscopy education, and a description is provided of an online depository of spectral data to support NMR instruction. Finally, there is a chapter devoted to an overview of available NSF funding resources that support the purchase of NMR instrumentation for undergraduate institutions and strategies for being a successful applicant.

Future Directions

Twenty five years ago the majority of superconducting NMR spectrometers were almost exclusively housed at research universities or industrial facilities that had sufficiently large budgets to purchase these instruments and employ specialized personnel to manage them. Technological limitations at the time meant that these older instruments needed a significant amount of floor space due to bulky electronic components and to minimize contact with stray magnetic fields. User interaction was through simplistic graphical user interfaces where experiments were loaded using an alphabet soup of commands, and each sample required manually shimming, meaning that the quality of a spectrum was highly dependent upon user skill. Detecting other nuclei required trained personnel to change cables and retune the probe.

As we approach the 70th anniversary of the discovery of NMR spectroscopy, NMR spectrometers have moved firmly from the realm of industry and research universities into standard instrumentation found at many undergraduate institutions. Several decades of increased access to private and public funding have made the purchase of these instruments more manageable for numerous undergraduate institutions, though these opportunities are becoming increasingly limited. In many cases, these instruments can be operated and maintained by teaching faculty with only basic training. Instrument hardware has decreased in size and high-field magnets are now better shielded, meaning that these instruments can be situated in existing spaces with minimum renovations. Gradient shimming has negated the need to manually shim a sample, so that even the most inexperienced user can generate a quality spectrum for any sample each and every time. Probe auto-tune hardware also allows users to easily detect non-standard nuclei without changing cables or manually tuning the probe. Undergraduate institutions are also taking advantage of the availability of automation options, which dramatically increases throughput. Additionally, low-field NMR instruments continue to maintain an important and critical place at many undergraduate institutions, providing high-quality spectra at a significantly lower purchase cost and with minimal upkeep costs. These instruments are often housed in existing instrumentation laboratories, or even on the bench alongside an infrared (IR) spectrometer, gas chromatograph (GC) or high performance liquid chromatograph (HPLC). Even with low-field instruments, automation may eventually be desirable on these instruments as a way to increase throughput. With either type of instrument, students and faculty now use graphical interfaces to interact with the spectrometer. New instruments no longer use analog buttons, switches and two letter commands, and in their place are icons and buttons that represent preset experiments optimized to provide the best results for a typical sample. Finally, the ubiquity of the Internet means that not only can data be easily moved from one location to the other, but that the instrument can also be controlled remotely. Clearly, the past 25 years had heralded significant advances and innovations in the field, but what does the future hold for NMR spectroscopy, specifically as it relates to undergraduate institutions?

For institutions with high-field magnets, a dependable and affordable supply of liquid helium remains a problem. Since nearly all of our current liquid helium comes from natural gas deposits (*16*), when issues with supply arise this can become a significant issue (*17*). In the short-term, new cost-effective methods for generating and maintaining a reliable supply of liquid helium needs to be developed so that magnets for NMR spectrometers and MRI instruments can be maintained. In the longer-term, the development of new materials that allow for superconductivity at liquid nitrogen temperatures would relieve many of the current pressures on our natural resources and will also significantly decrease the cost of maintaining these instruments.

Nearly all modern NMR instruments are now configured to access and use the Internet. Though not yet standard, built-in remote access to the instrument by the user will not only mean that data can be acquired from a remote site, but that NMR managers and NMR manufacturers will be able to diagnose and correct problems when they occur. Most consumers are used to updating software and firmware on their modern electronics, and this could become standard on NMR instruments, with updates recognizing and adapting to the configuration of a particular instrument. Finally, the Internet also allows users to take advantage of cloud storage/computing as a means to remotely store and manipulate data. These virtual storage solutions can also serve as places for students to collaborate. When combined with an NMR instrument capable of automation, these options provide powerful ways to deploy NMR spectral data to students in a variety of learning environments. Together, these advances in NMR hardware and software technology can significantly contribute to an increase in the overall impact of NMR spectroscopy in the undergraduate curriculum.

Conclusions

The use of NMR spectroscopy is becoming more widespread at the undergraduate level. As the following chapters demonstrate, faculty innovation at the intersection of chemical education and NMR spectroscopy, through a combination of pedagogical advances, research, and improvements in instrumentation, all serve to teach our students about this powerful spectroscopic technique. The NMR spectroscopy skills that students learn in the classroom not only enrich their education, but ultimately transfer to post-graduate careers in STEM fields.

References

1. Gorter, C. J. *Physica* **1936**, *3*, 503. Gorter, C. J. *Physica* **1936**, *3*, 995.
2. Bloch, F; Hanson, W. W.; Packard, M. *Phys. Rev.* **1946**, *69*, 127.
3. Purcell, E. M.; Torrey, H. C.; Pound, R. V. *Phys. Rev.* **1946**, *69*, 37.
4. http://www.nobelprize.org/nobel_prizes/physics/laureates/1952/ (accessed January 2013).
5. Ernst, R. R.; Anderson, W. A. *Rev. Sci. Instrum.* **1966**, *37*, 93.
6. *Encyclopedia of Nuclear Magnetic Resonance*; Grant, D. M., Harris, R. K., Eds.; Wiley: Chichester, 1996; Vol. 1.
7. Feeney, J. *Educ. Chem.* **1996**, *33*, 96.
8. Pfeifer, H. *Magn. Reson. Chem.* **1999**, *37*, S154.
9. Becker, E. D. *Anal. Chem.* **1993**, *65*, 295A.
10. Knight, J. M.; Shaw, I. C.; Jones, A. P. *Chem. Br.* **1996**, *32*, 37.
11. Lankin, D. C.; Ferraro, J. R.; Jarnutowski, R. *Spectroscopy* **1992**, *7*, 18.
12. Goldman, M. *L'actual. Chim.* **2004**, *273*, 57.
13. *SciFinder*; Chemical Abstracts Service: Columbus, OH; https://scifinder.cas.org (accessed January 2013).
14. *ACS Guidelines and Evaluation Procedures for Bachelor's Degree Programs*; American Chemical Society: Washington DC., 2008; p 6.
15. Haberle, F. personal communication, September 21, 2012.
16. http://www.blm.gov/nm/st/en/prog/energy/helium.html (accessed December 2012).
17. Reisch, M. S. *C&EN* **2012**, *90*, 32–34.

Chapter 2

Modern NMR Experiments: Applications in the Undergraduate Curriculum

David Soulsby*

Department of Chemistry, University of Redlands, Redlands, California 92374
***E-mail: David_Soulsby@redlands.edu**

Modern NMR spectrometers are capable of performing a dizzying array of experiments that can provide an enormous amount of information at the molecular level. Undergraduate curricula are increasingly taking advantage of these powerful instruments to provide valuable educational experiences. However, for those with limited NMR spectroscopy experience, the sheer number of experiments and their applications to relevant chemical problems can be overwhelming. This review will describe many of the experiments that are available on modern NMR spectrometers and couple these descriptions with recent applications from the undergraduate chemical literature.

Introduction

Nuclear Magnetic Resonance (NMR) spectroscopy is a technique that provides information about chemical systems at the molecular level. Discovered by Bloch and Purcell in the mid-1940's (*1*, *2*) it was not long before the first commercially successful instrument was introduced by Varian Associates in 1960 (*3*). Since then NMR spectroscopy has revolutionized the work of scientists across a broad array of disciplines. Current applications include structure determination, molecular dynamics, protein folding, kinetic analysis, and medical imaging. Since NMR spectroscopy has broad applicability to a vast array of problems it is important to effectively integrate the subject across the undergraduate curriculum so as to best prepare students for the numerous post-graduate opportunities where students will use this technique.

NMR spectrometers fall into two categories, those that use permanent or room-temperature electromagnets to provide magnetic flux densities up to approximately 2.35 T, and those that use superconducting electromagnets to achieve magnetic flux densities above 2.35 T. Permanent and room temperature electromagnet NMR spectrometers have undergone a renaissance over the past two decades. These rugged instruments now come equipped with modern electronics and software that allow students access to a range of powerful experiments. They are also very affordable, requiring little to no upkeep and maintenance. Higher magnetic flux density NMR spectrometers use superconducting magnets and offer higher resolution and better signal-to-noise ratios. These instruments are more expensive and require liquid nitrogen and liquid helium to maintain the superconductivity of the magnet. Improvements in hardware mean that many of these instruments incorporate technology such as gradient, tunable, and variable temperature probes, as well as automation capabilities. New software interfaces also make these instruments easier to operate by faculty and students, helping to increase their incorporation across many undergraduate curricula. NMR spectrometer manufacturers are beginning to recognize the growing importance of the undergraduate market and are designing instruments that specifically fit the diverse curricular and financial needs of these institutions.

Despite commercial instruments being available for over 50 years, NMR spectroscopy is unique among most spectroscopic techniques in that experiments (or pulse sequences) continue to be developed and optimized. Often driven by advances in magnet technology and hardware, these new experiments can result in shorter acquisition times, incorporation of new information into existing experiments, and even providing new ways to probe molecular connectivity. Since these experiments are software-based, NMR manufacturers can easily add new experiments and optimize old ones through software updates provided on CDs, DVDs, or the Internet. On most NMR spectrometers it is not unusual to be presented with a vast array of NMR experiments. Indeed, it can be a daunting challenge to figure out how these NMR experiments connect to pertinent chemical problems because much of the NMR literature is at an advanced level and is focused on those seeking a thorough understanding of NMR theory and to develop structural analysis skills (*4–10*). With the exception of the recent ACS symposium series volume *Modern NMR Spectroscopy in Education* (*11*), there have been few organized approaches that connect the experiments available on a modern NMR spectrometer to the investigation of chemical problems that are applicable to an undergraduate curriculum.

This review combines practical but non-theoretical explanations of the various liquid sample experiments that are available on many modern NMR spectrometers. Nearly every experiment is paired with relevant examples from the recent undergraduate chemical literature. This review begins with a description of various one-dimensional ^{13}C NMR experiments that include the proton broadband decoupled, inversion recovery, inverse-gated, gated, off-resonance, attached proton test, DEPT, and INADEQUATE experiments. These experiments are applied to problems as diverse as the *in vivo* studies of glycolysis through to the quantitative analysis of the components of Extra Strength Excedrin™ tablets. The one-dimensional ^{1}H NMR experiment is then introduced. Since this experiment is already discussed in great depth in most undergraduate curriculums, the discussion here is focused on less common topics such as the importance of coupling constants and the Karplus relationship to deduce the orientation of nuclei to each other in sugars. This section also includes discussions of the TOCSY and NOESY experiments and how they can be selectively applied to the analysis of more complex systems to aid structural assignments. The one-dimensional NMR experiment section concludes by discussing the ability to observe directly and indirectly nuclei other than ^{1}H or ^{13}C with applications that include the analysis of a mixture of phosphonates and the ability to calculate the isotopic ratio of boron in sodium borohydride. Two-dimensional NMR experiments introduced are those that can correlate carbon and hydrogen atoms over one, two or multiple bonds (HETCOR, HMQC, HSQC, HMBC, H2BC), carbon and carbon atoms over one bond (2D INADEQUATE), and hydrogen and hydrogen atoms over three, and sometimes multiple bonds (HH-COSY). Examples include the applications of these experiments to the structural analysis of Diels-Alder, Fischer esterification, and terpene products. Finally, the utility of two-dimensional versions of the TOCSY and NOESY experiments are illustrated with examples of the analysis of an octapeptide and the conformation of an aldol product. This selective review is not meant to be a comprehensive survey of all NMR experiments, or of the entire undergraduate NMR spectroscopy literature, but instead highlight relevant NMR experiments with pertinent examples from the literature. Because of the limited nature of this review, readers who want to explore the topic more fully are encouraged to explore this volume, the aforementioned *Modern NMR Spectroscopy in Education* (*11*), and publications such as the *Journal of Chemical Education or the Chemical Educator*.

Introduction to One-Dimensional ^{13}C NMR Experiments

Since NMR spectroscopy plays such an important role in organic chemistry, most organic chemistry textbooks justifiably spend an entire chapter dedicated to the subject. Topics normally include NMR theory, instrumentation, and a limited selection of one-dimensional NMR experiments. For ^{13}C NMR spectroscopy these experiments usually include the proton broadband decoupled, DEPT, and sometimes the off-resonance ^{13}C NMR experiment. Experiments that are typically absent at this stage include the inversion recovery, inverse-gated, gated, attached proton test, and 1D INADEQUATE experiments. During the discussion of what a typical one-dimensional NMR spectrum looks like, students are told that the abscissa of the spectrum displays the chemical shift scale in ppm and the ordinate of the spectrum displays signal intensity. In ^{13}C NMR spectroscopy the chemical shift scale is over 220 ppm and signals are usually well dispersed, even when the ^{13}C atoms are in very similar chemical environments. The chemical shift of the ^{13}C signal also correlates to the functionality surrounding the carbon, providing important structural information. Unfortunately, ^{13}C atoms are 60 times less sensitive (a function of the energy difference between the spin populations) than ^{1}H atoms and only 1.1% abundant, so acquisition times for a ^{13}C NMR experiment are normally significantly longer than those of a ^{1}H NMR experiment with a comparable signal-to-noise ratio. Despite these significant disadvantages, one-dimensional ^{13}C NMR experiments are regularly acquired across the undergraduate curriculum, either on their own or as part of a sequence of experiments, to solve a variety of chemical problems.

The Proton Broadband Decoupled ^{13}C Experiment

The proton broadband decoupled ^{13}C NMR experiment is easily the most common ^{13}C experiment taken at the undergraduate level because it is straightforward to analyze and provides information about the number of non-equivalent carbon atoms, molecular symmetry, and the functionality of each carbon atom. The data from this experiment can be used to corroborate molecular connectivity since carbon chemical shifts can be calculated either manually or computationally using established algorithms (*12*). In a proton broadband decoupled spectrum each non-equivalent carbon atom appears as a singlet because all *indirect spin-spin couplings* that provide information about the local environment of that carbon have been removed.

Magnetically active nuclei interact with each other via *direct* and *indirect spin-spin* coupling mechanisms. The direct spin-spin coupling mechanism is commonly seen in viscous liquids or solids where molecular rotation is hindered. However, when molecules are allowed to rotate freely, for example in dilute liquid samples, direct coupling averages to zero and only indirect spin-spin coupling is observed. This type of coupling is related to the distance of the coupled nuclei from each other and occurs over 1, 2, and 3 bonds, though longer range coupling is possible in rigid systems. Since the majority of molecules that are of interest to chemists contain adjacent magnetically active carbon and hydrogen atoms, indirect spin-spin coupling occurs to give spectra with multiple signals and poor signal to noise. This is because each signal is now spread out over multiple peaks. If this coupling information can be removed, or attenuated, then a spectrum would be simplified and would have enhanced signal intensities.

To remove unwanted signals caused by indirect spin-spin coupling the ^{1}H resonance region is irradiated with sufficient power to saturate all of the proton resonances, a process called broadband irradiation. Unfortunately, at higher magnetic flux densities broadband irradiation causes unwanted sample heating that can be particularly fatal to biological samples, so alternative pulse sequences are used that have the same effect. In either case all indirect spin-spin couplings resulting from neighboring protons are removed. Since broadband irradiation is only occurring over the proton resonance region, indirect spin-spin couplings caused by adjacent ^{13}C atoms remain. However, the probability of having adjacent ^{13}C atoms is so low that this coupling is not normally observed except in specific experiments such as the 1D and 2D INADEQUATE experiments where acquisition times can be very, very long. In the proton broadband decoupled ^{13}C NMR experiment, irradiation of the ^{1}H resonance region occurs over the entire pulse sequence (observation pulse, acquisition time, and the relaxation delay) leading to a significant enhancement in signal intensity of up to 200%. This is due to the nuclear Overhauser effect (NOE) which will be discussed later. In the literature, this experiment is commonly referred to as just a ^{13}C NMR experiment, with no indication that broadband decoupling has occurred. Of course it is more accurate when describing this experiment to use the full description and call it the proton broadband decoupled ^{13}C NMR experiment which can be abbreviated $^{13}C\{^1H\}$ where the $\{^1H\}$ notation signifies that broadband decoupling of the ^1H atoms has occurred.

Draanen and Page (*13*) use the $^{13}C\{^1H\}$ NMR experiment in an introductory organic chemistry laboratory experiment where students identify a series of benzene-containing C_9H_{12} structural isomers through prediction and experiment. Students begin by drawing and naming the eight benzene-containing isomers, paying careful attention to the symmetry of each molecule. By identifying the number of non-equivalent carbon atoms they can predict the number of signals each compound would give in a $^{13}C\{^1H\}$ NMR experiment. Students are then told that carbon atoms with directly attached hydrogen atoms experience a significant enhancement in signal intensity and that this can be explained via the nuclear Overhauser effect (NOE). Though further explanation may not be needed at this level, this signal intensity enhancement arises because irradiated nuclei bring about changes in the Boltzmann population distributions of non-irradiated nuclei that are close in space. Importantly, these nuclei need not be directly bonded to each other (see NOE experiment). By having a preliminary understanding of the NOE, students can now predict the intensities for each signal. Finally, signal position on the chemical shift scale is predicted using established algorithms. With all of this information in hand students are then assigned one of the isomers and acquire a $^{13}C\{^1H\}$ NMR spectrum on a 60 MHz NMR spectrometer. Even at lower magnetic field strengths the large spectral width of ^{13}C NMR spectroscopy provides sufficient signal dispersion that students are able to identify their unknown by correlating their predicted spectra with the spectrum of their unknown sample.

The large spectral width of ^{13}C NMR spectroscopy also provides significant advantages in the study of more complex systems, such as in the analysis of biological processes. Unfortunately, since ^{13}C atoms are only 1.1% abundant and acquisition times are long, ^{13}C NMR experiments that require frequent accumulations of data are not normally feasible unless the kinetics of the process being followed are slow enough that multiple scans can be taken to generate spectra with sufficient signal-to-noise ratios. However, this problem can be alleviated by using samples that have been enriched with ^{13}C in one or more positions. This results in significantly shorter experiment acquisition times and an ability to easily discriminate between enriched and unenriched carbon atoms within the sample. Particularly in biological samples, the use of isotopically enriched samples can be a valuable tool to observe the formation of intermediates and products in complex processes. However, care should be taken when samples are enriched at multiple or adjacent sites since indirect spin-spin coupling from adjacent atoms will now occur, leading to a more complex analysis and a reduction in signal intensity.

Mega et al. and Giles et al. describe related experiments that use the $^{13}C\{^1H\}$ NMR experiment to study the metabolic process of glycolysis using ^{13}C enriched 1-^{13}C-glucose in a biochemistry class (*14*, *15*). Despite the complexity of the sample which included yeast, buffer, glucose, and the products of glycolysis, students were able to observe and measure decreases in signal intensity for the C-1 carbon of glucose (92 ppm for the α-anomer and 96 ppm for the β-anomer) and increases in signal intensity for the C-2 carbon of ethanol and the C-1 carbon of glycerol. Figure 1 is a $^{13}C\{^1H\}$ NMR spectrum that shows all of these components.

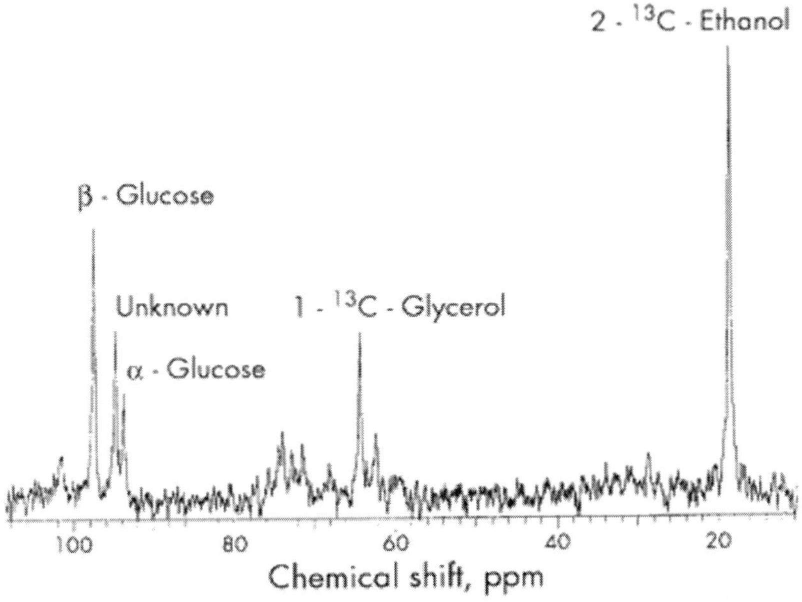

Figure 1. $^{13}C\{^1H\}$ NMR spectrum during glycolysis showing the presence of the α- and β-anomers of glucose, 1-^{13}C-glycerol and 2-^{13}C-ethanol. The unknown signal was attributed to an impurity in the glucose. Reprinted with permission from reference (15). Copyright 1999 American Chemical Society.

Students were also able to use the same data to explore more subtle effects, such as measuring the difference in conversion rates between the α- and β-anomers of glucose, and how the concentration of glycerol and ethanol varied according to the salinity of the buffer.

The Inversion Recovery ^{13}C Experiment

The *inversion recovery* experiment is used to determine the time it takes for the spin population to return to equilibrium. After a sample absorbs energy from a pulse, the sample can transfer that energy back to its surroundings, or lattice, through a variety of processes. Specifically, the spin-lattice relaxation time T_1 is the rate at which the z-component of the magnetization returns to equilibrium after a 180° pulse, and this is measured using the inversion recovery experiment.

Depending upon the nature of the sample, the time (T_1) for this process to occur can vary dramatically, with primary carbon atoms having T_1 values of 7-10 s and quaternary carbon atoms having T_1 values in excess of 60 s. The T_1 values derived from the inversion recovery experiment are particularly useful for experiments where sufficient delays between pulses are necessary in order to allow each nucleus to fully relax during the free induction decay so as to maximize signal intensity (see inverse-gated and 2D INADEQUATE experiments).

In the physical chemistry laboratory, Gasyna and Jurkiewicz describe the use of the inversion-recovery NMR experiment to measure the T_1 values for each carbon atom in *n*-hexanol (*16*). Excellent dispersion of the signals from each carbon makes analysis straightforward and students find that the T_1 values increase as the distance from the –OH group increases. Spin-lattice relaxation is typically dominated by dipole-dipole interactions, which depend upon the distance and orientation of the interacting dipoles, and the motion of the molecule. Surprisingly, the T_1 value for the terminal methyl group in *n*-hexanol is only slightly longer than that for the methylene groups and this is attributed to methyl group rotation which slows down the dipole-dipole relaxation pathway and is particular to this type of system where intermolecular hydrogen bonds are present.

The Inverse-Gated ^{13}C Experiment

The inverse-gated ^{13}C experiment is an experiment that quantitatively measures ^{13}C signals. In most ^{13}C NMR experiments, signal intensities can vary for a variety of reasons that include 1) decreased signal intensities when long T_1 values are present in the molecule resulting in some carbon atoms not returning to equilibrium before the next pulse occurs, 2) increased signal intensities for carbon atoms with attached hydrogen atoms that arise from NOE enhancements, and 3) variable signal intensities when the excitation pulse is of insufficient power to excite all nuclei equally across the spectral width of interest. The *inverse-gated* ^{13}C NMR experiment only irradiates the ^1H resonances during the observation pulse and the acquisition time. Since NOE signal enhancements build significantly during the relaxation delay, turning off the irradiation during that time mostly suppresses the NOE. Additionally, if the delay between pulses is equal to at least five times the longest T_1 value for the system, measured using the inversion recovery experiment, then the system has a chance to return to equilibrium before the next pulse. Unfortunately, five times the longest T_1 value can be in excess of tens of minutes, leading to unrealistically long acquisition times. Fortunately, the addition of paramagnetic compounds such as $Cr(acac)_3$ or TEMPO can induce relaxation and dramatically shorten the acquisition time. However, the addition of these compounds can cause changes in signal location and line broadening so caution should be used. Finally, experimental parameters are often easily adjusted so that the transmitter pulse is of sufficient strength so that all carbon atoms are irradiated equally.

To illustrate the issues surrounding NOE enhancements and the effect of long T_1 values LeFevre and Silveira describe the use of the $^{13}C\{^1H\}$ and the inverse-gated ^{13}C NMR experiment to quantitatively analyze a mixture of cyclohexanones in an advanced laboratory course (*17*). Students began by identifying and assigning signals for each compound separately before integrating and measuring these same signals in a mixture of unknown composition using the $^{13}C\{^1H\}$ NMR and inverse-gated experiments, Table 1.

Table 1. Illustrating the effect on the measured percentages of a cyclohexanone mixture using $^{13}C\{^1H\}$ NMR and inverse-gated experiments. The carbon atoms that were analyzed are highlighted in each structure. Adapted with permission from reference (*17*). Copyright 2000 American Chemical Society

Experiment	Acquisition Time (min)	Percentage			
		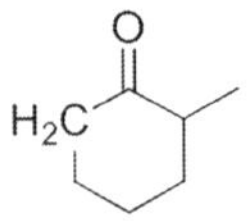 O, H_2C	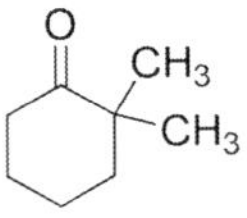 O, CH_3, CH_3	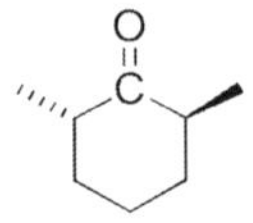 O, C	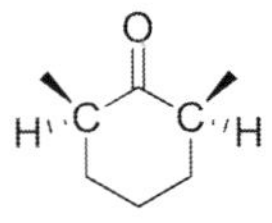 O, H, C, C, H
$^{13}C\{^1H\}$	7	57.6	20.7	7.3	14.4
Inverse-gated	270	46.8	16.5	6.4	30.3
Inverse-gated + $Cr(acac)_3$	26.2	46.3	17.1	6.0	30.6
Actual Composition		45.5	17.0	6.3	31.1

As expected the $^{13}C\{^1H\}$ NMR experiment gave percent compositions that, for the most part, varied from the actual composition values because of NOE signal enhancements. Upon switching to the inverse-gated ^{13}C NMR experiment combined with a delay between pulses of five times the longest T_1 value, quantitative analysis of the same carbon signals showed near perfect agreement with the actual composition values. Unfortunately, the long delay between pulses meant a lengthy acquisition time, something that can be impractical for an undergraduate experiment. However, the addition of $Cr(acac)_3$ dramatically shortened the relaxation rate and gave comparable results in a fraction of the time. LeFevre and Silveira noted that when students were allowed to choose any carbon to measure, most chose the carbonyl carbon since these are well dispersed in that area of the spectrum. Using this approach both the $^{13}C\{^1H\}$ and inverse-gated experiments surprisingly gave the same quantitative data as the actual composition. However, upon closer examination this is less surprising because each carbon atom is in a very similar environment and is expected to have essentially the same magnitude of NOE and very similar T_1 values.

In an instrumental analysis class, Schmedake and Welch describe how students explored the effects that different ^{13}C NMR experiments have on the ability to quantitatively analyze the ^{13}C atoms of acenaphthene (*18*). After optimizing experimental conditions, students quantitatively determined the amount of aspirin and acetaminophen in an analgesic tablet. Quantitative analysis of acenaphthene was best accomplished using the inverse-gated experiment. Despite the addition of $Cr(acac)_3$ to induce relaxation, Schmedake and Welch noted the importance of the delay time between pulses, with shorter delay times leading to greater deviation from the expected values. After optimal experimental conditions were found, students were then tasked with finding the amounts of aspirin and acetaminophen in an Extra Strength Excedrin ™ tablet using an internal standard. Unfortunately, solubility problems occurred with $Cr(acac)_3$, so 2,2,6,6-tetramethyl-1-piperidinyloxy (TEMPO) was substituted as the relaxation agent. For their analysis students chose ^{13}C signals from the alkane and carbonyl regions of the aspirin and acetaminophen since these could be easily compared to the carbonyl resonance from an acetophenone internal standard. By correlating their data with the internal standard, the amount of each analgesic was calculated and these values were found to be in good agreement with the reported literature values.

The Gated ^{13}C Experiment

Both the $^{13}C\{^1H\}$ NMR experiment and the inverse-gated ^{13}C NMR experiment provide valuable information about the general environment of the carbon based on the chemical shift, and in the latter case, a way to quantitatively analyze these signals. However, information that could arise from indirect spin-spin coupling to adjacent protons is lost because of when the irradiation to the ^1H resonance region is applied. Of course the decoupler could be turned off entirely which would retain all of the coupling information, but NOE enhancements would be lost and longer acquisition times would be required. Fortunately, indirect spin-spin coupling occurs instantaneously in a sample whereas NOE signal enhancements take time to build up. So, if the decoupling irradiation is turned on only during the relaxation delay between pulses, then coupling information is retained and signals still experience limited NOE signal enhancement. This experiment is called *gated decoupling*.

Delagrange and Nepveu use gated decoupling to investigate the regiospecific ring-opening of a monosubstituted anhydride in an organic chemistry laboratory experiment (*19*). Students begin by reacting ethanol with 2-(*S*)-acetoxysuccinic anhydride to generate ethyl 2-(*S*)-acetoxy-3-carboxypropionate. The identity of the product is not known at this stage but analysis of the product using ^1H and $^{13}C\{^1H\}$ NMR experiments establish that only one isomer has been formed and many of the signal assignments can be made at this point. An HMBC experiment (described later) allows the assignment of the remaining three carbonyl resonances. At this stage the students still need to distinguish between one of two possible regioisomers. An examination of the carboxylic acid region in the ^{13}C gated decoupling NMR experiment revealed a triplet of doublets. Since the carboxylic acid carbon of each isomeric product resides in a slightly different environment, it was predicted that they would undergo different indirect spin-spin couplings with a two-bond coupling constant that was larger than the three-bond coupling constant, Figure 2.

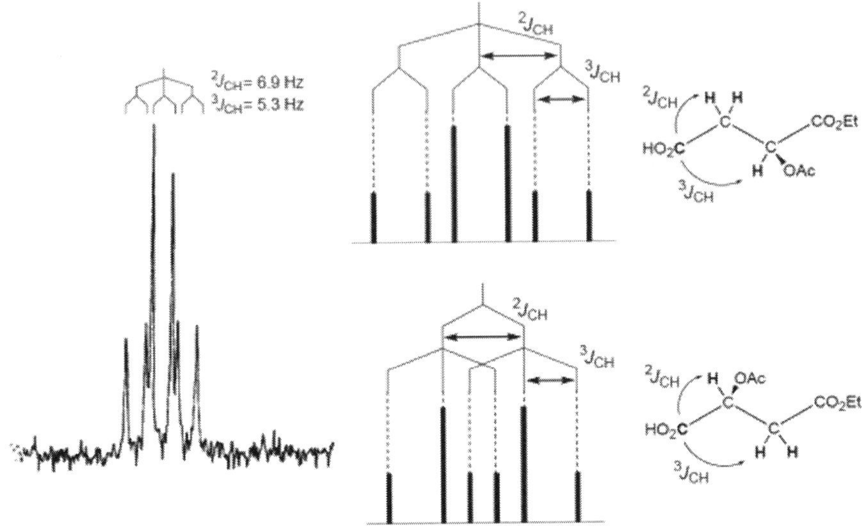

Figure 2. Proton-coupled ^{13}C NMR of the carboxylic acid resonance alongside predicted splitting patterns for the carboxylic acid carbon of ethyl 2-(S)-acetoxy-3-carboxypropionate and ethyl 3-(S)-acetoxy-3-carboxypropionate. Adapted with permission from reference (19). Copyright 2000 American Chemical Society.

Indeed, analysis confirmed that the product arising from the regiospecific ring-opening of 2-(*S*)-acetoxysuccinic anhydride is ethyl 2-(*S*)-acetoxy-3-carboxypropionate, illustrating the utility of the ^{13}C gated decoupling NMR experiment as a method for distinguishing between regioisomeric products.

The Off-Resonance ^{13}C Decoupling

The off-resonance ^{13}C NMR experiment uses less powerful decoupling irradiation that is centered away from the middle of the ^1H resonance region. When the appropriate experimental parameters are used, this has the effect of removing all indirect spin-spin couplings that are further than one bond away. Signals are still enhanced by the NOE because the irradiation is on during the entire pulse sequence, but each signal now gives a characteristic n+1 splitting pattern that reflects the number of directly attached hydrogen atoms. For simple molecules, this provides a straightforward method for determining whether carbon atoms are methyl, methylene, methine, or non-protonated. However, since each signal is now split over more peaks, longer acquisition times may be needed to achieve comparable signal-to-noise ratios when compared to the $^{13}C\{^1H\}$

NMR experiment. Additionally, more complex compounds will give spectra with overlapping signals, making interpretation more challenging. By varying the power of the irradiating transmitter and the distance from the ^{1}H resonance region, the off-resonance experiment can also be used to observe the evolution of complex splitting patterns. Hersh provides an example of the utility of the off-resonance ^{13}C NMR decoupling experiment in distinguishing between AA'X and ABX spin systems (*20*). Despite the utility of the off-resonance experiment, it is now rarely used having been replaced by more powerful experiments that provide the same information.

The ^{13}C Attached Proton Test

The *attached proton test* (APT) experiment, sometimes called the *J*-modulated spin-echo experiment, uses a pulse sequence that gives a spectrum with signals of different phases depending upon how many hydrogen atoms are attached to that carbon atom. For carbon atoms with 1 or 3 attached hydrogen atoms, the signals will have an opposite phase to those with 0 or 2 attached hydrogen atoms. Proton broadband decoupling occurs during different parts of the pulse sequence which allows NOE signal enhancements to build and also removes all of the coupling information meaning that each signal now becomes a singlet. However, care should be taken in assigning signals, particularly since methyl and methine carbons often give signals of similar intensity.

Cooke, Henderson, and Lightbody report on the use of the APT experiment to characterize the unusual product of a zeolite-catalyzed etherification of diphenylmethanol (*21*). The highly symmetric product shows very few signals in the APT spectrum, Figure 3.

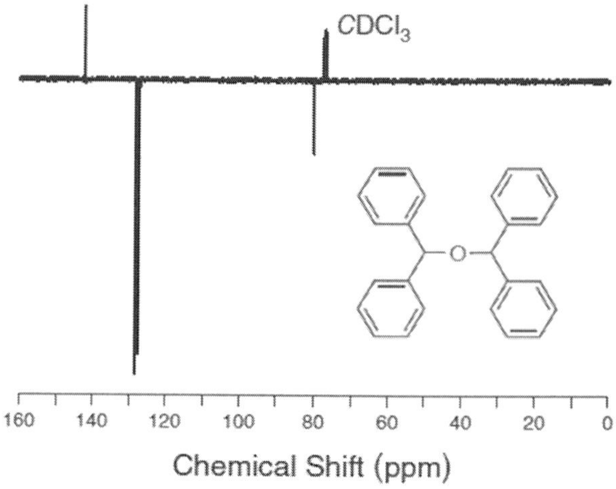

Figure 3. APT spectrum of $(C_6H_5)_2CHOCH(C_6H_5)_2$. Carbon atoms with 0 or 2 attached hydrogen atoms give signals with positive phase. Carbon atoms with 1 or 3 attached hydrogen atoms give signals with negative phase. Adapted with permission from reference (21).Copyright 2009 American Chemical Society.

The aromatic and benzyl CH atoms are easily distinguishable, having signals with negative phase, while the aromatic carbon atoms with no attached hydrogen atoms are distinguished with a positive phase signal. Though not always a reliable indicator, the relative intensities of these signals are also reflective of the number of carbon atoms that contribute to that signal. The APT spectrum in Figure 3 also provides an example of accidental chemical equivalency since there should be three non-equivalent aromatic methine carbons yet only two are clearly visible.

The DEPT Experiment

The Distortionless Enhancement by Polarization Transfer (DEPT) experiment has become the standard experiment to use when determining the local environment of a carbon atom since it does not have any of the disadvantages present in the off-resonance decoupling, gated decoupling, or APT experiments. The DEPT experiment is actually a series of experiments that differ with respect to an angle of 45°, 90°, or 135° that is incorporated into the pulse sequence. Each experiment generates a spectrum in which the substitution pattern of the carbon governs whether the signal will have a positive phase, a negative phase, or will be absent. In the DEPT experiment the 45° spectra show all protonated carbon atoms with positive phase signals, the 90° spectra show only methine carbon atoms as signals with a positive phase, with all other signals being absent, and the 135° spectra shows both methine and methyl groups as signals with a positive phase and methylene groups as signals with a negative phase. Quaternary carbon atoms can be readily identified by comparing the $^{13}C\{^{1}H\}$ NMR experiment to either the 45° or 135° DEPT spectrum. All three DEPT spectra can be shown, or just the 90° or 135° spectra, since these last two spectra contain all of the necessary information to make assignments. However, a more visually appealing set of spectra can be constructed by adding and subtracting these spectra from each other to generate new spectra that show only positive signals for each carbon atom type. A recent modification to the DEPT experiment, called the DEPTQ experiment, allows for the detection of non-protonated or quaternary carbon atoms, thus removing the need for a separate $^{13}C\{^{1}H\}$ NMR spectrum.

Reeves and Chaney describe a laboratory experiment that comes early in their organic chemistry curriculum and that uses the DEPT experiment to aid in the identification of C_6 or C_7 isomeric alkanes (*22*). Students are assigned an unknown isomer and get hands-on access to the instrument so that they can take $^{13}C\{^{1}H\}$ NMR and DEPT spectra. Upon returning to the lab they then use their knowledge of bonding, symmetry, and constitutional isomerism to predict the number and type of signals for all possible isomers of their C_6 or C_7 unknown. By comparing their predicted data with their spectra, identification of their unknown becomes straightforward particularly since the DEPT experiment allows students the ability to distinguish between isomers that have the same number of signals but differ in the immediate environment of the carbon atom. Other examples include a biochemistry laboratory developed by Ivey and Smith who use low-field NMR and the DEPT and COSY experiments to identify several amino acids (*23*), and the integration by Walsh et al. of the DEPT experiment into a longer experimental sequence that includes ^{1}H, $^{13}C\{^{1}H\}$, HH-COSY, HMQC, and HMBC experiments to elucidate the structure of parthenolide, a compound isolated from *Tanacetum parenthium* (*24*).

The 1D INADEQUATE Experiment

The one-dimensional Incredible Natural Abundance Double Quantum Transition (INADEQUATE) experiment provides information about the coupling constants of adjacent ^{13}C atoms. Coupling constants can provide important structural information, as evidenced in the gated ^{13}C NMR experiment where two- and three-bond carbon-hydrogen coupling was used to distinguish between isomers. Unfortunately, the low natural abundance of ^{13}C means that the probability of having two adjacent ^{13}C atoms in an unenriched sample is highly unlikely (about 0.01%). This means that in order to attain signals of sufficient intensity to observe carbon-carbon coupling, acquisition times will be very, very long. Though some structural information can be determined from this experiment, and since carbon-carbon coupling constants do not vary much, it is nearly always more straightforward to use other experiments such as HH-COSY, HSQC, and HMBC when analyzing structure. As a result, the 1D INADEQUATE experiment is rarely, if ever, used. However, the 2D INADEQUATE experiment, which will be discussed later, is a significantly more powerful experiment providing structural information that is comparable to X-ray crystallography data.

The One-Dimensional ^{1}H NMR Experiment

The one-dimensional ^{1}H NMR experiment provides information about the number of non-equivalent protons, their electronic environment, the number of adjacent protons, and the number of protons contributing to a given signal. Historically, protons were the first nuclei to be regularly observed because they are ubiquitous and have high natural abundance and sensitivity. Both of these factors contribute to very short acquisition times. Many organic chemistry textbooks start with a discussion of the ^{1}H NMR experiment since it provides such a wealth of structural information. Much like ^{13}C NMR spectroscopy, the number of signals and their position provide important information about the number and functionality of non-equivalent protons. ^{1}H NMR signals can also be quantitatively analyzed (integrated) with these values providing the number of protons contributing to a given signal. Finally, indirect spin-spin coupling readily occurs between non-equivalent protons that are 2 and 3 bonds away, with longer-range couplings sometimes occurring. The nature of the splitting pattern provides information about the number of adjacent protons. Furthermore, coupling constant values combined with the Karplus relationship can also provide structural detail, such as the geometry of an alkene or the relative orientation of adjacent atoms (*25*).

Adesoye et al. examined a glucopyranoside and used the Karplus relationship to study an S_N2 reaction at the anomeric center in an advanced organic synthesis laboratory experiment (*26*). Beginning with the commercially available

2,3,4,6-tetra-O-acetyl-α-D-glucopyranosyl bromide students synthesized the corresponding β-D-glucopyranosyl azide and analyzed it by ^{1}H NMR spectroscopy, Figure 4.

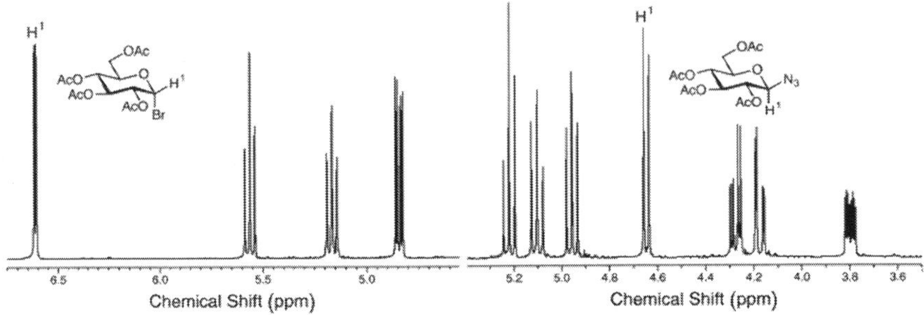

Figure 4. Partial ^{1}H NMR spectra of the 2,3,4,6-tetra-O-acetyl-α-D-glucopyranosyl bromide and the corresponding azide derivative. Reproduced with permission from reference (26). Copyright 2012 American Chemical Society.

Since the doublet of the anomeric proton atom is easily recognized (labeled H^1 in Figure 4) and the signals are well-dispersed, chemical shift changes resulting from the reaction do not hinder analysis. The S_N2 nature of the mechanism is confirmed by observing the 3-bond coupling constant of the anomeric proton increasing from 4.0 Hz to 8.8 Hz. This is consistent with a change from an equatorial-axial to axial-axial relationship and conforms to the Karplus relationship for such systems.

Sorensen, Witherell, and Browne describe a similar approach where students use ^{1}H NMR spectroscopy to identify a sugar from two unknowns, α-methylglucopyranoside or α-methylgalactopyranoside (*27*). These sugars differ only in the orientation of the H-4 proton. After synthesizing the peracetyl derivatives of their unknown α-methyl pyranoside, students obtain ^{1}H, ^{13}C and COSY spectra and assign the signals to the relevant proton and carbon atoms. Recognizing that the anomeric proton is the only ^{1}H NMR signal that gives a doublet, students worked their way around the molecule identifying the H-4 proton and measuring its 3-bond coupling constant to its adjacent protons. Using the Karplus relationship to identify that the coupling constant for an axial-equatorial relationship is less than 3 Hz, compared to an axial-axial relationship of about 10 Hz, students determined the orientation of the H-4 proton and subsequently identified their sugar.

The 1D NOESY Experiment

The one-dimensional Nuclear Overhauser Effect Spectroscopy experiment provides information about the proximity of nuclei to each other. NOE enhancements result from population changes that occur when an irradiated nucleus is close to a non-irradiated nucleus. This enhancement is not dependent upon indirect spin-spin coupling. As a result NOE signal enhancements are observed for any nuclei that are close enough together in space. With a distance dependence of $1/r^6$ this means that NOE enhancements are only observed at distances of less than 5 Å. Unlike the NOE signal enhancements detected in the $^{13}C\{^1H\}$ NMR experiment, where the maximum possible enhancement was 200%, NOE signal enhancements that result from interactions between the same nuclei result in maximum enhancements of only 50%. Since the NOE is solely dependent upon the proximity of the nuclei, one of the most important aspects of this technique is that NOE signal enhancements can be observed in nuclei that are close in space but many atoms apart, as might be found in complex biological systems.

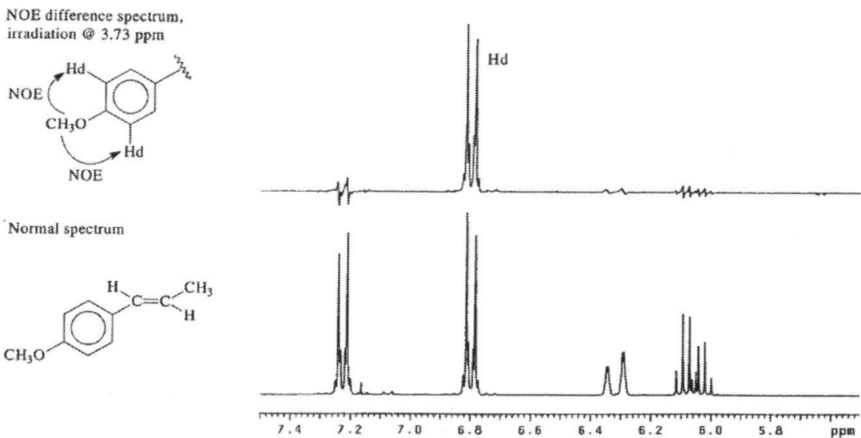

Figure 5. Partial ^1H NMR and NOE difference spectrum of trans-anethole. Adapted with permission from reference (28). Copyright 2000 American Chemical Society.

When run in one dimension, the NOE experiment is typically run as a difference experiment where a regular ^1H spectrum is subtracted from the 1D NOESY experiment that has undergone irradiation of a known signal. The resulting difference spectrum will then only show NOE enhancements for the spatially proximal atoms. It is worth noting that both positive and negative NOEs can be observed since the effect is also dependent upon the relative arrangement of atoms. In some cases, careful analysis of the magnitude of the NOE signal enhancements can lead to conclusions about the 3D arrangement of the atoms. LeFevre describes an advanced laboratory experiment where students isolate and determine the structure of *trans*-anethole, using the 1D NOESY experiment to finalize signal assignments (*28*). After performing a steam distillation of anise seeds followed by column chromatography of the product, each student takes ^1H and ^{13}C NMR spectra of their samples. Because of time constraints DEPT, COSY and HETCOR spectra are taken from a representative sample. With all of this information most of the assignments of *trans*-anethole can readily be made. The ^1H NMR spectrum shows two distinctive doublets that are indicative of a *para*-substituted benzene ring. Assignment of each doublet signal can be made if a student understands the effect of benzene substituents on chemical shifts. However, a more straightforward method that allows definitive assignment of each signal uses a 1D NOESY experiment, Figure 5.

Since the methoxy group and the aromatic protons (Hd) of *trans*-anethole are proximal in space, a significant NOE would be expected. Indeed when the protons associated with the methoxy group are irradiated, a strong NOE is observed in the protons at 6.8 ppm which correspond to the neighboring protons.

McDaniel and Weekly used NOE difference spectroscopy to unambiguously assign diastereotopic protons in a Diels-Alder experiment involving the reaction of 2,4-hexadien-1-ol with maleic anhydride in an organic chemistry laboratory (*29*). Students began by carrying out the synthesis of the isobenzofuranone Diels-Alder product which, after crystallization, generates good yields of only one isomer, Figure 6.

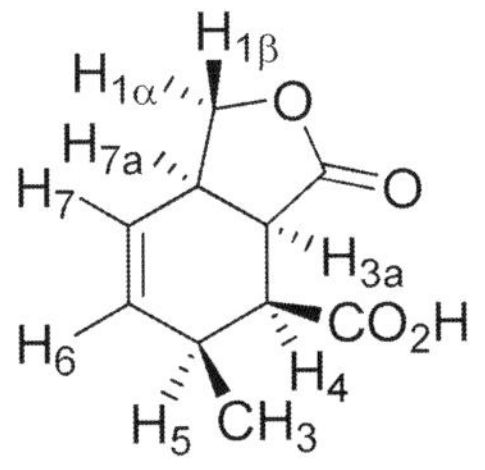

Figure 6. The isobenzofuranone Diels-Alder product arising from the reaction of 2,4-hexadien-1-ol with maleic anhydride. Reproduced with permission from reference (29). Copyright 1997American Chemical Society.

Much of the structure can be determined from ^1H and HH-COSY NMR experiments, but definitive assignment of the methylene hydrogen atoms $H_{1\alpha}$ and $H_{1\beta}$ can only be achieved using an NOE difference experiment. Indeed, irradiation of the H_{7a} proton shows NOE enhancements for the previously assigned H_{3a}, H_4, H_5 signals as well as a strong NOE enhancement for the signal at 4.43 ppm, attributed to $H_{1\alpha}$. Overall, the one-dimensional NOE difference experiment provides an elegant, and most importantly selective method, for assigning signals via through-space interactions.

The 1D TOCSY Experiment

The one-dimensional Total Correlation Spectroscopy (TOCSY) experiment provides a spectrum that only shows signals for a given spin system. In a typical ^1H NMR experiment, splitting patterns are dominated by 2- and 3-bond indirect spin-spin coupling. The TOCSY experiment allows indirect spin-spin coupling information to be relayed along an entire spin system so that the spectrum only contains signals that belong to that spin system. For the one-dimensional experiment a distinct proton resonance is irradiated and the spin-lock mixing time part of the pulse sequence is optimized so that the energy from the irradiated proton is transferred to the entire spin system of that proton. Shorter spin-lock mixing times decrease the distance over which this energy is transferred resulting in fewer signals. This can simplify analysis when overlapping signals are present. The TOCSY experiment has greatly benefited from improved hardware and electronics that allow for the formation of very specific pulses that can narrowly target the desired resonance. The TOCSY experiment is very useful in analyzing compounds that contain isolated spin systems, such as oligosaccharides and proteins, and can even be used to analyze reaction mixtures as long as signals are sufficiently resolved.

Sereda provides an example of using the 1D TOCSY experiment to analyze the reaction mixture arising from the iodochlorination of 1-hexene (*30*). Students carry out the iodochlorination of 1-hexene and generate two regioisomers, 2-chloro-1-iodohexane and 1-chloro-2-iodohexane. The ^{1}H NMR spectrum of this mixture is complex and difficult to analyze. However, the 1D TOCSY experiment can be used to observe each compound separately since distinct resonances exist for each. Students first explore the 1-chloro-2-iodohexane product by irradiating the resonance associated with the proton on the iodine-bonded methine group. By changing the spin-lock mixing time they observe how the coupling information propagates through the molecule. With shorter spin-lock mixing times only geminal and vicinal couplings are observed, whereas with longer spin-lock mixing times the entire spin-system can be seen. A similar analysis can be carried out for the 2-chloro-1-iodohexane product.

1D NMR Experiments of Other Nuclides

Most modern NMR spectrometers can be tuned to observe nuclides other than ^{1}H or ^{13}C. This significantly broadens the applicability of this powerful spectroscopic technique. With an appropriately configurable probe, tuning can occur using built-in hardware and software or by manually changing out cables and adding filters. The ability to observe nuclides is governed by a variety of factors, including the spin of the nuclide, the sensitivity, and the nuclides natural abundance. When nuclides have spins where $I \neq ½$ the observed splitting patterns are governed by the $2nI + 1$ rule, where n is the number of neighboring atoms and I is the nuclear spin of the coupled atom. This results in more complex splitting patterns which can complicate analysis. Nuclides with spins $I > ½$ are also quadrupolar and generally have much shorter relaxation times. This leads to line-broadening, meaning that the fine structure associated with coupling may not be visible. The sensitivity of nuclei to the NMR experiment is also important. Sensitivity is essentially a measure of the difference in energy between the spin states for that nucleus. The most sensitive nuclei that are most commonly observed, in decreasing order, are ^{1}H, ^{19}F, ^{59}Co, and ^{7}Li, with all other nuclei being at least an order of magnitude less sensitive. Finally, the natural abundance of the nuclei is very important; 100% abundant nuclides will mean shorter acquisition times, whereas lower abundant nuclei, such as ^{13}C, will require longer acquisition times.

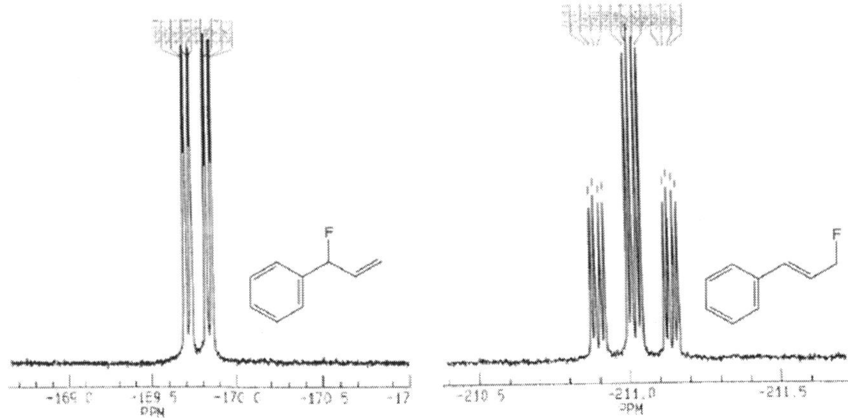

Figure 7. ^{19}F spectra recorded at 376.5 MHz show two signals that belong to 3-fluoro-3-phenylpropene and 3-fluoro-1-phenylpropene. Adapted with permission from reference (31).Copyright 1995 American Chemical Society.

Fluorine is rarely found in nature but has found many applications in materials and pharmaceuticals. While the ^{19}F nuclide is only slightly less sensitive than a proton, it is 100% abundant and has a spectral width of nearly 400 ppm that ranges from +50 ppm to -300 ppm. Note that atoms such as fluorine have spectral widths that are very different from those observed with ^{13}C and ^{1}H atoms. Fluorine is particularly sensitive to neighboring functional groups which results in excellent spectral dispersion. All of these characteristics make ^{19}F NMR spectroscopy an important addition to the more commonly observed nuclides. deMendonca et al. use ^{19}F NMR spectroscopy in an experiment that could be performed in an honors-level or advanced organic chemistry course (*31*). In this experiment students are asked to analyze the diethylaminosulfur trifluoride (DAST) fluorination of cinnamyl alcohol which can occur via S_N2, S_N2', or as a combination of both mechanisms. Analysis of the ^{1}H NMR spectrum of the product is difficult because the ^{19}F undergoes indirect spin-spin coupling with the ^{1}H atoms resulting in a complex spectrum with overlapping splitting patterns. A more straightforward approach, using ^{19}F NMR spectroscopy yields a spectrum with two multiplets at -169.77 ppm and -211.02 ppm, Figure 7.

In this case, the most up-field signal, at -169.77 ppm, shows a splitting pattern consistent with two-bond coupling with the benzylic proton, three-bond coupling with the vinyl proton, and four-bond coupling with the aromatic methine. This indicates that the product is 3-fluoro-3-phenylpropene which could only occur via the S_N2' mechanism. The second signal shows a splitting pattern consistent with two-bond coupling with the allylic methylene, and three- and four-bond coupling with the vinyl protons. This indicates that this product is 3-fluoro-1-phenylpropene which occurred via the S_N2 mechanism. Integration of each signal gives a 1.75/1.00 ratio illustrating that both mechanisms are active in the reaction, but that the S_N2' mechanism predominates.

Jenson and O'Brien demonstrate another benefit of ^{19}F NMR spectroscopy in simplifying the analysis of an inorganic cobalt complex (*32*). In this preparatory inorganic chemistry experiment students prepare a cobalt complex using cobalt carbonate and 1,1,1-trifluoro-2,4-pentanedione. The resulting product is a mixture of two cobalt complexes that vary in the orientation of the 2,4-pentanedione moiety around the cobalt. After the synthesis, each isomer was isolated by preparatory thin layer chromatography and the samples were pooled so that there was sufficient sample for ^{1}H, ^{13}C, and ^{19}F NMR analysis. One of the cobalt complexes is highly symmetric, meaning that all three trifluoromethyl groups are equivalent, showing only one signal in the ^{19}F spectrum. The other cobalt isomer is not symmetric and all three trifluoromethyl groups are non-equivalent. This should result in three signals in the ^{19}F spectrum. Unfortunately, in this case two of the signals were accidentally chemical equivalent, resulting in only two signals in the spectrum. Since accidental chemical equivalency is dependent upon magnetic field strength and solvent, signal separation can be achieved by changing either of these parameters. Jenson and O'Brien demonstrated that upon switching the solvent from $CDCl_3$ to acetone-d_6 the accidentally chemically equivalent signals resolved into three signals from the non-equivalent trifluoromethyl groups.

The ubiquity of phosphorus in biological systems, coupled with its high natural abundance, means that ^{31}P NMR spectroscopy can be an important tool in determining a variety of biologically important processes. Similar to ^{19}F, a ^{31}P spectrum covers about 500 ppm, approximately from +250 ppm to -250 ppm. This means that signal separation usually occurs even when there are only subtle differences in the electronic environment of each phosphorus atom. Fenton and Sculimbrene describe an experiment that uses ^{31}P NMR spectroscopy to determine the stereochemistry of di-*sec*-phenethyl phosphonate in a second-semester or advanced organic chemistry laboratory (*33*). Students are divided into groups and use either racemic, (*R*)- or (*S*)-*sec*-phenethyl alcohol with phosphorous trichloride to generate the corresponding dialkyl phosphonate. Because of the nature of the synthesis, all of the phosphorus is consumed in the formation of the dialkyl phosphonate so analysis of the ^{31}P NMR spectra is not complicated by unreacted starting materials or byproducts. Students using the racemic alcohol generated a statistical mixture of diastereomers which gave a ^{31}P NMR spectrum of three signals in a 1:2:1 ratio. Students who started with either of the enantiopure alcohols observed only one signal in the ^{31}P NMR spectrum since both alkyl groups give signals at the same resonance frequency (or chemical shift) because of their enantiotopic relationship.

Since indirect spin-spin coupling occurs between NMR active nuclides, the direct observation of other nuclei is not always required. Rather, nuclides can be observed indirectly by observing splitting patterns in ^{1}H or ^{13}C NMR experiment. The most obvious example of this is the three-line splitting pattern at 77 ppm in a ^{13}C NMR spectrum that results from the ^{13}C atom in $CDCl_3$ coupling to the deuterium atom, which has a nuclear spin of 1. Zanger and Moyna provide another excellent example of this phenomenon using both high- and low-field NMR spectrometers to determine the isotopic ratio of boron in sodium borohydride in an organic chemistry experiment (*34*). Boron exists as two isotopes, ^{10}B and ^{11}B, with nuclear spins of I=3 and I=3/2 respectively, and will give splitting patterns that are governed by the equation $2nI + 1$. Analysis of the spectrum is simplified because all of the hydrogen atoms in sodium borohydride are equivalent, and collectively see only one boron atom. For example, $Na^{10}BH_4$ has a boron atom with a nuclear spin of I=3. Using the equation given earlier, the number of signals is equal to (2 x 1 x 3) + 1, or 7 signals. A similar calculation can be performed for the other isotope of boron. Figure 8 shows the ^1H NMR spectrum of a 5% solution of sodium borohydride in D_2O at 400 MHz.

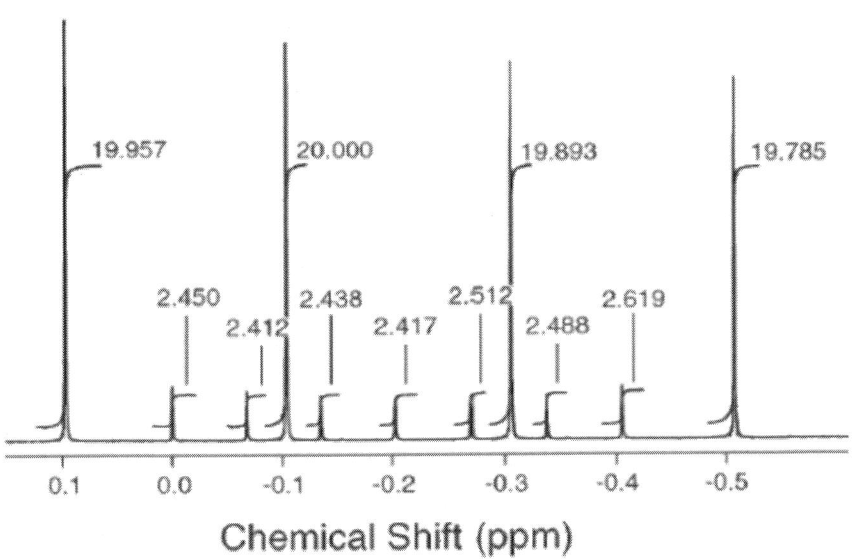

Figure 8. ^1H NMR spectrum of a 5% solution of sodium borohydride in D_2O at 400 MHz. Integral values associated with the signals is also shown. Reproduced with permission from reference (34). Copyright 2005 American Chemical Society.

After acquiring the data, students analyzed the number and intensity of the signals and determined that the ^{10}B isotope must be responsible for the seven smaller signals of equal height, while the ^{11}B isotope was responsible for the four signals of equal height. Even at lower magnetic flux densities (90 MHz NMR spectrometer) all 11 lines are still sufficiently resolved to allow for complete analysis. Students completed the experiment by integrating the signals to determine that the isotopic abundance is approximately 18-21% for ^{10}B and 82-79% for ^{11}B, values which is in excellent agreement with established literature values.

Introduction to 2D NMR Experiments

It is not surprising that one-dimensional NMR experiments predominate across the undergraduate curriculum. These experiments provide a tremendous amount of molecular information in a form that is usually easy for students to understand and analyze. However, two-dimensional experiments are becoming increasingly integrated at all levels of the undergraduate curriculum because they provide powerful ways of approaching chemical problems and are not that much harder to analyze than their one-dimensional cousins.

All two-dimensional experiments contain two time domains. One time domain is a result of the free induction decay that occurs at the end of every pulse sequence, while the second time domain arises from the addition of a mixing and evolution stage that occurs during the pulse sequence and which is incrementally changed over the course of the experiment. Fourier transforms of each time domain with respect to each other give two frequency domains that can then be plotted as a two-dimensional spectrum. Traditionally, two-dimensional experiments have long acquisition times because of the iterative nature of the pulse sequences. However, new variants of traditional 2D experiments have been developed that take advantage of gradient probe technology, resulting in significantly shorter acquisition times.

The HETCOR, HMQC, and HSQC Experiments

The family of experiments that include the Heteronuclear chemical-shift Correlation (HETCOR), Heteronuclear correlation through Multiple Quantum Coherence (HMQC), and Heteronuclear Single Quantum Correlation (HSQC) correlate protons that are directly attached to carbon atoms. Using only ^{1}H, ^{13}C, and DEPT NMR experiments to analyze molecular structure can lead to connectivity ambiguities, particularly in more complex molecules. The HETCOR, HMQC, and HSQC experiments provide an excellent method for making these assignments with certainty. A typical spectrum in this family of experiments will display an abscissa and ordinate that correspond to the chemical shifts for ^{1}H and ^{13}C. Cross peaks are most commonly represented as contour plots that correlate each characteristic signal. The major difference between the experiments is that the HETCOR experiment is designed to directly detect the far less sensitive and less abundant ^{13}C nuclei. This results in longer acquisition times and reduced sensitivity. Alternatively, the HMQC and HSQC experiments detect the more sensitive and essentially 100% abundant ^{1}H nuclei through a technique called inverse detection. In this technique the ^{13}C responses are transferred and observed in the more sensitive ^{1}H nuclei. This results in shorter acquisition times and increased sensitivity. Out of the three experiments, the HSQC experiment is now becoming the experiment of choice. The pulse sequence that is used provides increased sensitivity and peak shape and is less dependent upon instrumental parameters. Notably, the HSQC experiment can also be adapted to include multiplicity-editing, where the phase of the cross peak indicates whether that signal is a CH_3/CH or CH_2/non-protonated carbon. In fact a multiplicity-edited

HSQC spectrum provides information comparable to individually taken ^{1}H, ^{13}C, and DEPT spectra, highlighting the powerful nature of these types of experiments.

Shaw et al. describes the use of HMQC experiments to provide structural confirmation for a series of microwave-promoted Diels-Alder reactions in an organic chemistry laboratory (*35*). Using both cyclopentadiene and cyclohexadiene in the reaction with *trans*-1,2-benzoylethylene, students prepared various Diels-Alder products. The products were analyzed using ^{1}H NMR spectroscopy, with the HMQC experiment providing verification for several key assignments, Figure 9.

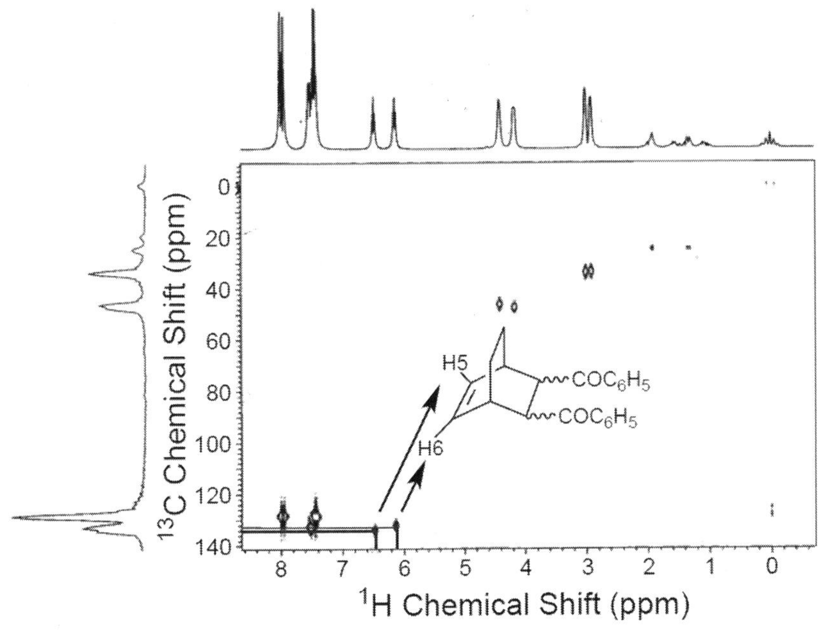

Figure 9. ^{1}H-^{13}C HMQC spectrum of racemic exo,endo-2,3-dibenzoylbicyclo[2.2.2]octa-5-ene. Stereochemistry of the benzoyl groups is not implied. Adapted with permission from reference (35). Copyright 2005 American Chemical Society.

The HMQC spectrum shows correlations between the H5 and H6 vinyl protons and the two alkene carbon atoms. Since the product is not symmetric, both signals are observed. Cross peaks correlating the carbon atoms adjacent to the carbonyl groups (approx. 42 ppm) and the allylic carbons (approx. 30 ppm) are also easily distinguishable. Notice that the ^{13}C signals are not as well-resolved as the ^{1}H signals. This is due to the nature of the experiment. If high-resolution 1D ^{1}H and ^{13}C NMR spectra have been taken and are available, most software packages will readily substitute these spectra onto the 2D spectrum which can simplify analysis.

Clausen describes an experiment that introduces the concept of diastereotopic protons via the synthesis of 2-methylbutyl acetate using standard Fischer esterification protocols in an organic chemistry laboratory experiment (*36*). Since the product of this esterification reaction is chiral, the methylene protons are diastereotopic and show up at different chemical shifts giving appropriate, but more complicated, splitting patterns. Since students at this level may not be aware of diastereotopicity, an analysis of the splitting pattern is challenging. However, the HSQC experiment easily identifies these protons as diastereotopic since the spectrum clearly shows that both of these protons are correlated to the same carbon atom.

The HMBC and H2BC Experiments

The Heteronuclear Multiple Bond Correlation (HMBC) experiment provides carbon-hydrogen correlations over 2 and 3 bonds, with longer range correlations being observed occasionally. Three-bond correlations are often stronger, since the magnitude of the coupling constant is larger, but care should be taken when analyzing an HMBC spectrum because this isn't always the case. The HMBC experiment is very similar to the HMQC or HSQC experiments but unlike those experiments that only show one cross peak for each carbon and hydrogen atom correlation, an HMBC experiment will have several cross peaks for each carbon atom.

Caes and Jensen describe the synthesis of 9-hydroxyphenalenone in an experiment that is suitable for a range of laboratory courses (*37*). The synthesis starts with commercially available 2-methoxynaphthalene, cinnamoyl chloride, and aluminum trichloride. The initial Friedel-Crafts acylation product quickly undergoes a series of reactions to generate the C_{2v} symmetric 9-hydroxyphenalenone product. The symmetric nature of the product simplifies analysis since the number of observed carbon signals is reduced from 13 to 8 and the number of proton signals from 8 to 5. ^{1}H, ^{13}C, and HH-COSY experiments provide much of the structural information, but an HMBC experiment is needed to finalize the assignments by correlating carbon atoms with more distant hydrogen atoms, Figure 10.

Indeed, analysis of the HMBC spectrum shows that the quaternary carbon at 3a shows 3-bond correlations with the H-5 and H-2. Similarly, carbon 4 shows 3-bond correlations with H-6 and H-3. While analyzing an entire HMBC spectrum is daunting, particularly when it is unclear whether 2- or 3-bond correlations are being observed, the HMBC experiment provides an excellent method for resolving structural ambiguities that remain after ^{1}H, ^{13}C, DEPT, HSQC, and HH-COSY experiments have been fully analyzed.

An alternative to the HMBC experiment is the Heteronuclear 2 Bond Correlation (H2BC) experiment which overcomes many of the ambiguities present in the analysis of an HMBC spectrum. The H2BC experiment is optimized to correlate carbon atoms to the proton of protonated carbons that are two bonds away. When combined with the HMBC experiment, the H2BC experiment provides an easy method for distinguishing between 2- and 3-bond couplings.

Alternatively, when H2BC data is combined with HSQC data it becomes possible to "walk" around the molecule and determine structural connectivity. This can provide comparable information to that of a 2D INADEQUATE experiment.

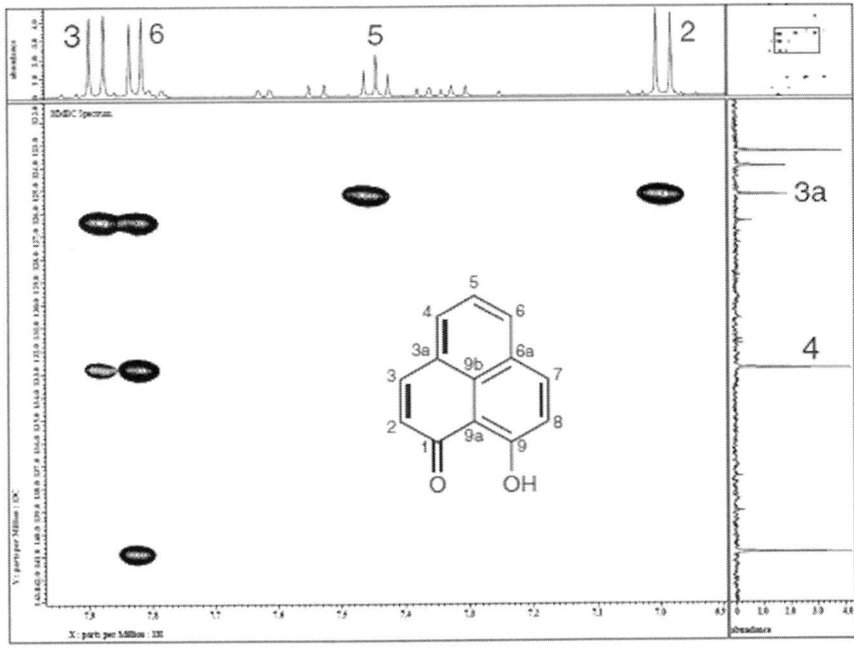

Figure 10. Expanded HMBC spectrum of 9-hydroxyphenalenone. Adapted with permission from the supplemental information provided in reference (37). Copyright 2008 American Chemical Society.

The 2D INADEQUATE Experiment

The two-dimensional INADEQUATE experiment, sometimes called the CC-COSY experiment, correlates carbon atoms with carbon atoms and provides a powerful method for structural analysis. In this experiment the abscissa is the ^{13}C NMR scale while the ordinate is a frequency domain that is centered on 0 Hz. Cross peaks exist as horizontal pairs that correlate one ^{13}C atom with an adjacent ^{13}C atom. Analysis of a 2D INADEQUATE spectrum begins by identifying all of the signals from the DEPT and $^{13}C\{^1H\}$ spectra as methyl, methylene, methine, or non-protonated carbon atoms.. Then starting with a known carbon atom, a cross peak is traced horizontally to its neighbor providing connectivity information. Vertical correlations from this cross peak are then traced to a new signal and new horizontal correlations are mapped out. If more than one vertical cross peak exists then this signifies that the carbon atom is bonded to more than one carbon atom. By continuing to trace out these correlations, the entire carbon-carbon skeleton of the molecule can be elucidated.

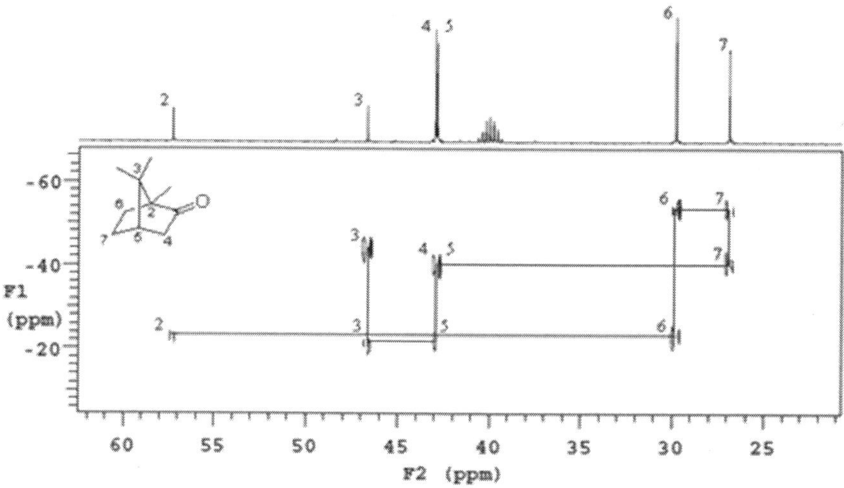

Figure 11. Partial INADEQUATE spectrum of camphor in DMSO-d_6 (signal at 40 ppm) illustrating the connectivites between carbon atoms. 250 mg of camphor, 8 hour acquisition time.

The spectral width of the ^{13}C NMR spectrum means that signals are typically well dispersed and even the structure of complicated compounds can be easily elucidated. The information generated in the 2D INADEQUATE experiment is comparable to that of X-ray crystallography. Of course the major disadvantage of this experiment is the low abundance of ^{13}C atoms and the very low probability that ^{13}C atoms are adjacent. This means that acquisition times will be very, very long. When performing a 2D INADEQUATE experiment, it is important to maximize the effectiveness of the experiment by measuring the T_1 values of all of the carbon atoms so that a suitable delay between pulses can be added. This ensures that all of the atoms have fully relaxed before the next pulse begins and that signal intensity is maximized. Much like the HSQC and HMQC experiments, a variation of this experiment, called the ADEQUATE experiment, exists that uses inverse detection to transfer energy from the ^{13}C atoms to the more sensitive ^{1}H atoms. However, since this experiment relies upon carbons with attached protons, carbons with no attached hydrogen atoms will not show correlations. Despite the powerful curricular opportunities provided by these experiments, the long acquisition times and other experimental difficulties means that they have yet to find applications in the undergraduate curriculum.

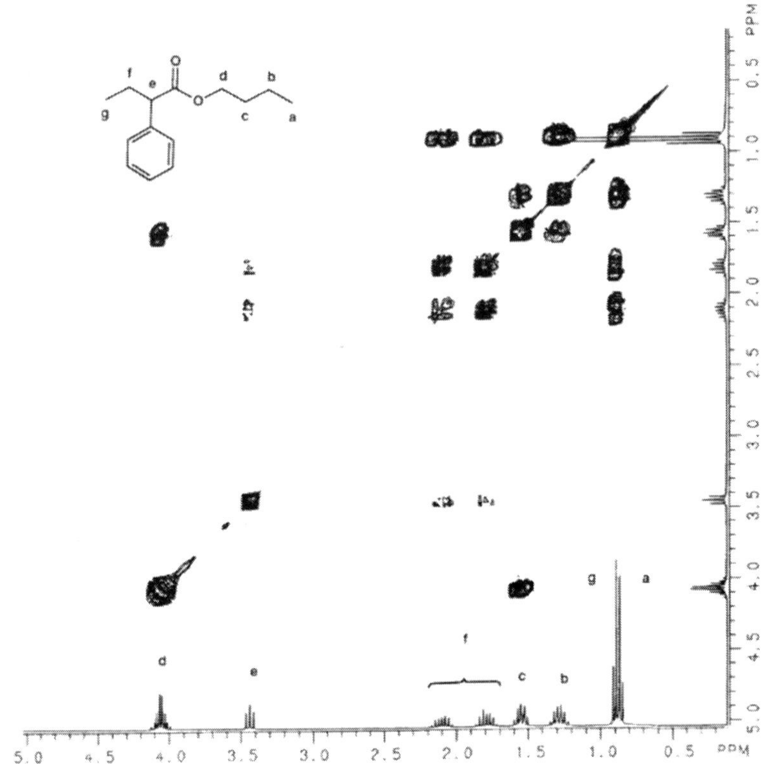

Figure 12. COSY spectrum of butyl 2-phenylbutyrate. Reproduced with permission from reference (38). Copyright 1995 American Chemical Society.

The HH-COSY Experiment

The Correlated Spectroscopy (COSY), or HH-COSY experiment, displays correlations between protons that are 2, 3, and rarely more bonds away. Using a traditional one-dimensional ^{1}H spectrum it can be difficult to determine molecular connectivity by observing coupled protons, particularly if signals overlap or splitting patterns are complex. The HH-COSY experiment measures proton resonances in each dimension with the diagonal corresponding to the ^{1}H spectrum. Cross peaks that occur off the diagonal correspond to protons that are coupled to each other, making the determination of molecular connectivity much more straightforward. Unfortunately, cross peaks arising from non-coupled protons, such as *t*-butyl or isolated methyl groups, are often large and can obscure off-diagonal cross peaks particularly when there is little spectral separation. If this is problematic the Double Quantum filtered COSY, or DQCOSY experiment, can be used since this experiment effectively suppresses all singlets and provides a spectrum that is easier to interpret. Similar to other two-dimensional experiments the HH-COSY experiment can require significant acquisition time. However, a gradient variant of the HH-COSY experiment exists which dramatically reduces the time needed and is preferable in nearly all cases that are relevant to the undergraduate curriculum.

Branz et al. described the utility of the HH-COSY experiment in a Fischer esterification experiment for the organic chemistry laboratory (*38*). In this experiment students are assigned an unknown $C_4H_{10}O$ alcohol and an unknown $C_{10}H_{12}O_2$ carboxylic acid, with the carboxylic acid partner containing a symmetrically substituted benzene ring. Using standard Fischer esterification conditions students synthesized the corresponding ester and analyzed it using IR, ^1H and HH-COSY NMR experiments. Figure 12 shows the COSY spectrum of butyl 2-phenylbutyrate.

By starting from known signals, such as $-OCH_2$ (d in Figure 11 at 4.05 ppm), correlations can be readily established by tracing horizontally or vertically from a diagonal cross peak to an off-diagonal cross peak. In this example signal d shows step-wise correlations to c, b and then to either a or g. The presence of this O-butyl spin system could be confirmed by using a 1D TOCSY experiment and irradiating the $-OCH_2$ group. Figure 11 also illustrates another introduction to the topic of diastereotopicity by showing that the methylene (f) has diastereotopic protons. Of course their presence could easily be confirmed via an HSQC spectrum, but in this case their presence can also be inferred from the integration of the ^1H NMR spectrum, and the fact that each proton couples to itself and with the methine (e) and methyl (g) groups.

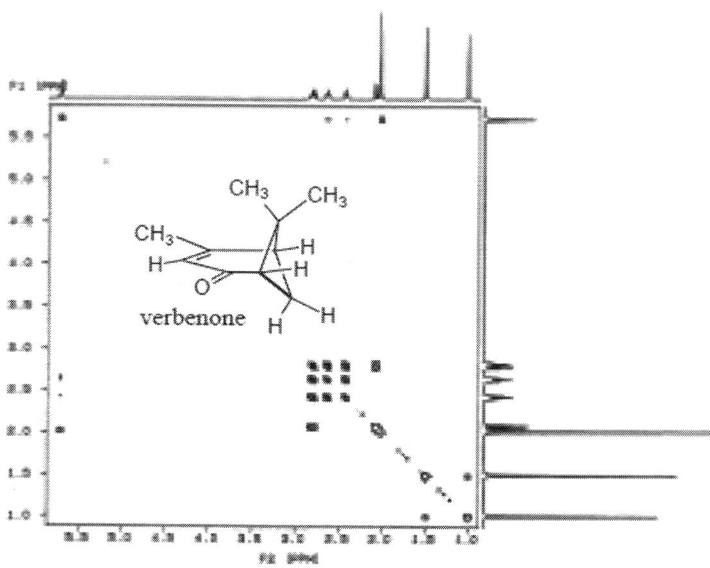

Figure 13. HH-COSY spectrum of verbenone. Adapted with permission from reference (39). Copyright 1996 American Chemical Society.

Mills provides a more involved HH-COSY experiment in the analysis of a series of bicyclic[3.1.1]terpenes in an instrumental analysis course (*39*). The rigid nature of the chosen bicyclic terpenes has significant ramifications on the observed coupling constants and so an understanding of the Karplus relationship is necessary. Students began by modeling the terpenes and measuring the dihedral angles between protons. Since the Karplus relationship predicts a strong dependence upon dihedral angle, students not only predict the magnitude of the coupling constant, but also identify those protons with dihedral angles close to 90° for which the Karplus relationship predicts that no coupling should be observed. Furthermore, the rigid nature of the system also allows for 4-bond, or W-coupling, to exist. Figure 13 shows the HH-COSY spectrum of verbenone, one of the bicyclic terpenes that were investigated.

The difficulties in analyzing this system are immediately apparent when looking at the vinyl proton signal at 5.8 ppm which shows correlations to three protons despite the apparent lack of neighboring protons. These cross peaks must therefore arise from longer range 4-bond coupling to the bridgehead and vinyl methyl group protons. Once these signals have been clearly identified, the remaining assignments are more straightforward. The analysis concludes by showing that the protons of the methylene group should show three bond coupling with the bridgehead protons. However, no cross peak exists because the dihedral angle is close to 90° and the Karplus relationship predicts that the coupling constant falls to zero in these instances.

The 2D NOESY Experiment

The two-dimensional NOESY experiment provides similar through-space proximity information to the one-dimensional NOESY experiment. However, unlike the one-dimensional form of this experiment, which requires selective irradiation of individual resonances, the two-dimensional NOESY experiment shows all of the NOEs in a single experiment. Huggins and Billimoria use the 2D NOESY experiment to analyze the alkene configuration of the products of the aldol reaction of various aldehydes with 3-ethyl-4-methyl-3-pyrrolin-2-one in a third-year instrumental analysis course (*40*). Students began with the aldol synthesis, then analyzed their products using ^{1}H, HSQC, and HMBC experiments. Correct assignment of the single vinyl proton is critical since it is required in order to determine the stereochemistry of the product using a 2D NOESY spectrum. Figure 14 shows a 2D NOESY spectrum of one of the aldol products.

Much like the HH-COSY experiment, 2D NOESY spectra have cross peaks that lie on the diagonal. These represent the one-dimensional spectrum. Cross peaks that reside off the diagonal correlate protons that are spatially proximal (less than 5Å). Since much of the structure has already been deduced from other experiments only a few key correlations are required in order to determine the stereochemistry of the product. In this case, the *Z* configuration of the product is confirmed by the existence of an NOE cross peak that exists between the vinyl proton (4) and the methyl group (3). Furthermore, the s-cis conformation of the product can be determined by an NOE cross peak between the vinyl proton (4) and the pyridine methine (5). Students concluded the experiment by showing that the *Z s*-cis conformation is the most stable via molecular modeling.

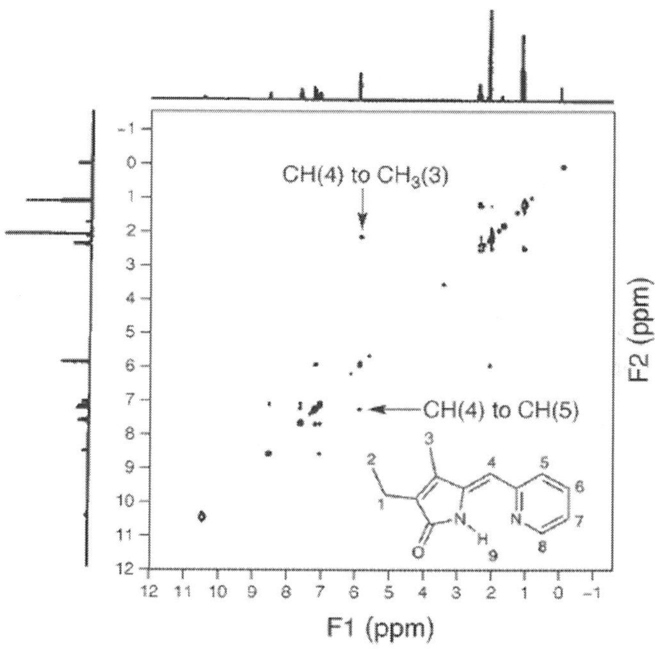

Figure 14. The 2D NOESY spectrum of 3-ethyl-4-methyl-5-pyridin-2-ylmethylene-1,5-dihydropyrrol-2-one. Adapted with permission from reference (40). Copyright 2007 American Chemical Society.

The 2D TOCSY Experiment

The two-dimensional TOCSY experiment correlates diagonal cross peaks with all coupled protons in that spin system. Unlike the one-dimensional TOCSY experiment, which relies upon irradiation of specific resonances, the 2D TOCSY samples all of the spin systems at once. This experiment is particularly useful for assigning resonances in complicated samples that have many isolated spin systems, such as biological molecules. Rehart and Gerig use the 2D TOCSY experiment in an introduction to NMR methods biochemistry laboratory. Students analyzed the biologically relevant octapeptide [Sar1]angiotensin II (Sar-Arg-Val-Tyr-Ile-His-Pro-Phe) which is believed to exist in a folded form in DMSO (*41*). Depending upon the approach, signal assignments can be made available so that students need only analyze the 2D NOESY spectrum to calculate the structure. When this experiment is used at a more advanced level, students are asked to use HH-COSY and 2D TOCSY to determine the assignments. Since the octapeptide's ^{1}H NMR spectrum contains many overlapping signals, the 2D TOCSY experiment plays a critical role in assigning these signals to the appropriate amino acid since each amino acid exists in an isolated spin system. Assignments can be made by focusing on the α-protons of each amino acid residue and looking for the appropriate cross peaks. Of course, running a series of 1D TOCSY experiments would provide similar information, but at the expense of longer acquisition times because each resonance must be irradiated separately. Students completed the experiment by analyzing their NMR data and used a molecular model kit and molecular modeling programs to determine the structure of the protein.

Conclusions

Modern NMR spectroscopy is a powerful tool that allows us to investigate a variety of chemical problems at the molecular level. The broad variety of NMR experiments that are available can provide information on molecular connectivity, three-dimensional structure, and valuable insights into the physical properties of molecules. Even in classes such as organic chemistry, where the focus has traditionally been on the use of one-dimensional ^{1}H and ^{13}C NMR experiments, the increasing incorporation of experiments such as the 1D TOCSY and NOESY, and 2D HSQC, HETCOR, HMQC, HH-COSY, HMBC, and TOCSY, demonstrates that these experiments can be appropriately and successfully utilized. Beyond organic chemistry, other undergraduate courses, such as physical chemistry, biochemistry, and instrumental analysis, are also benefiting from the integration of modern NMR experiments that allow students to tackle more relevant and challenging problems than ever before. Recent innovations such as the 2D ADEQUATE and H2BC experiments have yet to be adopted by the undergraduate community, but these experiments could make powerful contributions, particularly in the field of structural analysis. As aids to the reader, three tables are provided that summarize the experiments listed in this chapter along with brief descriptions of each experiment. Table 2 provides a list of the common NMR experiments, Table 3 lists less common one-dimensional ^{13}C experiments, and Table 4 lists less common NMR experiments that can be run in one or two dimensions. Finally, as technology becomes more available and affordable, NMR experiments continue to be developed and optimized, and software becomes more faculty- and student-friendly, we should expect our undergraduate curricula to continue to change and adapt, integrating, where appropriate, relevant chemical problems that better prepare our students for post-graduate studies in chemistry and related fields.

Table 2. Summary of the Most Common NMR Experiments

One-Dimensional Experiments	
$^{13}C\{^1H\}$	In a $^{13}C\{^1H\}$ spectrum, the number of signals represents the number of non-equivalent carbon atoms in a sample, and the signal location provides information about the electronic environment of the carbon atom. All indirect spin-spin couplings are suppressed via broadband decoupling. Quantitation (integration) of signals is unreliable because of long T_1 values, NOE signal enhancements, and potentially insufficiently powerful excitation pulses. Since ^{13}C is only 1% abundant and is less sensitive than ^1H, long acquisition times are needed.
^1H	In a ^1H spectrum, the number of signals provides information about the number of non-equivalent protons, the location of the signal provides information about the electronic environment of the proton, the splitting pattern of each signal provides information about the number and nature of adjacent magnetically active nuclei, and the integration of each signal provides information about the number of protons contributing to that signal. Since ^1H is very sensitive with near 100% abundance, acquisition times are short.
DEPT	A set of 1D ^{13}C experiments that give a series of spectra that display methine, methylene, and methyl carbons with different phases. In combination with a $^{13}C\{^1H\}$ spectrum, non-protonated carbon atoms can also be identified. The DEPTQ experiment provides the same data without the need for a separate $^{13}C\{^1H\}$ spectrum.
Two-Dimensional Experiments	
HSQC, HMQC, HETCOR	Provides spectra where the cross peaks correlate two directly bonded nuclei (usually ^1H and ^{13}C). Multiplicity-edited versions of these experiments exist that show cross peaks that correspond to C/CH_2, and CH/ CH_3 with different phases (providing similar information to a DEPT experiment). Also called CH-COSY.
COSY	Provides a spectrum where the off-diagonal cross peaks correlate proton atoms that are 2, 3, and sometimes more bonds away. Very sensitive, with relatively short acquisition times. Also called HH-COSY.

Table 3. Summary of Less Common One-Dimensional ^{13}C NMR Experiments

One-Dimensional Experiments	
Inverse-gated	Allows for the accurate quantitation (integration) of signals. Requires a delay between pulses of at least 5 times the longest T_1 value. A small amount of paramagnetic reagents such as $Cr(acac)_3$ or TEMPO can be added to greatly reduce the time needed between pulses but this can lead to line-broadening and signal shifts.
Gated	Displays indirect spin-spin coupling between magnetically active nuclei while still retaining some NOE signal enhancement. Provides information about the local environment of the atom at the expense of a more complicated spectrum and decreased signal intensity.
Off-resonance decoupling	Limits indirect spin-spin coupling to neighboring magnetically active nuclei resulting in a spectrum where the splitting pattern provides information about the local environment but at the expense of a more complex spectrum and decreased signal intensity.
APT	Provides a spectrum where C/CH_2 and CH/CH_3 signals have different phases. Similar to a DEPT experiment. Can be challenging to unambiguously assign CH and CH_3 signals. Also called the *J*-modulated spin-echo experiment.
Other Experiments	
Inversion Recovery (T_1)	Provides a way to measure the time required for the spin population to return to equilibrium. T_1 values are important to know when signal intensity must be maximized, such as in the inverse-gated ^{13}C or INADEQUATE experiments.

Table 4. Summary of Less Common NMR Experiments that can be run as One- or Two-Dimensional Experiments

One/Two-Dimensional Experiments	
INADEQUATE	The 1D experiment provides information about coupling constants of adjacent ^{13}C atoms. The 2D experiment provides information about the molecular connectivity of the entire carbon skeleton. In both cases very long acquisition times are needed since signals arise from ^{13}C-^{13}C coupling (0.01% abundant).
ADEQUATE	Similar to the INADEQUATE experiment but this 2D experiment relies on attached protons to increase the sensitivity and decrease the acquisition time of this ^{13}C experiment. Correlations between non-protonated carbon atoms are not observed.
NOE	A 1D NOE difference experiment provides information about the spatial proximity of magnetically active nuclei that are typically less than 5Å apart via irradiation of specific resonances. A 2D NOE experiment shows off-diagonal cross peaks for all nuclei that are less than 5Å apart.
TOCSY	The 1D TOCSY experiment provides information about all of the signals that belong to that resonance's spin system via irradiation of a specific resonance. The 2D TOCSY provides the same information but for all spin systems present in the molecule.
HMBC	Similar to the HETCOR, HMQC, and HSQC experiments except that cross peaks show correlations between two nuclei (usually ^{1}H and ^{13}C) over 2 or 3 bonds. Since it is difficult to distinguish between the different couplings these spectra can be challenging to analyze.
H2BC	Similar to the HMBC experiment except it is optimized to correlate carbon atoms to the proton of protonated carbon atoms that are two bonds away. When combined with the HSQC experiment, it can provide similar information to the INADEQUATE experiment but in a fraction of the acquisition time.

References

1. Bloch, F.; Hansen, W. W.; Packard, M. *Phys. Rev.* **1946**, *70*, 474–485.
2. Purcell, E. M.; Torrey, H. C.; Pound, R. V. *Phys. Rev.* **1946**, *69*, 37–38.
3. *NMR and MRI: Applications in Chemistry and Medicine*; American Chemical Society URL http://portal.acs.org/portal/PublicWebSite/education/whatischemistry/landmarks/about/CNBP_026831 (accessed June 2012).
4. Keeler, J. *Understanding NMR Spectroscopy*, 2nd ed.; John Wiley & Sons Ltd: Chichester, 2010.
5. Friebolin, H. *Basic One- and Two Dimensional NMR Spectroscopy*, 5th ed.; Wiley-VCH Verlag GmbH & Co. KGaA: Weinheim, 2011.
6. Sanders, J. K. M.; Hunter, B. K. *Modern NMR Spectroscopy: A Guide for Chemists*, 2nd ed.; Oxford University Press: New York, 1993.
7. Lambert, J. B.; Mazzola, E. P. *Nuclear Magnetic Resonance Spectroscopy: An Introduction to Principles, Applications, and Experimental Methods*, 1st ed.; Pearson Education, Inc.: Upper Saddle River, NJ, 2004.
8. Jacobsen, N. E. *NMR Spectroscopy Explained: Simplified Theory, Applications and Examples for Organic Chemistry and Structural Biology*, 2nd ed.; John Wiley & Sons, Inc.: Hoboken, NJ, 2007.
9. Macomber, R. S. *A Complete Introduction to Modern NMR Spectroscopy*, 1st ed.; John Wiley & Sons, Inc.: New York, 1998.
10. Simpson, J. H. *Organic Structure Determination Using 2-D NMR Spectroscopy, Second Edition: A Problem-Based Approach*, 2nd ed.; Elsevier: San Diego, CA, 2012.
11. *Modern NMR Spectroscopy in Education*; Rovnyak, D., Stockland Jr., R., Eds.; ACS Symposium Series 969; American Chemical Society: Washington, DC, 2007.
12. Pavia, D. L.; Lampman, G. M.; Kriz, G. S. *Introduction to Spectroscopy*, 2nd ed.; Saunders: Fort Worth, TX, 1996; pp 146–163
13. Draanen, N. A. V.; Page, R. *J. Chem. Educ.* **2009**, *86*, 849.
14. Mega, T. L.; Carlson, C. B.; Cleary, D. A. *J. Chem. Educ.* **1997**, *74*, 1474–1476.
15. Giles, B. J.; Matsche, Z.; Egeland, R. D.; Reed, R. A.; Morioka, S. S.; Taber, R. L. *J. Chem. Educ.* **1999**, *76*, 1564–1566.
16. Gasyna, Z. L.; Jurkiewicz, A. *J. Chem. Educ.* **2004**, *81*, 1038–1039.
17. LeFevre, J. W.; Silveira, A., Jr. *J. Chem. Educ.* **2000**, *77*, 83–85.
18. Schmedake, T. A.; Welch, L. E. *J. Chem. Educ.* **1996**, *73*, 1045–1048.
19. Delagrange, S.; Nepveu, F. *J. Chem. Educ.* **2000**, *77*, 895–897.
20. Hersh, W. H. *J. Chem. Educ.* **1997**, *74*, 1485–1488.
21. Cooke, J.; Henderson, E. J.; Lightbody, O. C. *J. Chem. Educ.* **2009**, *86*, 610–612.
22. Reeves, P. C.; Chaney, C. P. *J. Chem. Educ.* **1998**, *75*, 1006–1007.
23. Ivey, M. M.; Smith, E. T. *Chem. Educator* **2008**, *13*, 307–308.
24. Walsh, E. L.; Ashe, S.; Walsh, J. J. *J. Chem. Educ.* **2012**, *89*, 134–137.
25. Karplus, M. *J. Chem. Phys.* **1959**, *30*, 11.

26. Adesoye, O. G.; Mills, I. N.; Temelkoff, D. P.; Jackson, J. A.; Norris, P. *J. Chem. Educ.* **2012**, *89*, 943–945.
27. Sorenson, J. L.; Witherell, R.; Browne, L. M. *J. Chem. Educ.* **2006**, *83*, 785–787.
28. LeFevre, J. W. *J. Chem. Educ.* **2000**, *77*, 361–363.
29. McDaniel, K. F.; Weekly, R. M. *J. Chem. Educ.* **1997**, *74*, 1465–1467.
30. Sereda, G. A. *J. Chem. Educ.* **2006**, *83*, 931–933.
31. deMendonca, D. J.; Digits, C. A.; Navin, E. W.; Sanders, T. C.; Hammond, G. B. *J. Chem. Educ.* **1995**, *72*, 736–739.
32. Jensen, A. W.; O'Brien, B. A. J. *Chem. Educ.* **2001**, *78*, 954–955.
33. Fenton, O. S.; Sculimbrene, B. R. *J. Chem. Educ.* **2011**, *88*, 662–664.
34. Zanger, M.; Moyna, G. *J. Chem. Educ.* **2005**, *82*, 1390–1392.
35. Shaw, R.; Severin, A.; Balfour, M.; Nettles, C. *J. Chem. Educ.* **2005**, *82*, 625–629.
36. Clausen, T. P. *J. Chem. Educ.* **2011**, *88*, 1007–1009.
37. Caes, B.; Jensen, D., Jr. *J. Chem. Educ.* **2008**, *85*, 413–415.
38. Branz, S. E.; Miele, R. G.; Okuda, R. K.; Straus, D. A. *J. Chem. Educ.* **1995**, *72*, 659–661.
39. Mills, N. S. *J. Chem. Educ.* **1996**, *73*, 1190–1192.
40. Huggins, M. T.; Billimoria, F. *J. Chem. Educ.* **2007**, *84*, 471–474.
41. Rehart, A. M.; Gerig, J. T. *J. Chem. Educ.* **2000**, *77*, 892–894.

Organic Chemistry

Chapter 3

Data versus Dogma: Introducing NMR Early in Organic Chemistry To Reinforce Key Concepts

Paul A. Bonvallet* **and Judith C. Amburgey-Peters**

Department of Chemistry, The College of Wooster, 943 College Mall, Wooster, Ohio 44691
***E-mail: pbonvallet@wooster.edu**

NMR spectroscopy occupies a key position in the undergraduate curriculum because of the unparalleled structural information that it provides. Introducing this content early in an organic chemistry course helps to reinforce the key role that molecular structure plays in dictating physical and chemical properties. Furthermore, there are many ways to incorporate NMR techniques in support of foundational concepts of organic chemistry such as resonance, conformational analysis, stereochemistry, acid-base reactions, electrophilicity, and aromaticity. The use of data helps to portray these concepts as constructed scientific knowledge rather than simply a group of facts to be memorized.

Introduction

The scientific value of NMR spectroscopy is difficult to overstate due to the wealth of information that it conveys. As such, this technique has been a topic of central importance in the development and revision of undergraduate curricula, particularly in organic chemistry and integrated introductory courses (*1–4*). However, despite its significance, NMR spectroscopy does not appear until the middle portion of most organic chemistry textbooks (*1*). Furthermore, the historical approach of these texts has been to represent NMR spectroscopy solely as a technique for the structural elucidation of unknown samples and the validation of identity and purity in the laboratory. While these applications do indeed make NMR indispensable, we believe that presenting this key technique earlier and in a broader context helps students to better learn the other fundamentals of organic chemistry.

Chemical educators have highlighted the importance of teaching how scientists think rather than what they know (*5*). We emphasize to our students that science is a process of constructing knowledge through the collection and interpretation of laboratory data. It therefore feels inconsistent to present the foundational concepts of organic chemistry, such as resonance and stereochemistry, from a dogmatic memorize-the-facts perspective. In an environment of critical inquiry, we need to encourage both our students and ourselves to pose incisive questions like "How do we know this is true?" and "What is the evidence?" If the instructor has to resort to saying, "That's the way it is, and you just have to know it," then there is probably a better way of teaching it.

Thus, we have deliberately moved NMR spectroscopy and other forms of spectroscopy at the College of Wooster into the first month of our first-semester organic chemistry curriculum as a way to reinforce the structural basis from which we teach the material. We progress from describing structure (Lewis structures, conformational analysis, stereochemistry) to ways of knowing structure (IR and NMR spectroscopy, mass spectrometry), and ultimately to function in the context of transformation of structure (starting with nucleophilic substitution and elimination reactions). Placing spectroscopy early in the curriculum also highlights the central role of structure in dictating the physical and chemical properties of organic compounds. Most importantly, when students have a functional understanding of NMR spectroscopy, we can use this key method as an evidentiary tool to support the conceptual content of the course. We illustrate below some of the ways in which NMR can corroborate the foundations of organic chemistry, marking a triumph of data over dogma.

Curricular Examples

Evidence of Resonance

The concept of resonance in Lewis theory is profoundly important in understanding organic structure and reactivity. However, while it is simple enough to declare that alternative Lewis structures *can* be drawn for some compounds, the description of *what* a resonance hybrid is and *how* we know that they exist is oftentimes more challenging. For example, inspecting the condensed structure **1a** and major Lewis descriptor **1b** of N,N-dimethylformamide (DMF) could lead to the prediction that the two methyl groups are equivalent by rotation about the central carbon-nitrogen bond (Figure 1). The room-temperature ^{1}H and ^{13}C NMR spectra of DMF in Figure 1, however, show distinctly separate resonances for each methyl group.

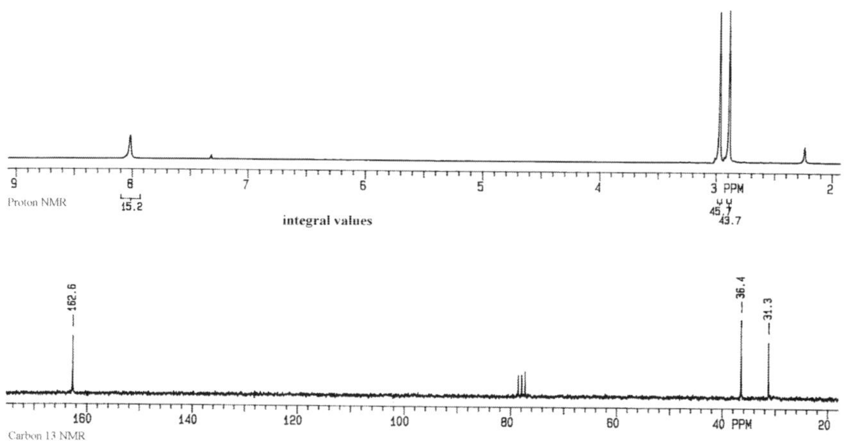

Figure 1. Structural depictions with ^{1}H (top) and ^{13}C NMR spectra (bottom) of DMF. Adapted with permission from reference (9). Copyright 1996 Sunbelt R & T Inc.

This experimental observation is explained through resonance contributor **1c**, which imparts some double bond character to the C—N bond and thus increases the rotational barrier beyond that of most ordinary C—N single bonds (Scheme 1). Structure **1c** is a resonance structure of DMF that contributes sufficiently to the hybrid structure **1d** that the interchange of methyl groups is slow enough to make them differentiable on the NMR timescale (*6*). Attributing this phenomenon to resonance is further reinforced by the observation that *N*-arylamines have a much higher C—N rotational barrier compared to saturated amines (*7, 8*).

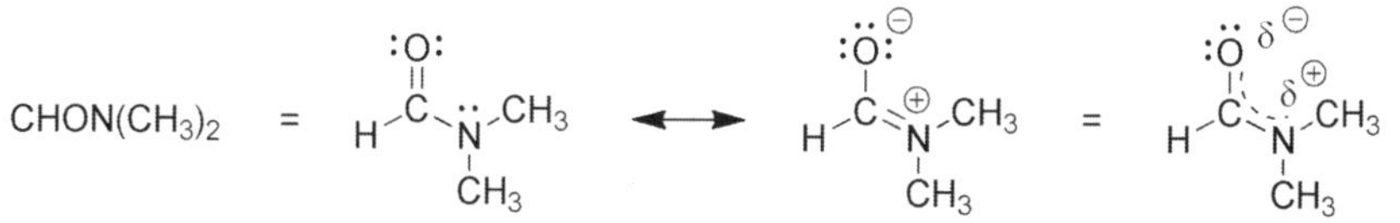

$CHON(CH_3)_2$ = [structure **1b**] ⟷ [structure **1c**] = [structure **1d**]

1a **1b** **1c** **1d**

Scheme 1. N,N-dimethylformamide (DMF)

This example illustrates the utility of NMR spectroscopy in reinforcing the foundational concept that the structure of an organic compound is a weighted blend of all possible resonance structures. A discussion of the NMR spectrum of DMF could be expanded to include the planar geometry of the nitrogen atom and the temperature dependence of the methyl group signals as they coalesce at elevated temperature (*10*). Resonance effects in DMF also appear in infrared spectroscopy, and the lower carbonyl stretching frequency (1672 cm^{-1}) is consistent with greater single bond character of the C—O bond (*7*). DMF is an attractive example because of its simple structure and relation to the amide bonds found in proteins and polypeptides.

Evidence for Conformational Preferences

Organic chemistry students learn that monosubstituted cyclohexanes preferentially adopt a chair conformation with the substituent in the equatorial position. Whereas some students are tempted to anthropomorphize the molecule and simply remember that the substituent "wants" to be equatorial, a more sophisticated learner would illustrate with molecular models or structural drawings that an equatorial substituent minimizes the number of energetically unfavorable gauche and 1,3-diaxial interactions in the molecule. Still, one might wonder how chemists know that conformational equilibrium even exists and how they are able to differentiate one conformer from the other.

One excellent example from the literature shows that the two conformers of iodocyclohexane are simultaneously present and "frozen out" on the NMR spectroscopy timescale at −80 °C (*11*). The equatorial α-proton in conformer I (Figure 2) should appear as a pentet (albeit unresolved at this field strength). By contrast, the axial α-proton in conformer II appears at an upfield chemical shift characteristic of axial protons (*12*) as a distinct triplet of triplets because of the greater magnitude of the axial-axial couplings relative to the axial-equatorial coupling constants (*13*).

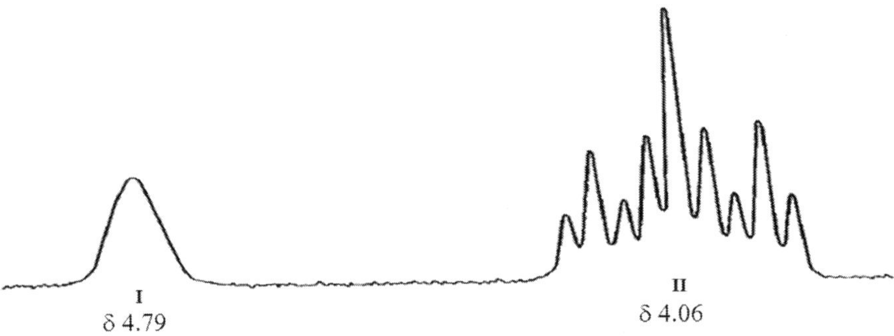

Figure 2. Newman projections (top) and partial 100 MHz ^1H NMR spectrum (bottom, showing the H_α signal) of the two conformers of iodocyclohexane). Adapted with permission from reference (11). Copyright 1969 American Chemical Society.

Using chair conformations and Newman projections, the differences in splitting patterns are easily rationalized on the basis of the different number and types of equivalent neighboring hydrogen atoms in each conformer (Scheme 2). The broadened pentet results from two sets of nearly equal H_α-H_a and H_α-H_e couplings (which leads to pseudo-first order coupling), whereas the triplet of triplets results from the larger H_α-H_a coupling relative to the H_α-H_e coupling.

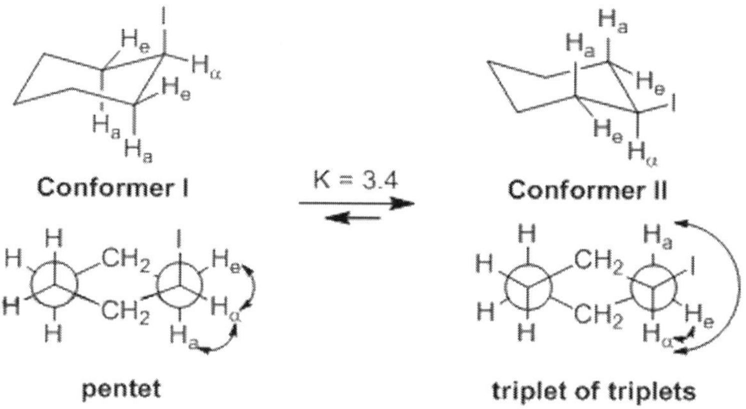

Scheme 2. Condensed, Lewis structure, and resonance contributor of N,N-dimethylformamide (DMF)

This example illustrates how splitting pattern analysis can be used to unambiguously assign the structure of two different conformers. Furthermore, the integration of each H_α signal at a known temperature enables one to determine the equilibrium constant (and thus the free energy difference) between the conformations, giving rise to a collection of commonly reported "A values" that describe the energetic penalty for a particular substituent adopting the axial conformation (*14*). This conceptual integration of spectroscopy, simple vs. complex splitting, and thermodynamics reinforces the key chemical ideas in conformational analysis.

Inequivalence of Diastereotopic Protons

Predicting or explaining the ^{1}H NMR spectrum of chiral organic molecules can be challenging because of the potential for stereochemically inequivalent protons. Two hydrogen atoms bonded to the same carbon atom can be anisochronous (have different chemical shifts) when they are close to an asymmetric center (*15*). Consequently, each chemically unique diastereotopic proton undergoes coupling with both its geminal and vicinal neighbors, making the coupling pattern more complicated than might have been initially expected.

To a novice learner, the OCH_2 protons in 2-methylbutyl acetate might be predicted to appear as a 2H doublet, whereas in the spectrum (Figure 3) they are actually two separate 1H signals, each appearing as a doublet of doublets (*16*).

A Newman projection of the *R*-enantiomer reveals that these two protons are in non-equivalent chemical environments, no matter how one rotates about the carbon-carbon bond, and thus logically should have different chemical shifts and undergo splitting independent of one another.

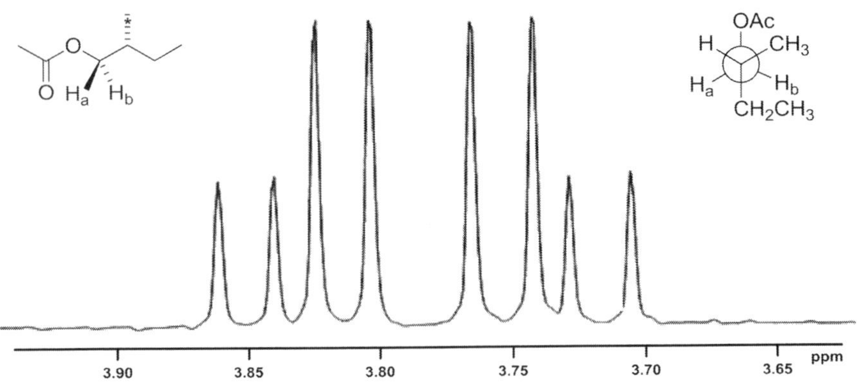

Figure 3. The structure of 2-methylbutyl acetate and a partial ¹H NMR spectrum showing the diastereotopic protons H_a and H_b. Adapted with permission from reference (16). Copyright 2011 American Chemical Society.

A variety of intellectual tools exists for the identification of diastereotopic protons, ranging from Newman and Fischer projections to the imaginary "Z replacement test" (*17*, *18*). As a supplement to these methods, it can be helpful to return to original experimental data demonstrating that diastereotopic protons actually are different from one another. Such evidence parallels our teaching that diastereomers are different compounds with different physical (and sometimes chemical) properties.

Acidic and Basic Functional Groups Undergo Chemical Exchange

The acid-base reactivity of organic functional groups is related to many of their physical and chemical properties. Yet how much class time do instructors dedicate to experimental evidence that proton exchange reactions really do take place? The characteristic broadening and loss of coupling for O—H and N—H signals in ^{1}H NMR spectroscopy is one piece of evidence for chemical exchange. A more dramatic example is the "D_2O shake" test, in which the spectrum of a sample is obtained before and after the addition of a few drops of heavy water (D_2O) (*19*). Exchangeable protons, such as those in alcohols, carboxylic acids, amines, and amides, have their signals diminish in intensity or disappear altogether as they become replaced with deuterium atoms. For example, among the many peaks in the ^{1}H NMR spectrum of menthol, the signal from the hydroxyl group is easily identified because it disappears when the sample is treated with D_2O (Scheme 3) (*20*).

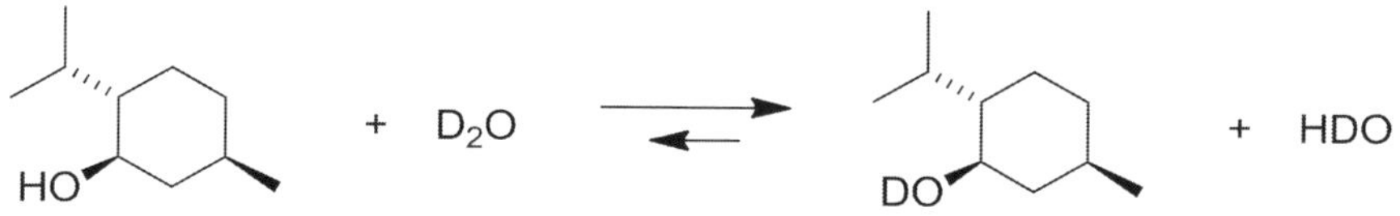

Scheme 3. Menthol undergoing a proton exchange reaction with D_2O

Such an example permits students to "see" the proton exchange of alcohols and illustrate a reaction that they have likely seen before but may not have examined the underlying experimental evidence for. More sophisticated experiments can be linked to mechanistic and kinetic studies, such as the exchange at the C-2 position of thiamine hydrochloride (Scheme 4).

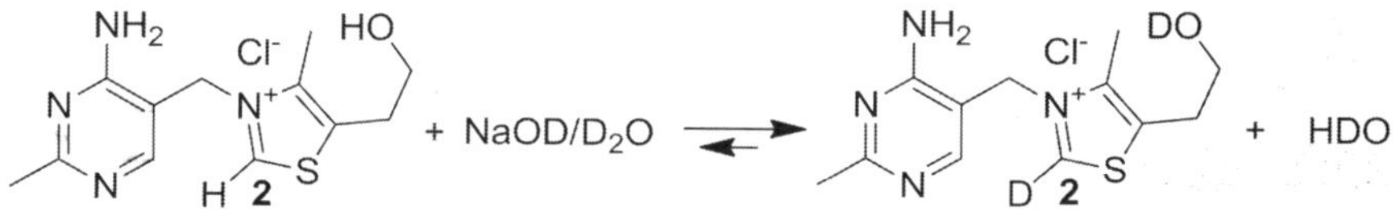

Scheme 4. Proton exchange at the C-2 position of thiamine hydrochloride

This example demonstrates the likely surprising acidity of thiazolium protons. Furthermore, it shows how NMR spectroscopy can be used to track a reaction as a function of time to not only reveal, but also measure, the rate of proton exchange (Figure 4) (*21*).

Electrophilicity Correlates with Deshielding

Identifying nucleophiles and electrophiles is a key skill that guides students' learning of organic chemical reactions. Although this talent may seem somewhat removed from the topic of NMR spectroscopy, the distribution of electron density that dictates differences in chemical environment also governs chemical reactivity within a molecule. Thus, those who understand the theory behind chemical shift differences are better equipped to explain and predict patterns in nucleophilic reactions.

One simplified approach is to correlate downfield chemical shifts with increasing electrophilicity. As shown in Figure 5, the partial positive (deshielded) character of the carbon atom in chloromethane can be correlated with the observation that alkyl halides are good S_N2 substrates whereas alkanes, which have much more highly shielded carbon atoms, are not.

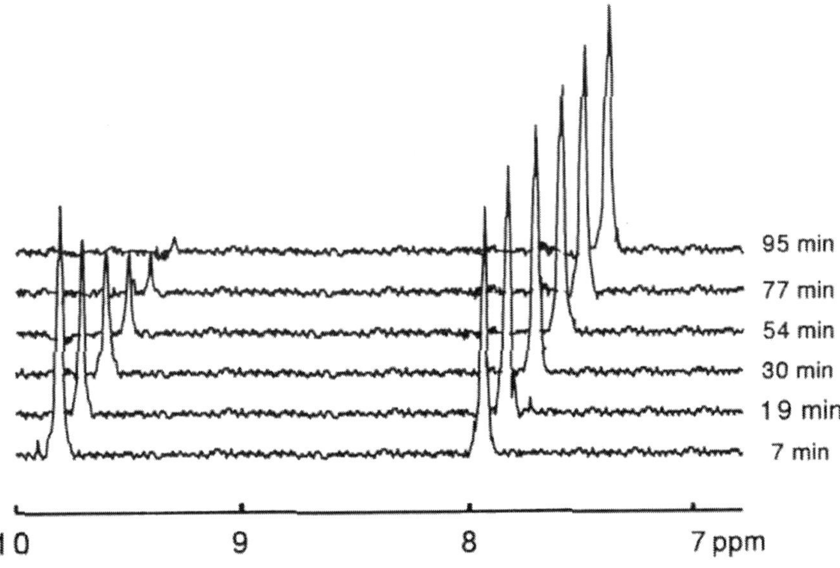

Figure 4. Sequential ^{1}H NMR spectra of thiamine, showing the disappearance of the C-2 proton signal (δ~10 ppm) as protium is exchanged with deuterium over time. Adapted with permission from reference (21). Copyright 1991 American Chemical Society.

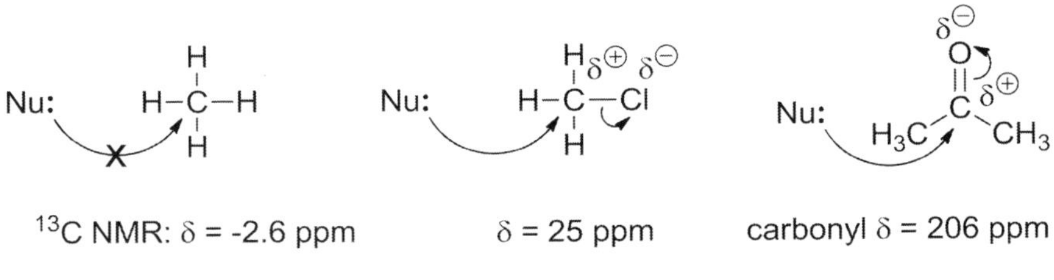

Nu: $H{-}CH_2{-}H$ Nu: $H{-}CH_2{-}Cl$ ($\delta^{\oplus}$, $\delta^{\ominus}$)   Nu: $H_3C{-}C(=O){-}CH_3$ ($\delta^{\oplus}$, $\delta^{\ominus}$)

^{13}C NMR: δ = -2.6 ppm   δ = 25 ppm   carbonyl δ = 206 ppm

Figure 5. The electrophilicity of a carbon atom correlates with its deshielding. Data are from reference (22).

Leaving group ability, steric bulk of the substrate, and other considerations are certainly important in enhancing the sophistication of the discussion, but fundamentally the deshielding effect of halogens can be applied equally well to the downfield shift of halides and to the electrophilicity of alkyl halides. A similar argument can be made for the electrophilicity of the carbonyl group, which is a key pattern in organic reactivity. The discussion can also be extended to explain inductive trends in the relative acidity of variously-substituted alcohols, carboxylic acids, and protons involved in β-elimination reactions, in which students can look for structural evidence consistent, or even potentially inconsistent, with the reported trends.

Aromaticity Is More than Resonance Alone

Although detailed explanations of aromaticity usually appear later in organic chemistry, NMR spectroscopy can illustrate the unique physical and chemical properties of aromatic systems. A compound like [18]annulene provides an interesting example to supplement the prototypical benzene molecule because of the dramatically different chemical shifts of protons inside and outside the ring (Figure 6) (*23*). Resonance cannot sufficiently explain this striking dissimilarity between what might initially appear to be very similar hydrogen atoms.

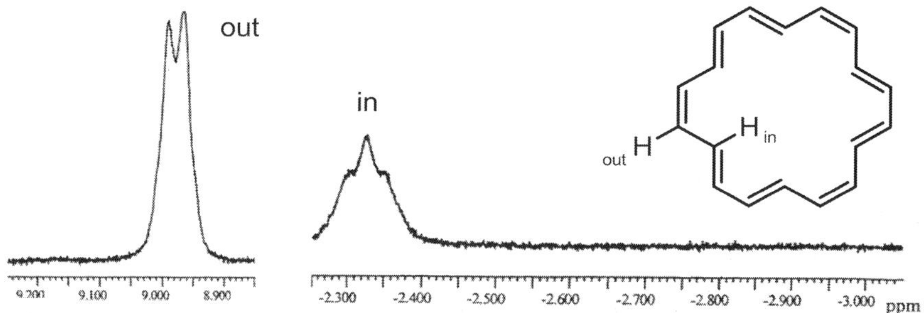

Figure 6. ¹H NMR spectrum of [18]annulene. Adapted with permission from reference (23). Copyright 2000 American Chemical Society.

At an early point in the curriculum it may be sufficient to point out that there is something "unusual" about aromatic compounds like benzene and [18]annulene because their ^{1}H NMR chemical shifts fall so far outside the normal range of other hydrocarbons. The conversation can be enhanced with a discussion of induced ring currents to explain the shielding and deshielding of atoms in aromatic systems (*24*), and ultimately culminate with information about Hückel's rule and the unique reactivity of aromatic compounds. As with the other examples, NMR spectroscopy can be used as a resource to drive a data-centered approach to learning about organic chemistry.

Conclusions

NMR spectroscopy is a powerful tool for structural elucidation in the classroom and laboratory, but it does not have to be limited to that role nor does it have to stand alone as a separate topic in the organic chemistry curriculum. Introducing this method early in the first semester enables the instructor to use concrete data to illustrate, integrate, and reinforce key concepts that build a framework for understanding chemical reactivity. Although many methods exist to use NMR spectroscopy in the teaching of organic chemistry, the above examples illustrate how scientific knowledge is built upon the collection and interpretation of laboratory data.

Acknowledgments

We are grateful to the National Science Foundation (grant CHE-9977546) for its support of a high-field NMR spectrometer at The College of Wooster.

References

1. Livengood, K.; Lewallen, D. W.; Leatherman, J.; Maxwell, J. L. *J. Chem. Educ.* **2012**, *89*, 1001–1006.
2. *Modern NMR Spectroscopy in Education*; Rovnyak, D., Stockland, R., Jr., Eds.; ACS Symposium Series 969; American Chemical Society: Washington, DC, 2007
3. Ball, D. B.; Miller, R. *J. Chem. Educ.* **2002**, *79*, 665–666.
4. Davis, D. S.; Moore, D. E. *J. Chem. Educ.* **1999**, *76*, 1617–1618.
5. Talanquer, V.; Pollard, J. *Chem. Educ. Res. Pract.* **2010**, *11*, 74–83.
6. Pavia, D.; Lampman, G. M.; Kriz, G. S.; Vyvyan, J. R. *Introduction to Spectroscopy*, 4th ed.; Brook/Cole Cengage Learning: Belmont, CA, 2009; pp 345–347.
7. Stewart, W. E.; Siddall, T. H. *Chem. Rev.* **1970**, *70*, 517–551.
8. Bushweller, C. H.; O'Neil, J. W.; Bilofsky, H. S. *Tetrahedron* **1971**, *27*, 5761–5766.
9. Thomasi, R. A. *A Spectrum of Spectra*, 2nd ed. [CD-ROM]; Sunbelt R&T: Broken Arrow, OK, 1996.
10. Rabinovitz, M.; Pines, A. *J. Am. Chem. Soc.* **1969**, *91*, 1585–1589.
11. Jensen, F. R.; Bushweller, C. H.; Beck, B. H. *J. Am. Chem. Soc.* **1969**, *91*, 344–351.
12. Silverstein, R. M.; Bassler, G. C.; Morrill, T. C. *Spectrometric Identification of Organic Compouds*, 5th ed.; John Wiley and Sons: New York, 1991; p 193.
13. Silverstein, R. M.; Bassler, G. C.; Morrill, T. C. *Spectrometric Identification of Organic Compouds*, 5th ed.; John Wiley and Sons: New York, 1991; p 243.
14. See, for example: Ansyln, E. V.; Dougherty, D. A. *Modern Physical Organic Chemistry*; University Science Books: Sausalito, CA, 2006; p 104.
15. Pavia, D.; Lampman, G. M.; Kriz, G. S.; Vyvyan, J. R. *Introduction to Spectroscopy*, 4th ed.; Brook/Cole Cengage Learning: Belmont, CA, 2009; pp 252–256.
16. Clausen, T. P. *J. Chem. Educ.* **2011**, *88*, 1007–1009.
17. See, for example: Wade, L. G.; *Organic Chemistry*, 8th ed.; Prentice Hall: Upper Saddle River, NJ, 2013; pp 591–593
18. Pavia, D.; Lampman, G. M.; Kriz, G. S.; Vyvyan, J. R. *Introduction to Spectroscopy*, 4th ed.; Brook/Cole Cengage Learning: Belmont, CA, 2009; p 252.
19. Pavia, D.; Lampman, G. M.; Kriz, G. S.; Vyvyan, J. R. *Introduction to Spectroscopy*, 4th ed.; Brook/Cole Cengage Learning: Belmont, CA, 2009; p 334.

20. Facey, G. *University of Ottawa NMR Facility Blog*; http://u-of-o-nmr-facility.blogspot.com/2007/10/proton-nmr-assignment-tools-d2o-shake.html (accessed March 2011).
21. Murray, C. J.; Duffin, K. L. *J. Chem. Educ.* **1991**, *68*, 683–684.
22. Reich, H. *Methyl Shifts*; http://www.chem.wisc.edu/areas/reich/handouts/nmr-c13/cdata.htm (accessed August 2012).
23. Stevenson, C. D.; Kurth, T. L. *J. Am. Chem. Soc.* **2000**, *122*, 722–72.
24. Pavia, D.; Lampman, G. M.; Kriz, G. S.; Vyvyan, J. R. *Introduction to Spectroscopy*, 4th ed.; Brook/Cole Cengage Learning: Belmont, CA, 2009; pp 128–129.

Chapter 4

Using NMR Spectroscopy To Promote Active Learning in Undergraduate Organic Laboratory Courses

John A. Cramer*

Department of Chemistry, Seton Hill University, 1 Seton Hill Drive, Greensburg, Pennsylvania 15601-1599
***E-mail: jcramer@setonhill.edu**

This chapter describes experiments where NMR spectroscopy has been employed in undergraduate organic laboratory courses at Seton Hill University to promote student active learning. A wide range of NMR-enabled concepts can be successfully taught as the result of the robust capability of NMR spectroscopy to connect students with the molecular world. In these experiments the need to answer significant scientific questions drives the work expected of students.

Introduction

Pedagogic research has demonstrated that chemistry students learn optimally when they are actively engaged in creating and interpreting knowledge as opposed to passively receiving information (*1*). This is especially true in laboratory courses. Active learning using NMR spectroscopy has been successfully employed in undergraduate organic laboratory courses at Seton Hill University, where students answer significant, realistic scientific questions as they engage a wide range of NMR-enabled concepts at multiple levels. One of the best strategies to engage students in the process and excitement of scientific discovery is to design laboratory experiments which are driven by the need to answer questions that students recognize as important.

NMR spectroscopy is an ideal engine for driving active learning. The vast array of fruitful applications of NMR spectroscopy that have been developed in a variety of scientific disciplines is mirrored by the diverse ways in which

NMR can be used to enhance the education of chemistry students (*2*). While it is crucially important to train students in the theory of NMR for its own sake, it is also important to recognize the pedagogic role that NMR spectroscopy can have in serving as a powerful hook to actively engage and connect students with the molecular world.

This chapter will describe how NMR spectroscopy has been used at Seton Hill University to promote active learning throughout a two-semester science major's organic laboratory sequence. The experiments, employing a 60 MHz Anasazi Eft permanent magnet instrument, are performed by students in laboratory sections of approximately sixteen students. In the first semester course NMR FID data are typically obtained by students working in small groups. Later in the second semester students collect NMR data individually.

Atom Equivalency and ^{13}C NMR Spectroscopy

Comprehending structural relationships in molecules is crucial to learning organic chemistry. An effective pedagogic response to this need is the early introduction of ^{13}C NMR spectroscopy to teach atom equivalency, and to thereby facilitate student understanding of molecular structure. In an example of such an exercise, students are asked early in the term to write structural formulas for the seventeen possible alkene isomers of molecular formula C_6H_{12}. After a brief introduction to structural equivalence in the context of ^{13}C NMR spectroscopy, including the practical aspects of ^1H decoupled ^{13}C NMR spectroscopy, students are presented with a spectrum of 2-ethyl-1-butene which consists of four singlet signals (*3*). Students are asked to determine which of their isomers could produce such a spectrum. Only the two C_6H_{12} alkene isomers in Figure 1 have four nonequivalent sets of carbon atoms.

Figure 1. C_6H_{12} alkene isomers with four nonequivalent sets of carbon atoms.

The task of distinguishing between these two isomers provides an impetus for students to have a deeper understanding of ^{13}C NMR spectroscopy, involving topics such as the dependence of chemical shift on chemical environment and the relatively shorter height of signals from carbon atoms with no attached hydrogen atoms. Alternatively, the need to distinguish between the two isomers can be an incentive for students to learn the principles of ^1H NMR spectroscopy. Through these exercises students gain an early appreciation for the power and relevance of NMR spectroscopy to the important task of determining molecular structure.

Use of Unknowns To Engage Student Learning

Incorporation of unknowns into synthetic experiments is an effective active learning strategy which serves to increase both interest and level of engagement. In one experiment, students are informed of the molecular identity of only one reactant in a synthesis that involves two reactants. Upon completion of the synthesis, students determine the molecular structure of their products by ^{1}H and ^{13}C NMR spectra. By inference they are then able to identify the unknown reactant.

A wide range of syntheses can be implemented which employ this strategy. One convenient example is the synthesis of aromatic tertiary alcohols by the reaction of phenylmagnesium bromide with unknown aliphatic ketones (Figure 2).

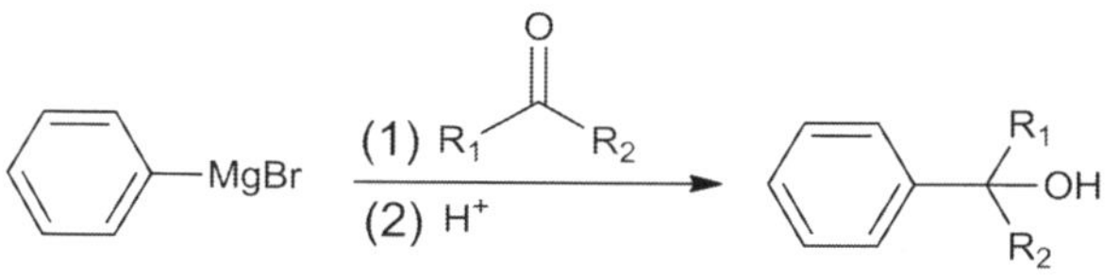

Figure 2. Synthesis of aromatic tertiary alcohols from the reaction of phenylmagnesium bromide with unknown ketones.

Structural analysis of the tertiary alcohol products by NMR spectroscopy permits students to identify the products formed, as well as the structures of the ketone reactants. In determining the identity of products from ^{1}H NMR spectra students set the integration of the aromatic region to five hydrogen atoms, which permits the determination of the number of remaining hydrogen atoms in the product molecule. Students then assign the structure of products and ketone reactants by working in small groups. The ketones employed in this experiment (Figure 3) provide a range of difficulty level, with acetone being a significantly less challenging ketone than the five-carbon ketones.

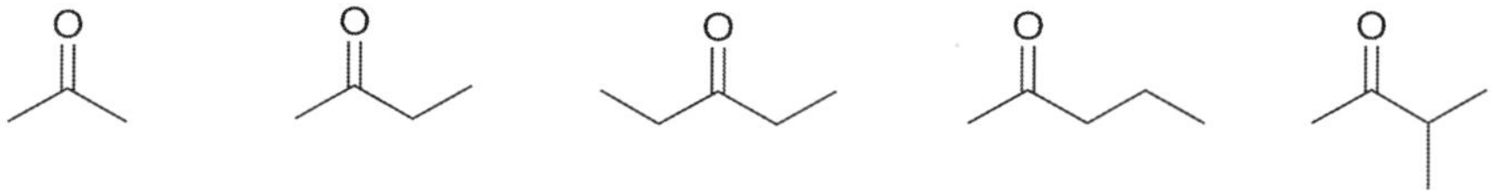

Figure 3. Unknown ketones used in tertiary alcohol syntheses.

Figure 4 shows the ^{1}H NMR spectrum of the alcohol product formed from the reaction of phenylmagnesium bromide with 3-methyl-2-butanone.

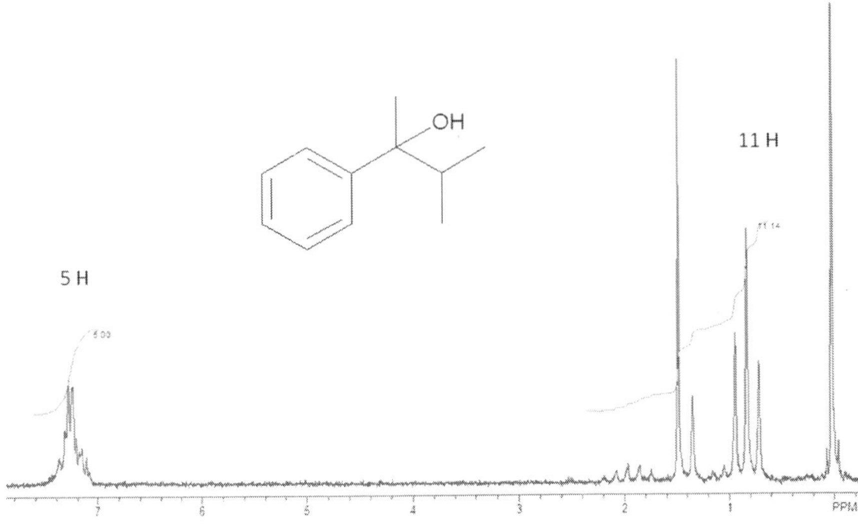

Figure 4. ^{1}H NMR spectrum of 3-methyl-2-phenyl-2-butanol.

In this example students initially determine that eleven aliphatic hydrogen atoms exist in the product molecule, in addition to the five aromatic hydrogen atoms. The major problem students have in assigning the spectrum is with the proper interpretation of what appears to be a triplet at 0.8 ppm. This provides an excellent opportunity to introduce the concept of diastereotopic atoms in a way that is logical, natural, and compelling for students. In this case, the apparent triplet is the result of two overlapping doublet resonances from the nonequivalent diastereotopic isopropyl methyl groups (*4*).

Molecular Structure of Limonene by NMR Spectroscopy

One of the experiments introduced early in the first semester of organic chemistry at Seton Hill is the isolation, characterization, and structural formula determination of limonene which students isolate from orange peels by steam distillation. Students calculate the molecular formula of limonene ($C_{10}H_{16}$) from elemental composition and vapor density data which are given. After an explanation of the concept of unsaturation index, students determine that the index of unsaturation for limonene is three. Students are then introduced to the six possible combinations of rings and/or pi bonds that are consistent with that index.

Students are next challenged with identifying which of the six combinations of pi bonds and/or rings is exhibited by the limonene molecule. As a response to the need to answer this question students are reintroduced to ^{13}C NMR spectroscopy at a deeper level than that of the earlier atom equivalency exercise. The observation of ten signals in the spectrum indicates that each of the ten carbon atoms in the limonene molecule has a unique chemical environment. After a discussion of how carbon chemical shift values correlate with various types of carbon atoms, students are able to identify the four alkene carbon peaks between 110 and 150 ppm in the ^{13}C NMR spectrum of limonene (Figure 5). This in turn leads students to the conclusion that limonene must contain two nonequivalent carbon-carbon double bonds and one ring.

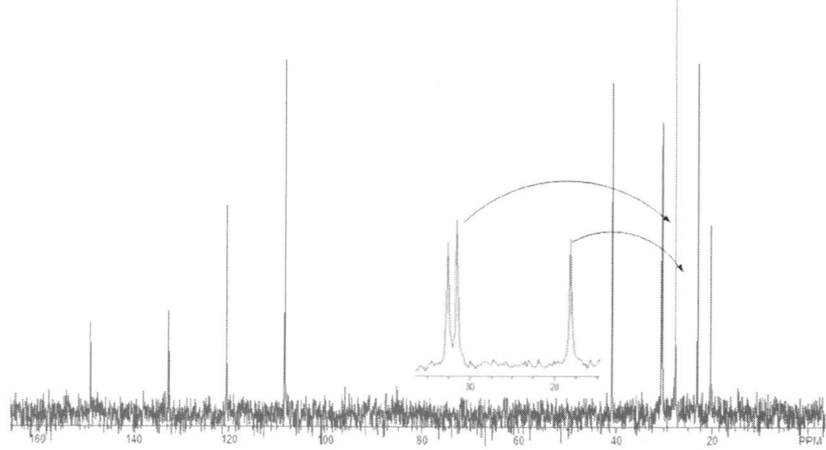

Figure 5. ^{13}C NMR spectrum of student isolated limonene ($C_{10}H_{16}$).

From the ozonolysis products of limonene (*5*), students are able to narrow the structure of limonene to the three possibilities (**1**, **2**, and **3**) shown in Figure 6, and ultimately choose the cyclic monoterpene structure **3** as the structure of limonene by comparing their experimental spectrum with spectra calculated for the three candidate molecules using ChemDraw.

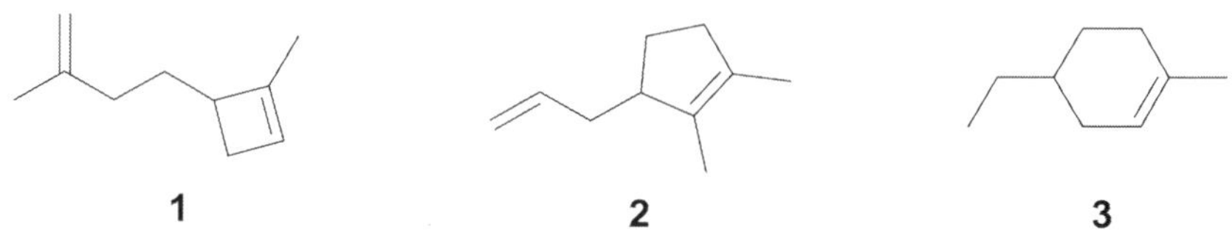

Figure 6. Possible structural formulas for limonene.

Analyses of Mixtures by ^{1}H NMR Spectroscopy

The linear relationship between the area of ^{1}H NMR signals and the number of hydrogen atoms causing that signal permits ^{1}H NMR spectroscopy to be used in the determination of the composition of mixtures. This type of analysis requires each component of interest in a mixture to have at least one ^{1}H NMR signal which is resolved from the signals produced by the other components. Such an analysis is easily included in experiments involving the fractional distillation of binary mixtures, where the composition of a distillation mixture are more traditionally determined from either gas chromatographic analysis, or from a distillation curve created by students from data collected during the distillation. However, if resolved ^{1}H NMR resonances exist for each component of the mixture, ^{1}H NMR spectroscopy can be used as an additional independent experimental method for assessing the composition of the mixture. A mixture consisting of acetone and propyl acetate is a good example of a fractional distillation mixture which can be conveniently analyzed by ^{1}H NMR spectroscopy. In this case the areas of the resolved methyl singlets at 2 ppm can be used to calculate the composition of the mixture. This analysis means that students must take into consideration the fact that the number of hydrogen atoms per molecule represented by the acetone peak is twice that of the methyl peak of propyl acetate. Requiring students to critically evaluate the accuracy and relative merits of multiple independent protocols is a great teaching exercise, since it requires students to think carefully about the assumptions and uncertainties of each type of analysis.

Another example of the application of the quantitative properties of ^{1}H NMR to assess the composition of a mixture is the alkene isomer analysis of the product mixture formed from the acid-catalyzed dehydration of an unsymmetrical alcohol such as 2-methylcyclohexanol. Determination of the isomer distribution with respect to 1-methylcyclohexene and 3-methylcyclohexene, as indicated in Figure 7, provides students with data that can be compared with the isomer distribution predicted by Zaitsev's Rule.

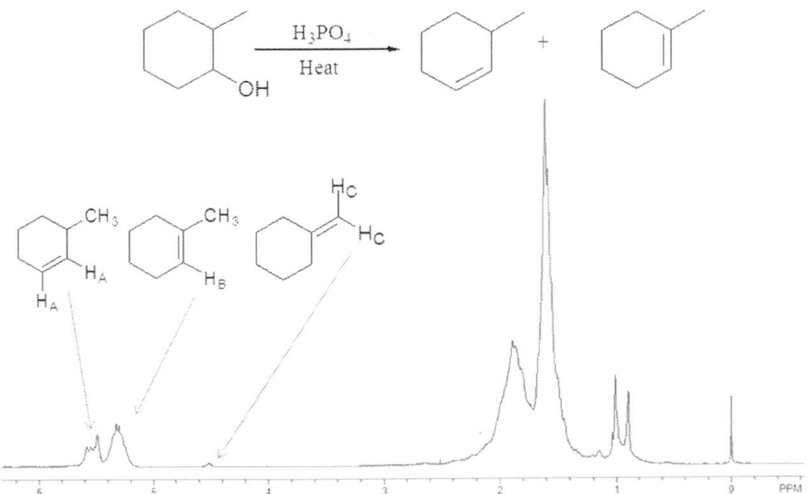

Figure 7. ^{1}H NMR spectrum of the alkene mixture formed from the acid-catalyzed dehydration of 2-methylcyclohexanol.

A further advantage of this analysis is the fact that student curiosity about the small peak at 4.5 ppm can serve as an impetus for the introduction of the concept of carbocation rearrangements.

Using NMR Spectroscopy To Teach Experimental Design

A guided inquiry multi-step synthesis project has been developed at Seton Hill in which NMR spectroscopy plays an important role in the structural determination of an unexpected product. This project involves the bromination of the dimethyl ester of *cis*-norbornene-5,6-*endo*-dicarboxylic acid **4** derived from the corresponding anhydride formed from the Diels-Alder addition reaction of cyclopentadiene and maleic anhydride as summarized in Figure 8.

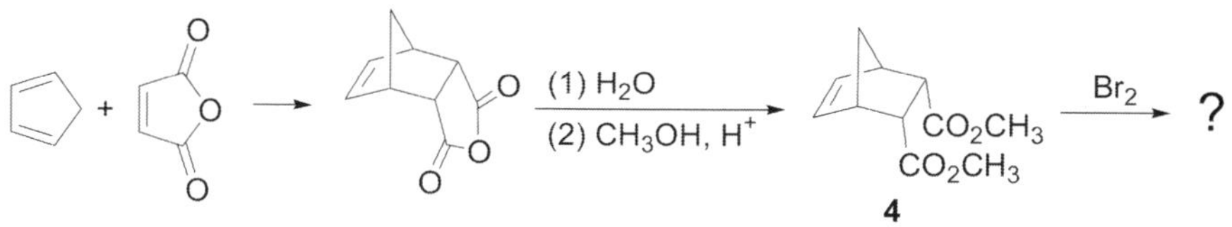

(1) H_2O
(2) CH_3OH, H^+
CO_2CH_3
CO_2CH_3
4
Br_2
?

*Figure 8. Synthesis of the dimethyl ester of cis-norbornene-5,6-endo-dicarboxylic acid (**4**).*

For the bromination reaction, students plausibly expect that anti addition of bromine to the carbon-carbon double bond of **4** will occur to give the corresponding *trans*-dibromide product in which the two methyl ester groups are retained. However, the integration data from the ^{1}H NMR spectrum of the bromination product of **4** (Figure 9) indicates the presence of just one methyl ester group.

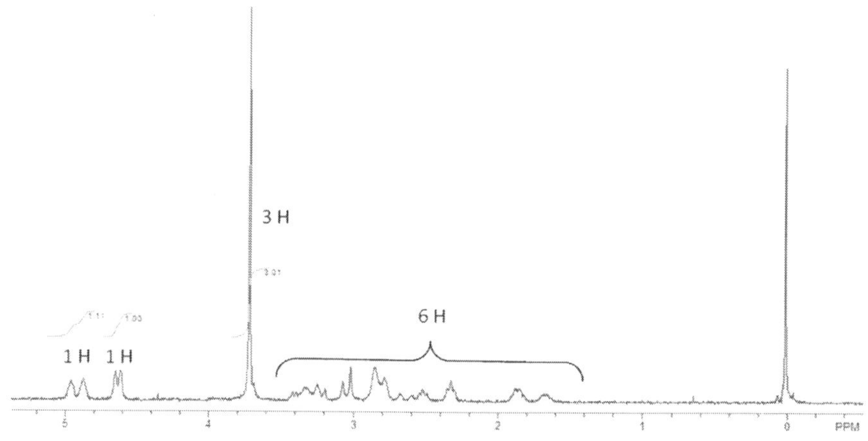

Figure 9. ^{1}H NMR spectrum of the reaction product formed from the bromination of the dimethyl ester of cis-norbornene-5,6-endo-dicarboxylic acid.

Infrared analysis of the crystallized product reveals the presence of a lactone carbonyl group and an ester carbonyl. This information, in conjunction with stereochemical insights gained from building a model of the assumed exo-bromonium ion intermediate, guides students to the conclusion that the bromolactonic ester, shown in Figure 10, is the product formed from the bromination of the *cis*-norbornene-5,6-*endo*-dicarboxylic acid (*6*).

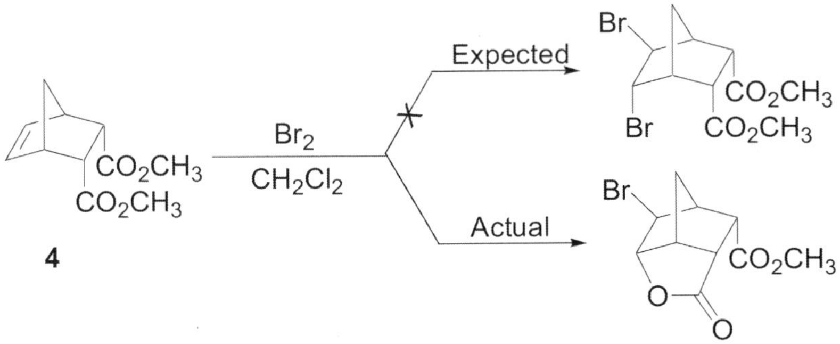

*Figure 10. Formation of bromolactonic ester from the bromination of the dimethyl ester of cis-norbornene-5,6-endo-dicarboxylic acid (**4**).*

A proposed mechanism for this reaction is discussed with students in which the *exo*-bromonium ion undergoes nucleophilic attack from the carbonyl of the methyl ester as opposed to bromide ion, Figure 11.

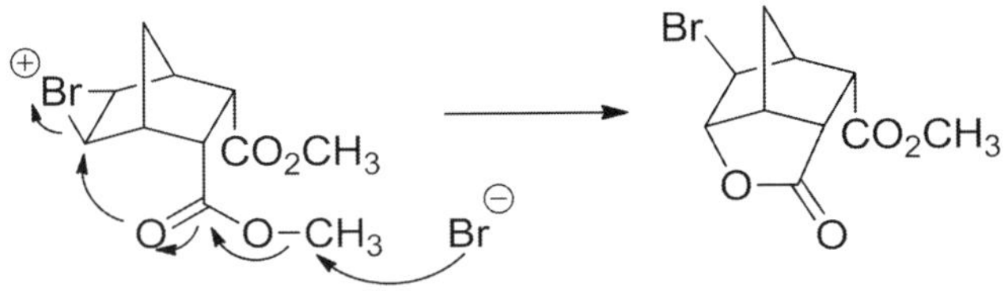

*Figure 11. Proposed mechanism for the formation of bromolactonic ester from the bromination of the dimethyl ester of cis-norbornene-5,6-endo-dicarboxylic acid **4**.*

Finally, students are asked to design an experiment which would represent a test of the proposed mechanism. Since the mechanism involves the formation of methyl bromide, which has a boiling point of 4 °C, students are encouraged to consider how the reaction could be conducted such that any methyl bromide formed could be observed by ^{1}H NMR spectroscopy. Through a process of guided inquiry involving leading questions, students are led to an experimental design where a solution of the dimethylester reactant (**4**) in carbon tetrachloride is brominated in an NMR tube fitted with the usual plastic cap to prevent the escape of methyl bromide. This microscale experiment can be conveniently performed by the addition of a 1 M solution of Br_2 in CCl_4 to an NMR tube containing the dimethyl ester dissolved in CCl_4. Upon addition of Br_2 the appearance of a sharp singlet at 2.6 ppm, as shown in the ^1H NMR spectrum in Figure 12, indicates that methyl bromide is indeed formed in the reaction. This observation provides support for the proposed mechanism and is particularly effective in teaching students how experiments can be carefully designed in order to answer scientific questions.

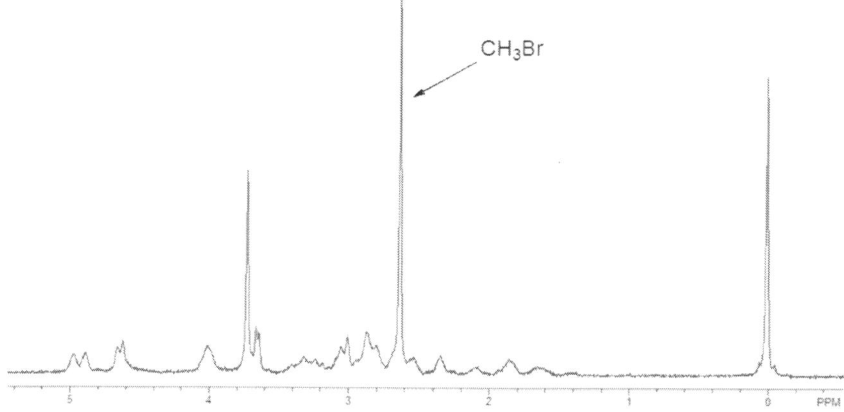

Figure 12. ^1H NMR spectrum of the product mixture formed from the bromination of the dimethyl ester of cis-norbornene-5,6-endo-dicarboxylic acid (4) in a closed NMR tube.

Reaction Kinetics by NMR Spectroscopy

Performing microscale reactions in NMR tubes positioned in the temperature controlled probe of an NMR spectrometer provides an especially convenient way of teaching students the principles of reaction kinetics. Monitoring the progress of such reactions can be accomplished by determining the area of a resolved resonance of either a reactant or a product with respect to time. Two examples of this type of experiment are presented.

The dimerization reaction of cyclopentadiene (Figure 13) is relatively slow, showing second order kinetics with a half-life at room temperature of approximately one day. Spectra of neat freshly distilled cyclopentadiene monomer can be taken serially and the progress of dimerization monitored by following the area of the resolved reactant or product peaks with respect to time. Plotting the reciprocal of cyclopentadiene concentration versus time gives a linear plot from which the rate constant for the reaction can be evaluated.

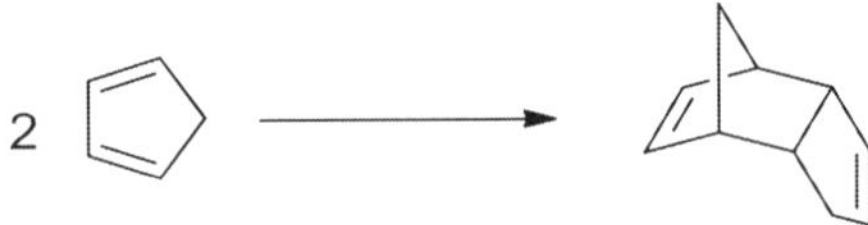

Figure 13. Dimerization of cyclopentadiene.

An example of a kinetic study, which is more amenable to a three or four hour undergraduate laboratory period, is the solvolysis of *t*-butyl bromide in methanol or aqueous methanol. Solvolysis kinetics of *t*-butyl halides can be conveniently studied by ^{1}H NMR spectroscopy in ordinary (protium) aqueous methanol solvent since the ^{1}H NMR chemical shifts of methanol and water are sufficiently removed from those of the reaction substrate and products. The fact that deuterated methanol and water solvents are not required significantly lowers the cost of these experiments.

The ^{1}H NMR spectrum shown in Figure 14 reveals that *t*-butyl bromide reacts with 80% aqueous methanol to form the corresponding *t*-butyl methyl ether and *t*-butyl alcohol products as a result of the S_N1 reaction of the *t*-butyl carbocation intermediate with methanol and water. Furthermore, evidence of the competing E1 reaction to form 2-methylpropene is also indicated by the spectrum.

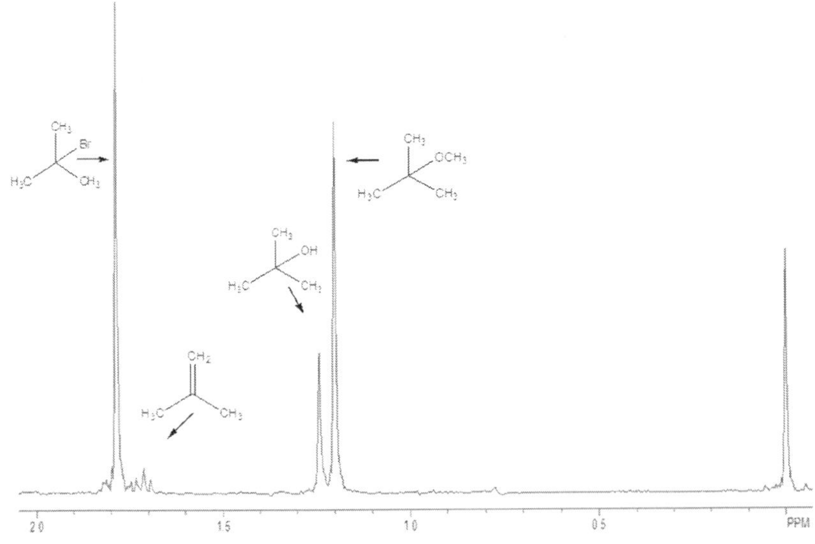

Figure 14. ^{1}H NMR spectrum of the solvolysis of 1 M t-butyl bromide in 80% aqueous methanol, t = 50 minutes. Scale is from 2.0 ppm to 0 ppm.

Figure 15 shows serial spectra collected at various times following the addition of *t*-butyl bromide to 80% aqueous methanol in an NMR tube. The plot of the natural logarithm of the area of the *t*-butyl bromide peak versus time gives a linear plot from which the first order rate constant can be calculated as

0.016 min^{-1} which corresponds to a half-life of 43 minutes. The time scale of this experiment is particularly well suited to the typical length of undergraduate laboratory periods.

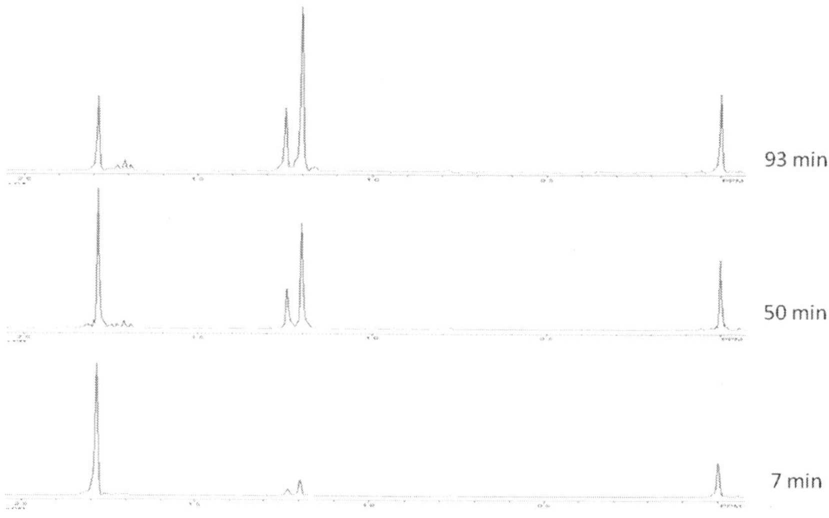

Figure 15. Solvolysis of 1 M t-butyl bromide in 80% aqueous methanol, t = 7 min, t = 50min, t = 93 min.

Since the ^1H NMR resonances of the three products shown in Figure 14 are resolved from each other and from the reactant, a wide range of experiments can be designed to teach students the important principles of nucleophilic substitution reactions. The leaving group effect on reaction rate constants can be easily studied by comparing the rate constants for the solvolysis of *t*-butyl bromide and *t*-butyl chloride. Varying the percentage of water in the aqueous methanol solvent permits the effect of solvent polarity on rate constants to be examined. Furthermore, factors affecting the ratio of substitution to elimination can be conveniently explored as well.

Conclusions

In this chapter experiments have been presented which illustrate how NMR spectroscopy can be employed in the undergraduate organic laboratory curriculum to promote active learning and to effectively introduce students to a wide range of important concepts and principles in organic chemistry. A major pedagogic strategy in this work has been to design experiments involving NMR spectroscopy where the laboratory work expected of students is in response to the need to answer important scientific questions. The significant degree to which this teaching approach has been successful in enhancing student learning is a reflection of the robust pedagogic power of NMR spectroscopy in connecting and engaging students with the molecular world.

References

1. Paulson, D. R. *J. Chem. Educ.* **1999**, *76*, 1136–1140.
2. *Modern NMR Spectroscopy in Education*; Rovnyak, D., Stockland, R., Eds.; ACS Symposium Series 969; American Chemical Society: Washington, DC, 2007.
3. http://riodb01.ibase.aist.go.jp/sdbs/cgi-bin/cre_index.cgi (accessed December 2012).
4. Kieboom, A. P. G.; Sinnema, A. *Tetrahedron* **1972**, *28*, 2527–2532.
5. Griesbaum, M.; Hilss, M.; Bosch, J. *Tetrahedron* **1996**, *52*, 14813–14826.
6. Fleming, I.; Michael, J. *J. Chem. Soc., Perkin Trans. 1* **1981**, *5*, 1549–1556.

Chapter 5

NMR Spectroscopy in Nondeuterated Solvents (No-D NMR): Applications in the Undergraduate Organic Laboratory

John E. Hanson*

University of Puget Sound, Department of Chemistry, Tacoma, Washington 98416
***E-mail: hanson@pugetsound.edu**

Expensive deuterated solvents have traditionally been used for NMR spectroscopy in order to facilitate locking and shimming, as well as to suppress the large solvent signal that would otherwise occur in the proton NMR spectrum. Advances in NMR instrumentation now make the routine use of deuterated solvents unnecessary. The use of nondeuterated solvents for NMR spectroscopy (No-D NMR) results in significant savings, both of time and money. No-D NMR can be used for identifying unknown compounds, confirming the structure and purity of compounds isolated from a reaction, determining the ratio of products in a crude reaction mixture, and following reaction progress. Practical issues relating to the implementation of No-D NMR experiments, along with applications in the undergraduate organic laboratory, are discussed.

Introduction

One of the first things that students are often taught about NMR spectroscopy is to prepare the sample by dissolving it in a deuterated solvent.

"The solvent molecules should have all hydrogen atoms replaced with deuterium atoms (^{2}H) for two reasons. First, if you are doing proton (^{1}H) NMR, you do not want the solvent resonance to dominate your spectrum. Solvent molecules typically outnumber solute molecules by 1000 to 1, so you would not really see your solute spectrum at all. Second, the spectrometer needs a deuterium (^{2}H) signal to "lock" the magnetic field strength and keep it from changing with time. Because the NMR experiment usually adds together a number of FIDs (scans), if the field changes during the experiment the frequency changes with it and the NMR peaks will not add together correctly. The deuterium NMR signal is used to monitor "drift" of the field and to correct it… (*1*)."

"If one uses the protonated solvent, then a ca. 0.5-1.0 ppm region will be obscured by the solvent peak….for proton FT studies, the deuterated solvent is vital, as the majority of pulse spectrometers use the solvent deuterium signal to stabilize and lock the spectrometer system. Also, the large signal of the protonated solvent in both ^{1}H and ^{13}C spectra causes digitization problems in FT spectrometers… (*2*)"

In demanding applications, particularly those in which the amount of sample is very limited or the complexity of the sample is very high, the use of deuterated solvents is important for many of the reasons articulated above. But in more routine applications, such as those typically encountered in the undergraduate teaching laboratory, NMR experiments performed in nondeuterated solvents (No-D NMR) normally work well (Figure 1). We were inspired to implement No-D NMR experiments into our second-year undergraduate organic laboratory class by a series of papers published in 2004-2005 by Professor Thomas Hoye and co-workers demonstrating the usefulness of No-D NMR for a wide variety of applications (*3–6*).

In early FT-NMR spectrometers, limitations on the analog-to-digital converters (ADC's) that were used made it difficult to observe small sample signals in the presence of large solvent peaks. This is much less of an issue in modern instruments. In applications where relatively large amounts (>20 mg) of relatively simple compounds (MW <300 g/mol) are being analyzed, the dynamic range of the ADC converters are more than sufficient to accurately detect the sample peaks without the need for solvent suppression techniques. Another reason for using deuterated solvents is to provide a lock signal that can be used to correct magnetic field drift. In most applications in the undergraduate lab, NMR experiments are typically completed in a few minutes, and rarely take more than 30 minutes. Modern superconducting magnets are stable enough that over this period of time the amount of drift is so small that it does not significantly degrade the quality of the spectrum.

A major reason for using deuterated solvents in ^{1}H NMR spectroscopy is a concern that solvent peaks will overlap some of the signals of the sample. In modern, relatively high field, spectrometers (^{1}H resonance ≥200 MHz) the chemical shift dispersion is large enough that this is a relatively rare occurrence. For a nonviscous solvent with a single peak, the area actually obscured by the solvent peak is usually less than 0.2-0.3 ppm (Figure 1). And in cases where the structure of the compound to be analyzed is known or partially known, and hence the expected positions of sample peaks can be estimated, a solvent can usually be chosen that will not interfere with the sample peaks.

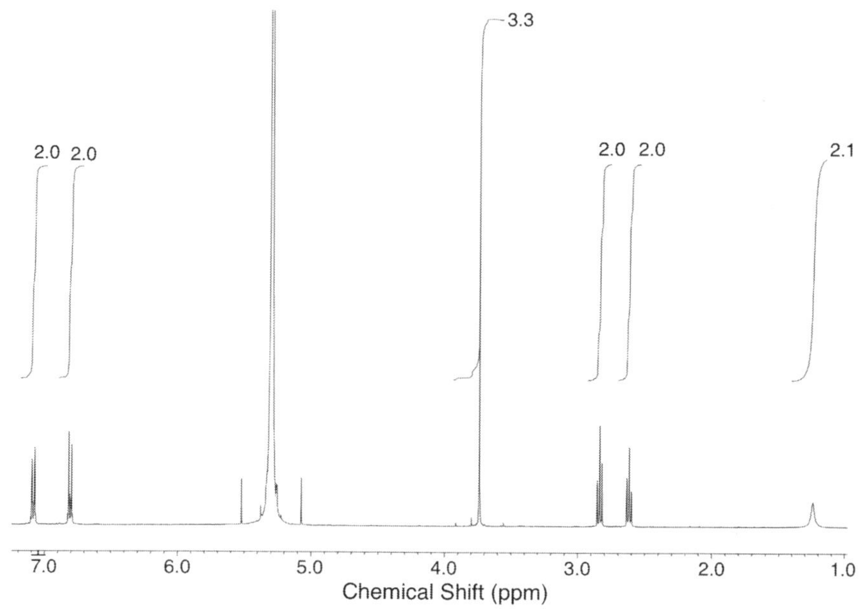

Figure 1. NMR spectrum of 2-3 drops (40-60 mg) of 4-methoxyphenethylamine in approximately 1 mL of dichloromethane.

Advantages of No-D NMR

The main motivation for using No-D NMR in the undergraduate teaching laboratory is the high cost of deuterated solvents relative to their nondeuterated congeners (Table 1). While high-quality nondeuterated solvents cost between 2 and 10 cents a gram, the deuterated analogs range from $0.23 to $32.55 per gram.

Since deuterochloroform is the least expensive deuterated solvent, it is the most commonly used for NMR analysis of organic compounds, and the only practical one (from a cost standpoint) for routine use in the teaching laboratory. Even so, $CDCl_3$ is approximately ten times more expensive than its nondeuterated form, and thus, in a large teaching laboratory, significant savings can be realized by switching to No-D NMR. Maybe not as obviously, but even more importantly, by switching to No-D NMR there is no longer a strong financial impetus to use chloroform, and thus safer, less toxic solvents such as dichloromethane or acetone may be substituted. In fact, the relatively low cost of the nondeuterated solvents means that samples that are not soluble in deuterochloroform can now be cost effectively analyzed by NMR in the teaching laboratory.

Table 1. Cost of common NMR solvents[a]

Solvent	*Nondeuterated List Price ($/g)*	*Perdeuterated List Price ($/g)*
Acetone	0.04	2.23
Acetonitrile	0.10	3.69
Benzene	0.06	2.62
Chloroform	0.03	0.23
Dichloromethane	0.03	12.10
Dimethyl formamide	0.06	32.55
Dimethyl sulfoxide	0.07	3.12
Methanol	0.03	7.46
Tetrahydrofuran	0.07	29.40
Toluene	0.04	5.68
Water	0.02	0.60

[a] All prices from the 2012-2014 Aldrich Handbook of Fine Chemicals. The nondeuterated price is for a 2-4 L bottle of a high purity grade of the solvent. The perdeuterated price is for the largest quantity listed of the lowest atom D grade ≥95%.

Another advantage of No-D NMR is that reaction progress can be routinely monitored by NMR spectroscopy without having to perform them in expensive deuterated solvents (*3*, *5*). In addition, if a reaction is being run for non-synthetic reasons, for example to determine the ratio of products obtained under different conditions, then there may be no need to evaporate the reaction solvent and switch to a deuterated solvent; the crude reaction mixture can be analyzed directly by No-D NMR resulting in a significant savings in time (*7*).

Finally, a somewhat trivial, but not insignificant, advantage of No-D NMR is that NMR tubes do not need to be thoroughly dried between applications. In the past it was a constant struggle to get NMR tubes cleaned and dried between uses in the teaching laboratory. But now, students just rinse the tube with the solvent in which they plan to record their next NMR spectrum and it is ready to use.

Solvent Selection and Sample Preparation

In our experience, dichloromethane is a good general purpose No-D NMR solvent for the undergraduate teaching laboratory. It has many desirable properties that include (i) while not as polar as chloroform, it dissolves a wide range of organic compounds, (ii) it has a single, sharp resonance (at 5.3 ppm for ^{1}H spectra and 53.5 ppm for ^{13}C spectra) that normally does not obscure peaks of the analyte, (iii) it is readily available in pure form and typically is not contaminated with water, stabilizers, or NMR-active impurities, (iv) it is nonflammable and less toxic than chloroform, and (v) it is volatile and can be readily removed if one wishes to recover the sample. Nevertheless, dichloromethane, like most organic solvents, is not without its hazards and students should be instructed to dispense it in a hood and to avoid contact with the skin.

Another attractive alternative is to run the sample neat; but this is limited to situations in which relatively large amounts (~1 mL) of nonviscous liquids are being analyzed (*8*). Of course it would also be possible to use solvents that do not contain hydrogen atoms at all. For example, carbon tetrachloride and carbon disulfide were used in many early ^{1}H NMR studies. Many organic compounds are not soluble in these nonpolar solvents, and, even more importantly, they are relatively toxic and/or flammable, making them particularly undesirable for use in a teaching laboratory.

Most reagent grade solvents are pure enough to be satisfactory for use as No-D NMR solvents. However, users should be aware that some common solvents (for example chloroform, ether, and THF) are often distributed with stabilizers that may result in undesired peaks in the NMR. For example, chloroform is often sold with 0.5-1% ethanol added as a stabilizer. Although it is not usually necessary to use "anhydrous" grade solvents, water is a ^{1}H NMR-active impurity common in some solvents (for example, acetone and DMSO), and whether the level of water is acceptable for the planned use should be considered. In practice, when exploring the use of a new solvent, we typically open a relatively new reagent grade bottle of the solvent and make up an NMR sample using 20-40 mg of a test compound. In the vast majority of cases, we find that the solvent is sufficiently pure for the desired use.

We usually suggest that students make up solutions containing approximately 40-60 mg of sample in 1 mL of nondeuterated solvent for ^{1}H-NMR samples, and approximately double that if they plan to acquire a ^{13}C-NMR spectrum. For most situations in the teaching laboratory, we are not limited by the amount of compound, and using relatively large amounts of sample insure that good signal to noise will be obtained in a relatively short time. This is an especially important consideration for ^{13}C NMR spectroscopy. We also recommend that students use a full 1 mL of solvent so that the shim settings are relatively similar from sample to sample.

If necessary, good No-D NMR spectra can normally be obtained using sample amounts of approximately one-tenth that recommended above. The actual practical limit will depend on the field strength of the NMR, the solvent being used, the amount of acquisition time one can tolerate, and the complexity of the sample to be analyzed.

Instrument Setup and Shimming

We use the same standard pulse sequence as for deuterated samples. For a ^{1}H NMR spectrum we typically acquire 8-16 scans, for a total acquisition time of 1-2 minutes. More complex pulse sequences are available to suppress the solvent peaks, but they are not necessary with the relatively high concentrations we are using. For ^{13}C NMR spectra, it may be necessary to adjust the default decoupling parameters, depending upon the solvent used.

One instrumental parameter that typically needs to be adjusted (relative to what is used in deuterated solvents) is the receiver gain. Otherwise the large solvent signal will overload the receiver and result in a poor spectrum. In the teaching laboratory, where all the students are using the same solvent, the receiver gain may simply be set at the appropriate level once. Alternatively, if available, an automatic gain adjustment can be incorporated into the standard NMR procedure, at the expense of adding another minute to each run.

One of the advantages of use of a deuterated solvent is to provide a convenient signal to monitor while shimming, either manually or automatically. However, other approaches can be used effectively for No-D NMR samples ((*3*), footnote 8). Many modern NMR spectrometers are equipped with pulsed-field gradient hardware that makes it possible to quickly and automatically shim a sample – so called "gradient shimming" (*9, 10*). Gradient shimming techniques are often used with deuterated solvents, but they are equally effective in No-D NMR samples containing a strong proton signal from the solvent. All of the No-D spectra acquired by students in our undergraduate organic course (and hence all the spectra shown in this article) are acquired using automated gradient shimming, a process that takes about one minute per sample. Alternatively, the magnet may be manually shimmed (either on the FID or the spectrum) while the sample is rapidly pulsed. In the undergraduate teaching laboratory, where students are usually using the same solvent and similar samples and volumes, it is only necessary to touch up Z1 and Z2 between samples and this can be done fairly rapidly. In fact, acceptable results are often obtained under these conditions without shimming at all between samples.

Spectral Analysis

In No-D NMR there is typically no need to add a chemical shift standard, such as TMS, to the sample; the NMR chemical shift (either ^{1}H or ^{13}C) of the solvent is used as the reference. Since the solvent peak is the largest in the spectrum, it is easily identified by students. After setting the reference, students then y-scale the spectrum so that the NMR resonances of their sample are visible and the solvent peak is clipped. Students also learn to ignore the solvent peak when integrating the spectrum, producing a final spectrum like that in Figure 1.

One source of confusion for students using No-D NMR is the presence of satellite peaks on either side of the large solvent peak. These peaks are due to artifacts from spinning the sample ("spinning sidebands") and coupling of the protons to the 1% of carbon atoms that are ^{13}C in the sample ("^{13}C satellite peaks") (Figure 2). For the spinning sidebands, the distance between the peaks (in Hz) corresponds to the spin rate (or a multiple of the spin rate), typically 10-15 Hz. If desired, the spinning sidebands may be removed by simply turning off the spinner. In most cases, the slight decrease in resolution that results from turning off the spinner does not significantly affect the appearance or interpretation of the spectrum. In No-D NMR the intensity of the satellite peaks are often similar to the sample peaks. The satellite peaks are symmetrically disposed on either side of the main solvent peak so they can normally be readily identified. The distance between the ^{13}C satellite peaks corresponds to the C-H coupling constant, typically 110-210 Hz.

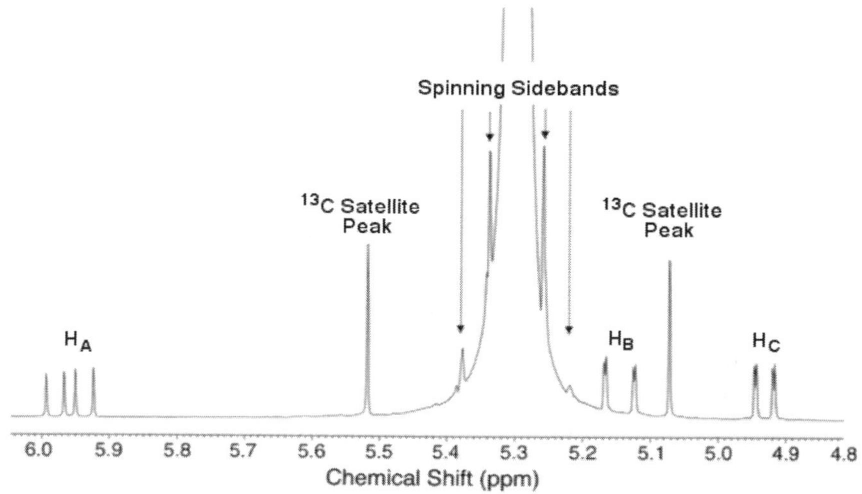

Figure 2. A portion of the ^{1}H-NMR spectrum of 2-methyl-3-buten-2-ol in dichloromethane. H_A, H_B, and H_C correspond to the vinylic protons of the sample.

Applications in the Undergraduate Organic Laboratory

From 2004 through 2005 Professor Thomas Hoye and coworkers published a series of papers describing the utility of No-D NMR for analyzing a variety of reaction and reagent solutions (*3–6*). While those papers emphasized the usefulness of No-D NMR in research applications, the same principles are equally applicable in the undergraduate teaching laboratory. However, relatively few examples describing the use of No-D NMR in the undergraduate teaching laboratory have appeared (*7, 8, 11, 12*). Alonso and Wong outlined an experiment for the general chemistry laboratory in which students identified an unknown organic compound by acquiring the ^{1}H decoupled ^{13}C NMR spectrum of neat samples of simple organic liquids (*8*). An experiment for the organic teaching laboratory used chloroform (not deuterochloroform) as the solvent for obtaining No-D ^{13}C NMR spectra of the alkene products from dehydration of unknown cyclohexanols and cyclopentanols (*11*). Students were then able to infer the identity of their starting alcohols. An interesting application of No-D ^{1}H NMR for the quantitative analysis of acetone in nail polish remover was developed for a second-year analytical chemistry course (*12*). In a similar vein, although not developed as an undergraduate laboratory experiment, Hoye and coworkers described the use of No-D NMR for the quantitative determination of ethanol in a variety of commercial alcoholic beverages (*3*).

We have incorporated No-D NMR throughout our second semester organic chemistry laboratory, and found that students quickly become proficient in the acquisition and analysis of No-D NMR spectra and find it no more challenging than "normal" NMR with deuterochloroform. All spectra shown in this paper were acquired by students in our organic teaching laboratory using No-D NMR on a JEOL ECA-400 spectrometer.

Identification of Unknowns

A common undergraduate organic experiment is the identification of unknown compounds on the basis of their physical and spectral properties. At the beginning of the second semester laboratory, students are assigned an unknown solid and unknown liquid compound. Students are not given any information about the identity of these compounds, except to note that they are organic compounds purchased from a major chemical supplier. The unknowns are selected from leftover chemicals purchased for other purposes, for example, starting materials from research labs. We rotate the unknowns as they get used up, or if we find they do not work well due to an impurity or some other problem with their analytical data. The unknown compounds typically have molecular weights less than 250 g/mol and contain one or two common functional groups (e.g., 4-isobutylacetophenone, 2-propylphenol, 3-phenyl-1-propanol, piperonal, 4-methoxyacetophenone, 4-methyl-2-pentanone, or 3,3-dimethylacrylic acid).

For both the solid and liquid unknown, students acquire ^{1}H NMR and IR spectra and do some simple solubility tests in water and aqueous NaOH, $NaHCO_3$, and HCl solutions. In addition, for the solid unknowns, students determine the melting point and acquire a ^{13}C NMR spectrum. For the liquid unknowns, students determine the index of refraction and perform a GC/MS analysis. For the past few years we have had students analyze these unknowns using No-D NMR with dichloromethane as the solvent. If they suspect that there is overlap between peaks in their compound and the dichloromethane peak, they can try acetone.

Figures 1 and 2 show No-D ^{1}H NMR spectra from two student unknowns made by adding 2-3 drops (40-60 mg) of sample in 1 mL of CH_2Cl_2. In Figure 1 the dichloromethane peak at 5.3 ppm is well-separated from other resonances so analysis of the spectrum is no more difficult than using deuterochloroform. While it is true that students must learn to recognize and ignore the large dichloromethane peak, we have found that this is no harder for students to do than learning to recognize and ignore the residual $CHCl_3$ peak in deuterochloroform.

In a few of the unknowns containing vinylic hydrogen atoms, some of the ^{1}H NMR resonances are very close to the dichloromethane peak (Figure 2). In nearly all cases it is still possible to analyze the splitting patterns, and, although accurate integration of peaks near the solvent peak (e.g., H_B in Figure 2) is problematic because it overlaps with the tail of the large dichloromethane resonance, it is clear from the peak height that the integral for H_B is similar to H_A and H_C. Of course, if overlap causes significant difficulties in interpretation, students may run the spectrum in a different solvent such as acetone.

Esterification – Product Purity

In another experiment that occurs during our second semester organic laboratory, students perform a simple esterification. Each pair of students is assigned a combination of an acid (acetic acid, propanoic acid, or butanoic acid) and an alcohol (methanol, ethanol, propanol, butanol) that produces an ester of between 5 and 8 carbons. That way, the product is water insoluble (which helps in the workup) and distills at a reasonable temperature (99-145°C). Although most students get reasonably pure material (Figure 3, Group A), the No-D NMR spectrum clearly indicates when they do not (Figure 3, Group B).

This experiment is done on a relatively large scale (~10 g of ester) to provide enough material for easy distillation. With acetic acid and propanoic acid, the alcohol is made the limiting reagent, while with butanoic acid the acid is the limiting reagent. The other reactant is used in a 3-fold molar excess. The acid and ester and ~1 mL of concentrated sulfuric acid are refluxed for about 1 hour. The reaction mixture is cooled, washed with water, 5% aqueous bicarbonate (carefully!), and saturated NaCl, and then dried (Na_2SO_4). After purifying the product by simple distillation, students analyze the product purity by ^1H and ^{13}C No-D NMR in dichloromethane. Hoye and coworkers have also shown that it is feasible to actually follow the progress of a similar Fischer esterification by No-D NMR of the reaction mixture (*3*).

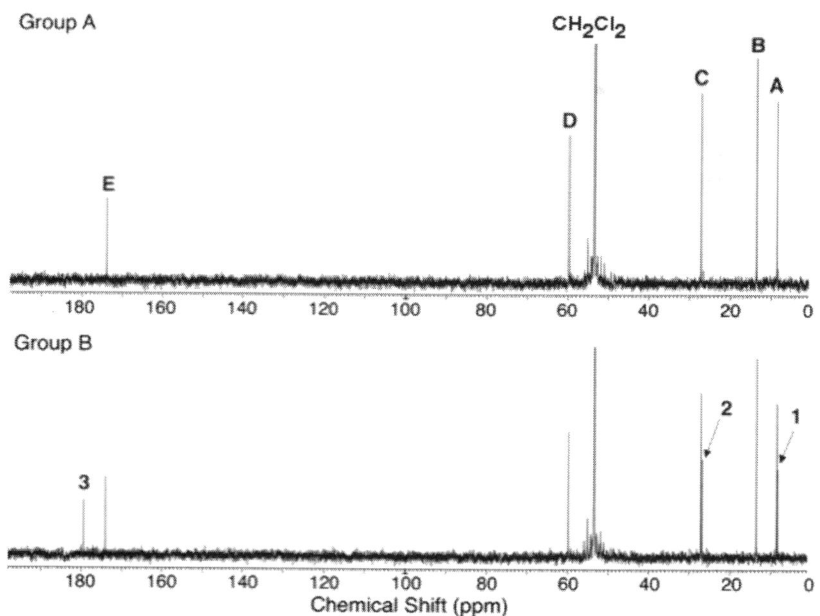

Figure 3. ^{13}C-NMR spectrum of product from esterification of propanoic acid and ethanol. Group A has prepared reasonably pure ethyl propanoate (carbons A-E). Group B's product is contaminated with propanoic acid (carbons 1-3).

Exploring the Stereochemistry of the Wittig Reaction

No-D NMR is a cost-effective method for characterizing purified compounds, but it is even more useful when analyzing crude reaction mixtures, as might be the case when one wants to determine the ratio of products produced under different reaction conditions. For example, we have described an organic laboratory experiment in which students determine the ratio of *E* and *Z* alkenes produced from a Wittig reaction run in the presence of different alkali metal salts (*7*). When we originally developed this experiment, students quenched the reaction (run in THF) with water, extracted with hexane, removed the solvent by rotary evaporation, then redissolved the mixture in $CDCl_3$. Now they simply pipet the hexane extract directly into an NMR tube and acquire a No-D NMR spectrum. Even though the NMR spectrum is dominated by large hexane and THF peaks, by zooming in on the vinylic region, students are readily able to determine the ratio of stereoisomers by integration and assign them as *E* or *Z* based on analysis of the coupling constants. Waiting to use the rotary evaporator used to be a bottleneck in the lab, so being able to omit this step has allowed this experiment to be performed more smoothly.

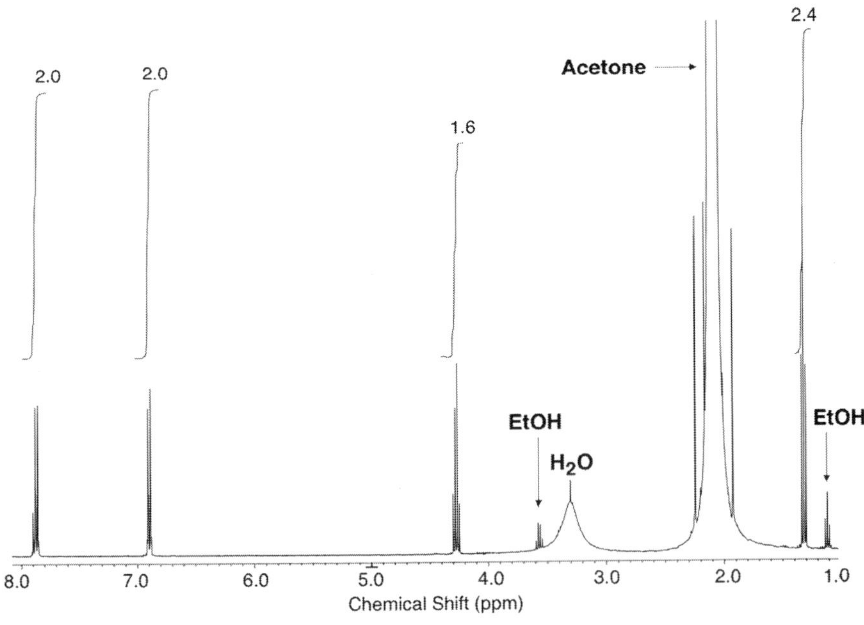

Figure 4. ¹H NMR spectrum following the progress of the esterification of 4-hydroxybenzoic acid with ethanol. Peaks at 6.9 and 7.9 ppm arise from overlapping resonances of both the starting 4-hydroxybenzoic acid and the product ester. The peaks at 1.3 and 4.3 ppm correspond to only the product ethyl ester. The low integration of these ethyl peaks relative to the aromatic peaks indicates that the reaction is not yet complete.

Synthetic Projects: Monitoring Reactions

As a capstone experience in our organic laboratory course, students spend the last five weeks of the semester working on a variety of extended synthetic projects. These projects are designed and supervised by faculty members teaching in the laboratory course. The project ideas are adapted from a variety of sources, including articles from *The Journal of Chemical Education*, the primary research literature, and the faculty members' research projects. Examples of synthetic targets include ibuprofen (*13*, *14*), frontalin (*15*), dihydroxyacetone phosphate (*16*), desmethoxyxanthohumol (*17*), coniferin (*18*), and sulotroban (*19*).

During these projects No-D NMR is routinely used to characterize products and reaction mixtures. For example, the esterification of 4-hydroxybenzoic acid with ethanol can be analyzed by removing small aliquots from the reaction, performing a quick workup, and taking the ^{1}H NMR spectrum in acetone, since there is limited solubility in dichloromethane and chloroform (Figure 4). Comparison of the integrations of the aromatic protons relative to those of the ethyl ester peaks shows that the 4-hydroxybenzoic acid is not yet completely esterified and that the reaction should be continued. Although acetone typically contains some water, which is seen as a broad signal in the NMR spectrum, it does not interfere with the analysis in this case.

In another project, based on a paper by Lee and Xia (*17*), the reaction shown below was performed in refluxing xylene, Scheme 1. At various times, an aliquot of the crude reaction mixture was added to an NMR tube and a No-D NMR was obtained. Integration of the phenolic hydrogen atoms was then used to follow the extent of reaction (Figure 5).

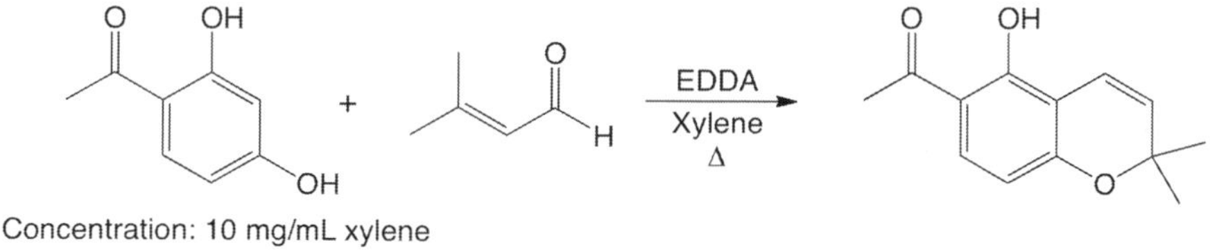

EDDA
Xylene
Δ

Concentration: 10 mg/mL xylene

Scheme 1. Synthesis of a pyranochalcone

During the synthetic projects we occasionally have the need to run NMR experiments on other nuclei, such as ^{31}P and ^{19}F. No-D NMR is even more convenient in these cases since there is no signal from the solvent. It is also feasible to use No-D NMR when conducting 2D NMR experiments (for example, COSY). We still use deuterated solvents from time to time, especially in situations where the sample is unusually complex or available in limited amounts, but No-D NMR works well in the majority of situations.

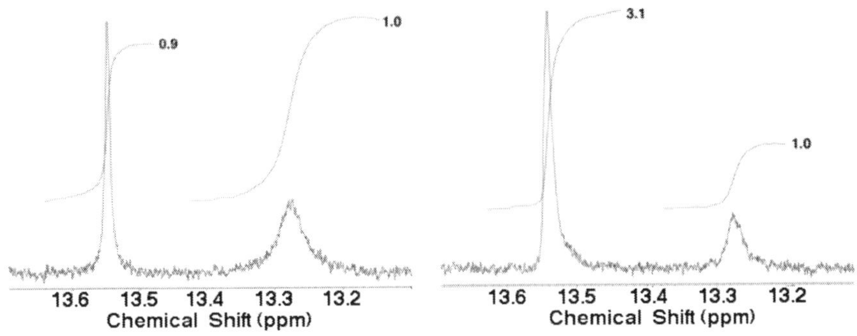

Figure 5. ¹H NMR spectra of the reaction mixture at the beginning of the reaction (left) and after 24 hrs (right). The continued presence after 24 hours of a peak slightly below 13.3 ppm indicates that the reaction is not yet complete.

Conclusions

No-D NMR is a convenient, versatile, and cost-effective method for characterizing organic compounds and monitoring the progress of organic reactions. No-D NMR can be adapted for use in almost any undergraduate organic laboratory experiment that uses NMR spectroscopy in a deuterated solvent. In addition, No-D NMR can make feasible NMR-based experiments that would be cost prohibitive if deuterated solvents were required.

Acknowledgments

I would like to thank Jim Kabrhel for bringing to my attention the usefulness of No-D NMR and Chip Detmer for helping to implement No-D NMR on our spectrometer. I am also grateful to my organic colleagues Bill Dasher, Eric Scharrer, and Tim Hoyt for being supportive in testing out new techniques and experiments in the organic teaching labs. In addition, I would like to thank all the organic students over the years who have been eager to learn new skills in the organic lab. Finally, I appreciate the generous support of William Canfield who helped make possible the acquisition of our NMR spectrometer.

References

1. Jacobsen, N. E. *NMR Spectroscopy Explained: Simplified Theory, Applications and Examples for Organic Chemistry and Structural Biology*; Wiley-Interscience: Hoboken, NJ, 2007.
2. Abraham, R. J.; Fisher, J.; Loftus, P. *Introduction to NMR spectroscopy*; Wiley: Chichester, NY, 1988.

3. Hoye, T. R.; Eklov, B. M.; Ryba, T. D.; Voloshin, M.; Yao, L. J. *Org. Lett.* **2004**, *6*, 953–956.
4. Hoye, T. R.; Eklov, B. M.; Voloshin, M. *Org. Lett.* **2004**, *6*, 2567–2570.
5. Hoye, T. R.; Kabrhel, J. E.; Hoye, R. C. *Org. Lett.* **2005**, *7*, 275–277.
6. Hoye, T. R.; Aspaas, A. W.; Eklov, B. M.; Ryba, T. D. *Org. Lett.* **2005**, *7*, 2205–2208.
7. Hanson, J.; Dasher, B.; Scharrer, E.; Hoyt, T. *J. Chem. Educ.* **2010**, *87*, 971–974.
8. Alonso, D. E.; Wong, P. A. *Chem. Educ.* **2008**, *13*, 234–235.
9. Prammer, M. G.; Haselgrove, J. D.; Shinnar, M.; Leigh, J. S. *J. Magn. Reson.* **1988**, *77*, 40–52.
10. Van Zijl, P. C. M.; Sukumar, S.; Johnson, M.; Webb, P.; Hurd, R. E. *J. Magn. Reson.* **1994**, *A111*, 203–207.
11. Dunlap, N. K.; Mergo, W.; Jones, J. M.; Martin, L. *Chem. Educ.* **2006**, *11*, 378–379.
12. Hoffmann, M. M.; Caccamis, J. T.; Heitz, M. P.; Schlecht, K. D. *J. Chem. Educ.* **2008**, *85*, 1421–1423.
13. Cleij, M.; Archelas, A.; Furstoss, R. *J. Org. Chem.* **1999**, *64*, 5029–5035.
14. Faigl, F.; Schlosser, M. *Tetrahedron Lett.* **1991**, *32*, 3369–3370.
15. Bartlett, P. A.; Marlowe, C. K.; Connolly, P. J.; Banks, K. M.; Chui, D. W. -H.; Dahlberg, P. S.; Haberman, A. M.; Kim, J. S.; Klassen, K. J.; Lee, R. W.; Lum, R. T.; Mebane, E. W.; Ng, J. A.; Ong, J .-C.; Sagheb, N.; Smith, B.; Yu, P. *J. Chem. Educ.* **1984**, *61*, 816.
16. Meyer, O.; Ponaire, S.; Rohmer, M.; Grosdemange-Billiard, C. *Org. Lett.* **2006**, *8*, 4347–4350.
17. Lee, Y. R.; Xia, L. *Synthesis* **2007**, *20*, 3240–3246.
18. Daubresse, N. R.; Francesch, C.; Mhamdi, F.; Rolando, C *Synthesis* **1998**, 157–161.
19. Nuhrich, A.; Varache-Lembége, M.; Lacan, F.; Devaux, G. *J. Chem. Educ.* **1996**, *73*, 1185–1187.

Chapter 6

Overcoming Problems Incorporating NMR into the Organic Chemistry Lab

Luke A. Kassekert and J. Thomas Ippoliti*

Department of Chemistry, University of St. Thomas, 2115 Summit Avenue, St. Paul, Minnesota 55105
***E-mail: jtippoliti@stthomas.edu**

Organic chemistry laboratory courses present exceptional opportunities to teach students how to utilize NMR to its fullest potential. When implementing experiments that use NMR to analyze products in the undergraduate teaching lab a number of practical problems arise, even with the use of an automated instrument. The pertinent issues in our situation involve sample submission timing, data transfer, and spectra processing. This chapter addresses what can be done to alleviate those problems using a newly designed experiment incorporating flow hydrogenation that has been incorporated into our organic laboratory sequence. Solutions include having a teaching assistant present by the NMR spectrometer room as students are submitting their samples, using Dropbox as a versatile approach to distributing and accessing data, and that iNMR reader is a flexible solution for students to process their data.

Introduction

In 1992 the Department of Chemistry at the University of St. Thomas purchased a Bruker AM300 NMR spectrometer with the help of an NSF- ILI grant (*1*). We fully incorporated this instrument into our curriculum, but by 2009 problems associated with the instrument were becoming increasingly time consuming. Therefore, in the summer of 2009, we wrote an NSF Major Research Instrumentation grant for a new NMR spectrometer. This proposal was funded in 2010 (*2*), and in the summer of that year a new JEOL ECS 400 MHz NMR spectrometer was installed. The most important factor considered when deciding on a new spectrometer was automation. Implementing NMR into our organic labs with our old instrument was problematic because taking spectra and working up data were complex. In particular, placing the samples manually in the magnet, shimming the magnet, and working up the data were sufficiently difficult that some of the faculty teaching our labs (which included many adjunct faculty) were hesitant to learn how to do these things. This made using the NMR in our organic teaching laboratories challenging. From this experience we knew that an automatic sample changer and an instrument capable of automated shimming were a necessity. The JEOL ECS 400 spectrometer we purchased has an auto sample changer that holds 24 samples and has fully automated tuning and shimming. The Delta™ Software (*3*) that controls the spectrometer runs on Apple's OSX system software and is capable of fully automated data acquisition. The instrument is also capable of running No-D NMR (*4*), which enables spectra to be recorded without the use of deuterated solvents. Use of the sample changer, gradient shimming and auto-tuning can automate the entire process of routine analysis, making it ideal for the undergraduate teaching laboratory.

Although the new NMR spectrometer is fully automated, there are still a number of practical problems associated with teaching the students how to use the new instrument, as well as how to access and work-up the associated data. To illustrate the strategies used to overcome these problems, we describe a new experiment, Hydrogenation of Eugenol, we designed and implemented in our Organic Chemistry II laboratory. Continuous-flow reactors are a modern solution to the inherent safety hazards of hydrogenation (*5–12*). Specifically, we used an instrument made by ThalesNano called the H-Cube Continuous-Flow Reactor (*13*). In this instrument, hydrogen gas is generated from the electrolysis of deionized water and is never stored in large quantities. It is immediately combined with the solution containing the unsaturated compound, pressurized, and passed through a pre-packed cartridge of catalyst that can be heated if required. The pressure, temperature, and flow rate are all digitally controlled. The pre-packed cartridges are especially convenient because they eliminate the need to weigh out flammable catalysts. This instrument is state-of-the-art and is starting to be used routinely in research labs around the world.

The incorporation of the H-Cube Continuous-Flow Reactor into the undergraduate curriculum is an ideal way to demonstrate how to safely carry out the hydrogenation of an alkene. To this end, we designed an experiment to perform the hydrogenation of eugenol (Scheme 1) in an undergraduate organic chemistry lab using the H-Cube Reactor.

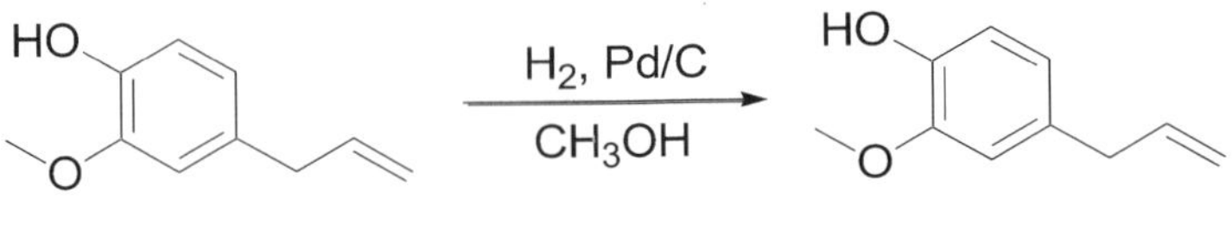

$\xrightarrow[CH_3OH]{H_2,\ Pd/C}$

Scheme 1. Hydrogenation of Eugenol

There are 16 students in each lab section and students work in pairs. Each pair of students hydrogenates an 80 mg sample of eugenol using the H-Cube hydrogenator. ^{1}H NMR spectroscopy is the primary tool for determining if, and to what extent, hydrogenation has occurred. This experiment exemplifies how important NMR spectroscopy is for structure determination, and in particular for following a functional group transformation. As with any experiment involving analysis by NMR spectroscopy, students must be taught how to prepare samples and how to properly insert NMR tubes into the spin collar. This is most easily accomplished using handouts, listing detailed steps with illustrative pictures. The practical problems of timing, data transfer, and data work-up are addressed below.

Experimental

Materials and Instruments

Eugenol was purchased from Sigma-Aldrich Corp. (St. Louis, MO, USA). The H-Cube® Continuous-flow Hydrogenation Reactor was purchased from ThalesNano (Budapest, Hungary).

Typical Experimental Procedure

80 mg of eugenol is added to a 25 mL Erlenmeyer flask. 5.0 mL of methanol is added to the same flask using a 10.0 mL graduated cylinder. A second 25 mL round-bottom flask is weighed and the weight of that flask recorded. This second flask serves as the collecting flask. The outlet tube from the H-Cube Continuous-Flow Reactor is placed into the 25 mL round-bottom collection flask. Within two seconds the sample inlet line is then placed into the 25 mL Erlenmeyer flask containing the eugenol seconds so that no air bubbles travel into the reactor. The sample is then pumped through the reactor and collected in the second round-bottomed flask. Throughout the collection period, the bottom disk located under the cylinder of the sample inlet line remains submerged in the sample solution. When about 1 mL is left in the Erlenmeyer flask, 3.0 mL of methanol is added to the flask using a plastic syringe. Once the solution has dropped to about 1 mL, the collecting capillary is transferred back to the clean methanol vial it was previously stored in. Methanol is transferred for several more minutes to wash out any remaining product in the system.The solvent is then removed from the round-bottomed flask using a rotary evaporator. Finally, the mass of the flask is recorded and the weight of the empty flask subtracted to obtain the actual yield in grams. Using the theoretical and actual yields, the percent yield for the reaction is determined.

For NMR analysis about 5 mm of product is sucked into a clean Pasteur pipette via capillary action. This is then placed into an NMR tube. Approximately 1 mL of deuterated chloroform is added to the top of the pipette containing the sample to wash the sample into the NMR tube.

Results and Discussion

Our initial approach to analyzing student samples was to bring the 24-sample carousel to the lab, wait until all of the students finished the experiment and loaded their tubes in the carousel, and then bring the fully loaded carousel back to the instrument. This approach was quickly discontinued because students finished the experiment at different times causing many to wait extended periods of time for their data. Our revised approach was to have each pair of students bring their freshly prepared NMR samples to the NMR room as soon as they finished. The hydrogenation reaction takes about 8 minutes total (5 minutes to run the eugenol solution through and a 3 minute rinse), and the solvent evaporation takes about 7 minutes. This means that all of the students (8 pairs) can finish the experimental part of the laboratory in about 2 hours. The actual time it takes to get the NMR data, from the time the tube is dropped into the magnet to the point the data file appears, is 7 minutes. Although we assumed the fully automated sample shimming and tuning would alleviate the need for a teaching assistant (TA), we soon found this not to be the case. In fact, we found it necessary for a TA to oversee the students in a number of steps, including placing their tubes in the spin collar, checking to see if the depth was properly adjusted, and placing the sample tube in an empty slot. One problem we did not anticipate was that the students would often lean against the magnet while placing their tube in the sample changer. Under the guidance of a TA, students would type in their filename and slot number, then click on the appropriate NMR experiment, in this case the ^{1}H NMR experiment.

At this point, data transfer and manipulation were our remaing concerns. The problem of data transfer is certainly platform and institution dependent. At St. Thomas, we use Apple computers. The JEOL instrument control software (Delta™) runs on an Apple iMac computer and the organic lab utilizes Apple MacBook Pro laptop computers. The iMac hard drive can be shared on a local network, but only five computers can connect at one time which is problematic. One solution we considered was to install Apple Server software on the computer that stores the data. Apple Server software is not expensive; however, one needs to erase the entire hard drive to install it. We chose not to do this because we would then need to reinstall the NMR software as well. If an institution is installing a new NMR spectrometer and using an Apple Macintosh system to run it, this is certainly something that should be considered. We recently discovered an innovative way around this issue that also allows access to the NMR data from anywhere on or off campus. We used a Dropbox account and created a symbolic link (sym link) for the lab NMR data folder. This is because aliases do not work with Dropbox (*14*). The sym link folder has the same name as the original folder and is placed in the Dropbox so that as soon as a file is created on the hard drive it is automatically copied to the Dropbox folder with the same name. Students can access the data from any type of computer as long as they have a Dropbox account and we have shared access to the folder with them. In our case, we put Dropbox folders on all of our lab computers, which provides a one click link to all of the NMR data.

Once the students have access to their data they need to be able to manipulate it using NMR processing software. For the Apple platform the choices are Delta™,

MestReNova (*15*), and iNMR (*16*). Out of these three we chose iNMR because, in our opinion, it has the cleanest and easiest interface to use. The version of iNMR we chose was iNMR Reader because site licenses are available at reasonable cost. Although iNMR Reader does not allow one to save the worked up spectrum in the program, one can print it or copy and paste it into a report. One can also save a pdf of the spectrum by choosing "Save as PDF" in the print menu. iNMR can also overlay multiple spectra and adjust the vertical offset to any value. This was important to us because it makes it very easy to see differences between starting material and product (see example in Figure 1). Overlaying spectra in this manner illustrates very clearly a functional group transformation. One can also overlay solvent spectra, and providing a folder containing spectra of various solvents is useful for this type of analysis. iNMR automatically processes the data (Fourier transform and phasing), and the students simply open the NMR data file to see the spectrum. There are only three operations the students need to learn: expansion, integration, and overlay. These spectra manipulations can be shown to the students in lab, using a computer attached to a projector, in less than 10 minutes at the beginning of the lab period. To help carry out the comparison of starting material to product and confirm functional group transformation, we provided them with the ^{1}H NMR data of the starting material (eugenol). Then we had the students integrate all the peaks of their spectrum and overlay the spectrum of the starting material. The students quickly learned how to carry out these operations and commented on the ease of use of the iNMR software. The experiment was successfully completed by all of the students in the four hour time frame allotted for the lab. The outcome of the experiment was that nearly all of the students achieved quantitative yields, and through NMR analysis of their products, found that complete hydrogenation had occurred at 20 bar pressure.

We have found that it is quite instructive to look at the change in splitting patterns of the benzylic methylene in the starting material and the product. In the starting material it is a broadened doublet with a coupling constant of 6.7 Hz (labeled S2 in Figure 1). It is interesting to note that the broadened nature is due to the long range coupling to the terminal olefin protons (see Figure 2). After the double bond is hydrogenated, the benzylic methylene is now a triplet with a coupling constant of 7.7 Hz (labeled P2 in Figure 1), typical for vicinal alkyl protons. This notable change in coupling, as well as the change in chemical shift of the methylene protons, is the subject of multiple questions in the pre-lab and lab report (*17*). The lab report also asks the students to compare data obtained from carrying out the hydrogenation at two different pressures (10 and 20 bar). The overlay given in the report of pure eugenol (starting material) with the product isolated from the reaction carried out at the two different pressures is shown in Figure 1. The overlay clearly shows how the olefin peaks (at 5.0 and 5.9 ppm; S4, S5 and S3 respectively) are not present when the reaction is carried out at 20 bar but are present at 10 bar, demonstrating that eugenol is only partially hydrogenated at 10 bar. In the spectrum obtained at 10 bar, the product peaks (benzylic CH_2 at 2.5 ppm; P2) are integrated and compared to the starting material (benzylic CH_2 at 3.3 ppm; S2) and the students are asked to determine the extent of hydrogenation from the integration data.

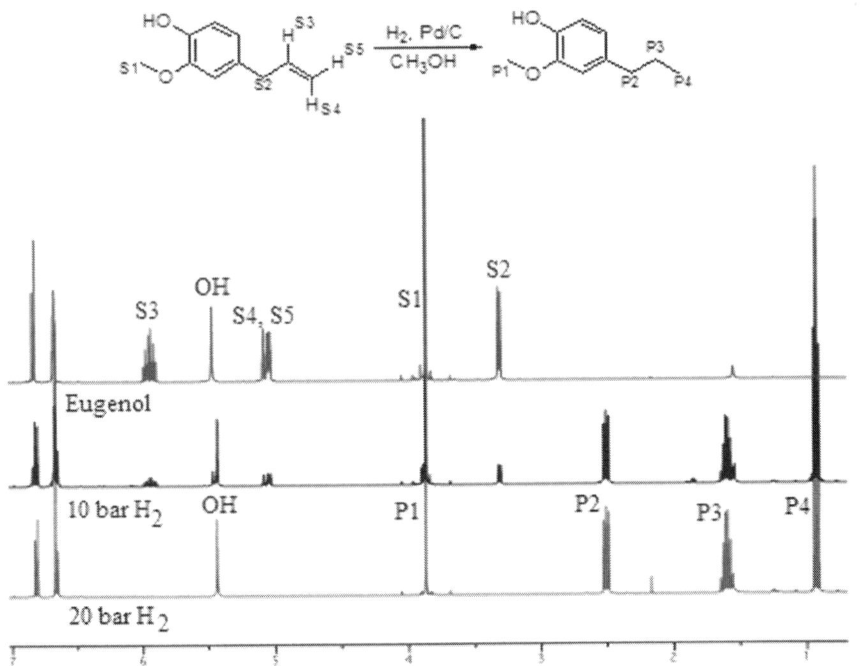

Figure 1. Overlay of ^{1}H NMR ($CDCl_3$) spectra of eugenol with isolated hydrogenation products(aliphatic protons assigned).

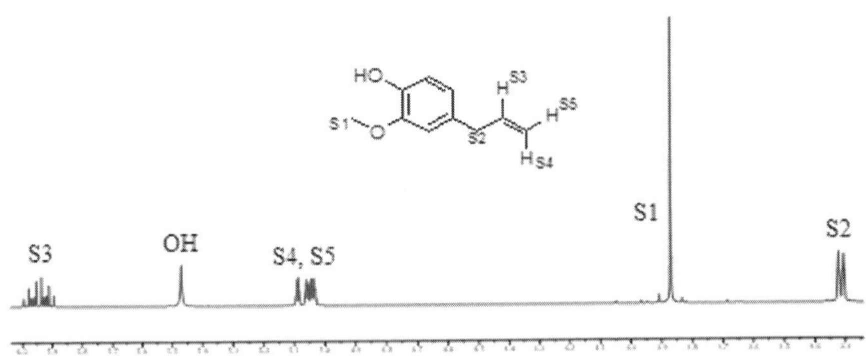

Figure 2. Partial expanded ^{1}H NMR ($CDCl_3$) spectrum of eugenol.

Several changes are being considered for future implementation in this lab. One change would be to utilize the No-D feature of the JEOL spectrometer so that the reaction can be directly analyzed in the solvent used to carry out the reaction (in this case methanol). This change would save time in that the solvent would not need to be removed by evaporation and would decrease costs since deuterated solvents would not be needed (*4*). Also, at 20 bar all of the students saw complete hydrogenation. In future experiments we plan on having groups of students carry out the hydrogenation at different pressures, 5, 10, 15 and 20 bar where we know complete hydrogenation does not occur. Students can then compare their results, allowing us to introduce a more "guided inquiry" approach to this experiment.

Conclusions

In summary, the three most pertinent problems of incorporating NMR with an automated instrument were timing, data transfer, and data work-up. Having a TA in the NMR spectrometer room allows students to run NMR experiments on their products as soon as they are finished. Using Dropbox to share data provided a flexible solution to NMR data transfer and access issues. A signficiant advantage of using Dropbox is that any computer connected to the Internet, even if it is removed from the campus network, can access the NMR data, giving the instructor the option of assigning data workup and analysis as homework. Finally, data workup is platform dependent. Students using a PC have the option of using the free NMR processing software by ACD labs or MestReNova (there is also a beta version of iNMR for the PC available). Students using a Mac have two choices: iNMR or MestReNova. Both of these software companies now offer site licenses at reasonable costs to educational institutions (*15*, *16*). We have found that the overlay feature in iNMR is most instructive for determining functional group transformation. In summary, all of these problem-solving strategies have been implemented in the newly developed Hydrogenation of Eugenol experiment described above and serve as an example of how to take full advantage of NMR spectroscopy in the undergraduate organic laboratory.

References

1. NSF Grant: 9251450 "Integration of High Field, Multinuclear NMR into the Undergraduate Chemistry Curriculum."
2. NSF Grant: CHE-0959322 "MRI-R2: Acquisition of 400 MHz Nuclear Magnetic Resonance (NMR) Spectrometer."
3. http://www.jeol.com/PRODUCTS/AnalyticalInstruments/NuclearMagneticResonance/DeltaNMRSoftware/tabid/395/Default.aspx (accessed November 2012).
4. Hoye, T. R.; Eklov, B. M.; Ryba, T. D.; Voloshin, M.; Yao, L. *Org. Lett.* **2004**, *6*, 953–956.
5. Amoa, K. *J. Chem. Educ.* **2007**, *84*, 1948–1950.
6. De, S.; Gambhir, G.; Krishnamurthy, H. G. *J. Chem. Educ.* **1994**, *71*, 992–993.

7. Hanson, R. W. *J. Chem. Educ.* **1997**, *74*, 430–431.
8. Mohrig, J. R.; Hammond, C. N.; Schatz, P. F.; Davidson, T. A. *J. Chem. Educ.* **2009**, *86*, 234–239.
9. Navarro, D. M.; Navarro, M. *J. Chem. Educ.* **2004**, *81*, 1350–1351.
10. O'Connor, K. J.; Zuspan, K.; Berry, L. *J. Chem. Educ.* **2011**, *88*, 325–327.
11. Plummer, B. *J. Chem. Educ.* **1989**, *66*, 518–519.
12. Wilen, S. H.; Kremer, C. B. *J. Chem. Educ.* **1962**, *39*, 209–210.
13. http://www.thalesnano.com/products/h-cube (accessed November 2012).
14. The two articles we used to understand symbolic links or sym links were: "A brief tutorial on symbolic links" http://hints.macworld.com/article.php?story=2001110610290643 and "Symbolic links made simple" http://www.macworld.com/article/1153437/symlinkservice.html (accessed November 2012).
15. http://mestrelab.com/ (accessed November 2012).
16. http://www.inmr.net/ (accessed November 2012).
17. For digital copies of all of the handouts and the report given to the students, email the principal author at jtippoliti@stthomas.edu.

Chapter 7

Using NMR To Investigate Products of Aldol Reactions: Identifying Aldol Addition versus Condensation Products or Conjugate Addition Products from Crossed Aldol Reactions of Aromatic Aldehydes and Ketones

Nanette M. Wachter*

Chemistry Department, Hofstra University,
Hempstead, New York 11549-1510
***E-mail: chmnwj@hofstra.edu**

Crossed aldol condensation of benzaldehydes with acetophenone is typically part of an undergraduate organic laboratory curriculum. The experiment involves the reaction of the enolate of acetophenone with an aromatic aldehyde generating benzalacetophenone (chalcone). The initially formed β−hydroxyketone rapidly dehydrates to generate an α,β-unsaturated ketone, even under the basic conditions usually employed for the acyl addition reaction. Elimination is favored because of resonance stabilization of the fully conjugated product. However, in the reaction of 2′-hydroxyacetophenone with benzaldehyde, the partially saturated β−hydroxyketone intermediate initially formed can be isolated in moderate to good yield. In a similar experiment, when an excess of acetophenone is treated with 2-pyridine carboxaldehyde, the initially formed α,β-unsaturated ketone undergoes rapid Michael-addition with a second equivalent of the enolate to yield a symmetric diketone. The unanticipated products of both of these reactions contains diastereotopic methylene hydrogen atoms that are easily identifiable and well resolved in the ^{1}H NMR spectrum. Moreover, a COSY experiment clearly demonstrates coupling between the methylene protons with one another and with the proton of the β-carbon. Each of these experiments highlights concepts taught in organic chemistry using techniques traditionally introduced in an introductory organic laboratory course.

Introduction

The Claisen-Schmidt Reaction

The aldol addition reaction is an important synthetic method for the formation of carbon-carbon bonds (*1–4*), and aldol condensations are typically covered in the second semester of the organic chemistry curriculum. An indication of the importance attributed to this reaction is the frequency with which it appears on standardized organic chemistry examinations. In fact, every popular organic chemistry laboratory text has a crossed aldol condensation experiment between an aromatic aldehyde and the enolate of an aromatic ketone that generates *trans*-chalcone (1,3-diphenyl-2-propen-1-one), or a substituted chalcone (*5–8*). The Claisen-Schmidt condensation is a crossed aldol reaction between an aromatic aldehyde and either an aliphatic or aryl ketone that produces an α,β–unsaturated ketone (*9, 10*). Aromatic carboxaldehydes are popular substrates for the reaction because they lack an α–hydrogen and therefore cannot form an enolate anion. While the enolate of the acetophenone could potentially undergo self-condensation, the anion preferentially attacks the carbonyl carbon of the aldehyde because it is more electrophilic than the carbonyl of acetophenone. The crossed aldol experiment that is most typically performed generates an α,β–unsaturated ketone due to rapid elimination of water from the β–hydroxyketone initially formed, as depicted in Scheme 1. The condensation often occurs readily at ambient conditions because the resulting enone is conjugated with both benzene rings (*11*).

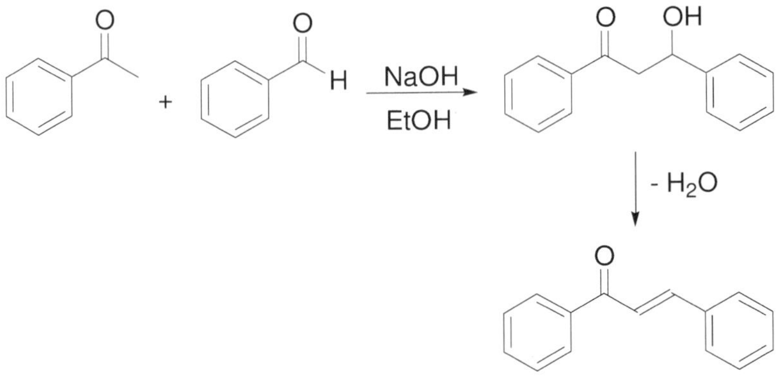

Scheme 1. Crossed aldol condensation of acetophenone with benzaldehyde

Surprisingly, when benzaldehyde is treated with an ethanolic solution of the sodium enolate of 2′-hydroxyacetophenone, as depicted in Scheme 2, the major product obtained after 30 minutes at room temperature is the β–hydroxyketone and not the expected α,β–unsaturated ketone (*12*). A simple ^{1}H NMR spectrum of the addition product demonstrates that the methylene protons on the alpha carbon are non-equivalent. Diastereotopic protons are not magnetically or chemically equivalent to one another and therefore can be readily observed with modern high-field magnets (*13*).

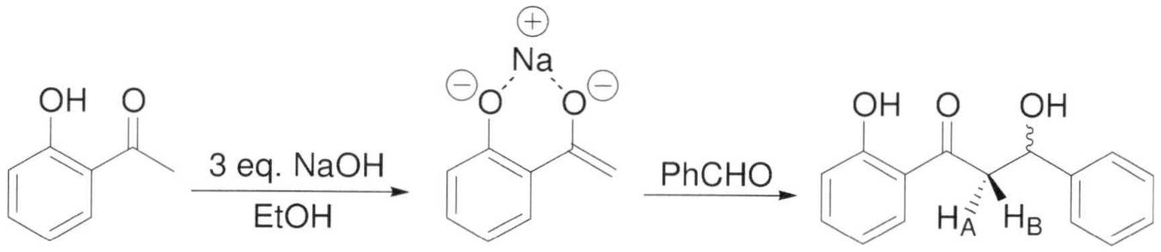

Scheme 2. Aldol condensation of 2'-hydroxyacetophenone with benzaldehyde

A more intriguing example of anisochronous methylene protons is observed in the product resulting from the coupling reaction between acetophenone and 2-pyridinecarboxaldehyde (*14*, *15*). As depicted in Scheme 3, when 2-pyridinecarboxaldehyde is treated with the enolate of acetophenone, the major product isolated is the Michael adduct resulting from the reaction of a second equivalent of the enolate with the initially formed chalcone. If two equivalents of acetophenone and sodium hydroxide are used, the conjugate addition product can be obtained in nearly quantitative yield. In this reaction, the initially formed α,β-unsaturated ketone produced from the condensation of the enolate of acetophenone with 2-pyridinecarboxaldehyde rapidly undergoes conjugate 1,4-addition with the excess enolate to produce a symmetrically substituted diketone. The product, 3-(2-pyridinyl)-1,5-diphenyl-1,5-pentanedione, is *achiral*, yet the methylene protons of the aliphatic chain are nonequivalent as defined by the symmetry criterion of diastereotopic ligands (*16*).

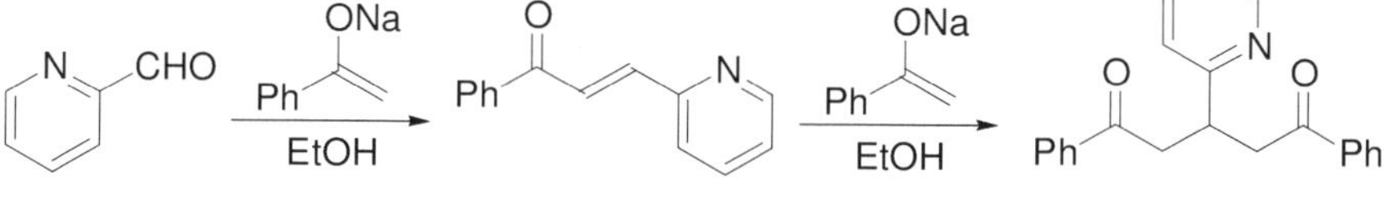

Scheme 3. Crossed aldol condensation and Michael addition of acetophenone with 2-pyridinecarboxaldehyde

Experimental Overview

The synthesis of either compound can be performed in one laboratory period. In the reaction of 2′-hydroxyacetophenone with benzaldehyde, three equivalents of hydroxide are added to ensure that there is sufficient base present, since one equivalent of base will immediately remove the phenolic proton on the acetophenone. Acetophenone, benzaldehyde, 2′-hydroxyacetophenone and 2-pyridinecarboxaldehyde were purchased from Aldrich Chemical Company. The sodium hydroxide solutions were prepared ahead of time.

Hazards

Sodium hydroxide is caustic and causes severe burns. The concentrated solution should be handled with gloves and appropriate eye protection. Acetophenone, 2′-hydroxyacetophenone, benzaldehyde and 2-pyridinecarboxaldehyde are irritating to the eyes, respiratory system and skin. They should be dispensed in a hood and suitable protective clothing should be worn. Acetophenone, ethanol and benzaldehyde are combustible and must be kept away from sources of ignition. No flames should be present in the laboratory when the experiment is being performed. Hydrochloric acid is corrosive and causes severe burns. Appropriate eye protection should be worn when handling the acid solution.

1-(2′-Hydroxyphenyl)-3-phenyl-2-propenone

0.106 mL (1 mmol) of benzaldehyde is dissolved in 1 mL of 95% ethanol in a 5-mL conical vial equipped with a magnetic spin vane. While stirring, 0.132 mL (1.1 mmol) of 2′-hydroxyacetophenone is added to the reaction vial. An air condenser is attached and 0.500 mL of 6 M aqueous sodium hydroxide is carefully added to the solution. A bright yellow color develops. The mixture is stirred at room temperature for 30 min. During the reaction a light orange precipitate forms. The reaction is quenched by adding 0.5 mL of water then 3 M aqueous HCl dropwise until the mixture is at neutral pH. The solid is collected using a Hirsch funnel and the filter cake is washed with cold ethanol. Recrystallization from 50% ethanol affords the product.

3-(2-Pyridinyl)-1,5-diphenyl-1,5-pentanedione

To a 5-mL conical vial equipped with a magnetic spin vane are added 0.10 mL (1.1 mmol) of 2-pyridinecarboxaldehyde, 0.26 mL (2.2 mmol) of acetophenone and 1 mL of 95% ethanol. The solution is stirred at room temperature and 1.0 mL of 2.5 M aqueous sodium hydroxide is added. The solution acquires a golden tint and a white precipitate develops. The mixture is stirred for 15 minutes at room temperature and then poured into a 10-mL beaker containing 2-mL of ice cold water. The alkaline reaction mixture is neutralized by dropwise addition of 3 M aqueous HCl. The solid is collected by vacuum filtration and the filter cake rinsed with small portions of cold ethanol. Recrystallization from 95% ethanol affords the product.

Discussion

Aldol Reaction of 2′-Hydroxyacetophenone with Benzaldehyde

The crossed aldol reaction between an aromatic aldehyde and ketone typically generates a β-hydroxyketone that readily dehydrates, even at room temperature, to produce a highly-conjugated α,β-unsaturated ketone. Base-promoted dehydration favors formation of a trans substituted double bond due to unfavorable steric interactions in the transition state for the formation of the cis alkene (Figure 1) (*11*).

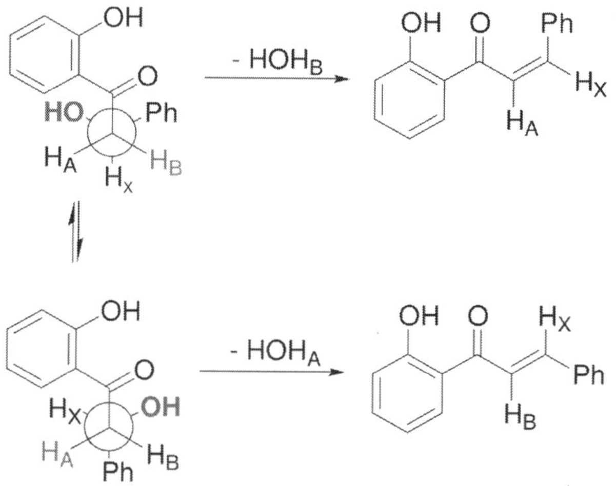

Figure 1. Newman projections of aldol adduct conformations for base-promoted dehydration to produce a cis- or trans-alkene.

Under alkaline conditions, the reaction proceeds through an E1cB pathway in which deprotonation of the aldol addition product precedes elimination of the beta hydroxyl group (Scheme 4).

In the reaction of 2′-hydroxyacetophenone with benzaldehyde, the ortho hydroxyl group of the addition product is believed to stabilize the enolate thereby slowing the elimination reaction (Figure 2). Since heat promotes dehydration, the chalcone can be obtained in good yield if the reaction is heated for 60 minutes. The two reaction products can be distinguished by their color; the chalcone is bright yellow, while the partially saturated β–hydroxyketone is colorless.

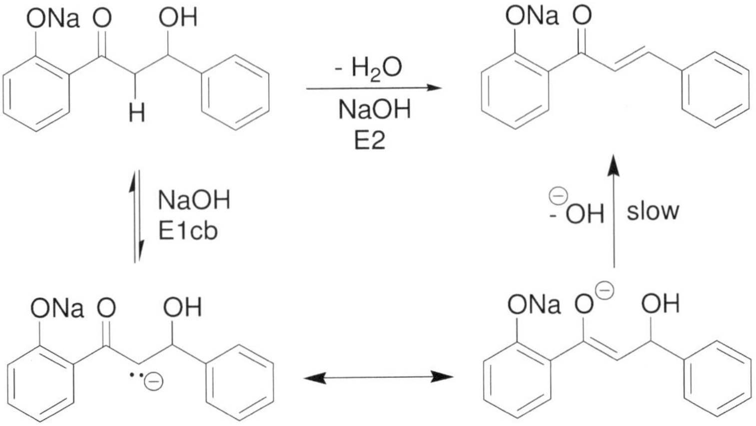

Scheme 4. E1cB pathway for the dehydration of a β-hydroxyketone

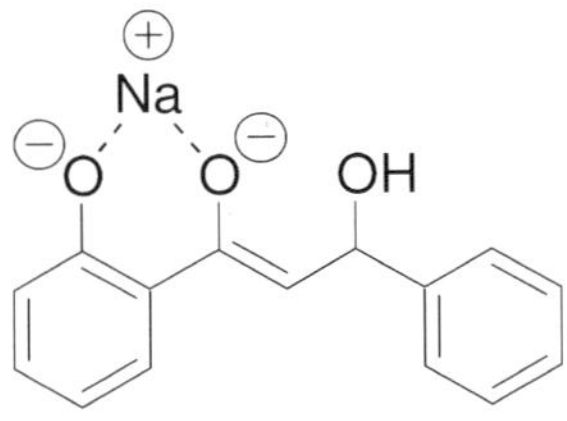

Figure 2. Stabilized enolate anion of β-hydroxyketone intermediate.

Perhaps of greater interest than the unexpected outcome of the reaction is the anisochrony of the methylene protons in the acyclic chain that connects the aromatic rings. ^1H NMR analysis of the white solid obtained in this experiment reveals three signals between 2 and 6 ppm produced by the diastereotopic methylene protons of the α carbon and the β hydrogen of the β–hydroxyketone which are all coupled to one another (Figure 3). This is a classic ABX pattern; each of the three protons on the aliphatic chain gives rise to a doublet of doublets. Typically, rotationally restricted or chiral molecules are employed to depict diastereomerism in undergraduate laboratories (*17–19*). Indeed, the β carbon of the ketone is a stereogenic center, conferring magnetic anisochrony to the α-methylene protons of the aliphatic chain. Not only are the chemical shifts of the α methylene hydrogen atoms appreciably different from one another, but a rather significant coupling constant for the geminal protons is observed. The relatively large J value is due to π electron donation by the benzoyl moiety (*13*). It should be pointed out that coupling over two bonds, (i.e., geminal coupling) results in parallel spin polarizations and J, by convention, has a negative sign so $^2J_{HCH}$ = -17 Hz. Geminal couplings are dependent on the H-C-H bond angle and typically range from -5 to -20 Hz for alkanes (*13*). The vicinal coupling constants between each of the methylene protons with the α-proton are also significantly different resulting in the multiplicities of the hydrogen atom's signal to appear as a doublet of doublets. Splitting diagrams that depict proton-proton coupling are a useful instructional tool to illustrate the effect of the coupling constant magnitude on the multiplicity of the signal. A sample tree diagram that uses the absolute values of the germinal and vicinal coupling constants for the aliphatic hydrogen atoms of the β-hydroxyketone is shown in Figure 4 below.

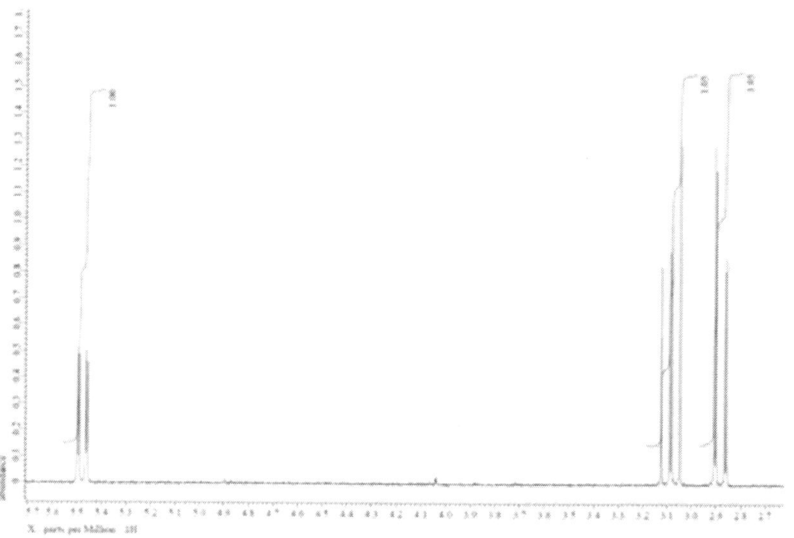

Figure 3. ^{1}H NMR spectrum of the alkyl region of 3-hydroxy-1-(2'-hydroxyphenyl)-3-phenyl-1-propenone depicting the alpha protons at 2.84 ppm (^{1}H, doublet of doublets, J = 17 and 3 Hz) and 3.15 ppm (1H, doublet of doublets, J = 17 and 13 Hz), and the beta proton at 5.62 ppm (1H, doublet of doublets, J = 13 and 3 Hz).

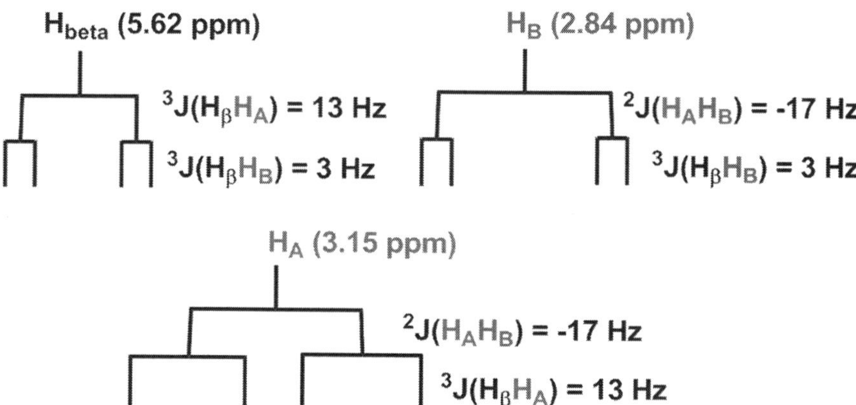

Figure 4. Tree diagrams depicting coupling between the aliphatic protons of 3-hydroxy-1-(2'-hydroxyphenyl)-3-phenyl-1-propenone.

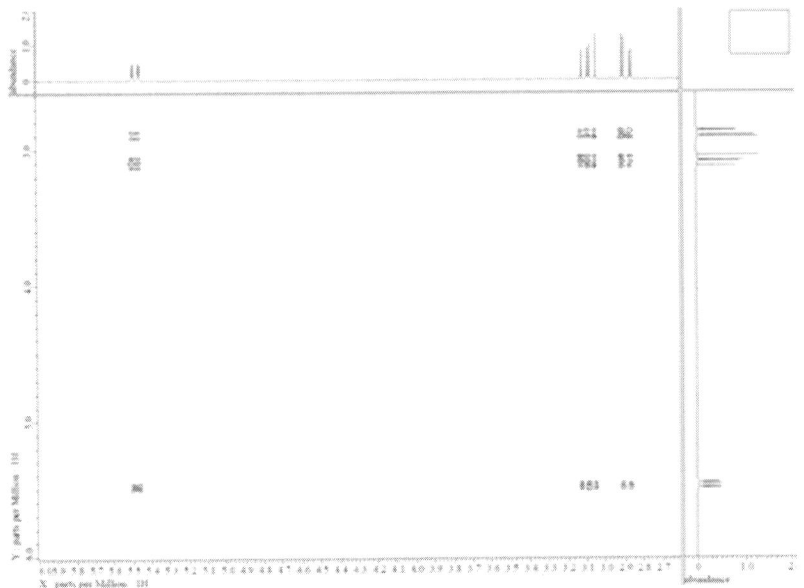

Figure 5. Aliphatic region of COSY of 3-hydroxy-1-(2'-hydroxyphenyl)-3-phenyl-1-propenone demonstrating coupling between the geminal α-protons and the β-proton.

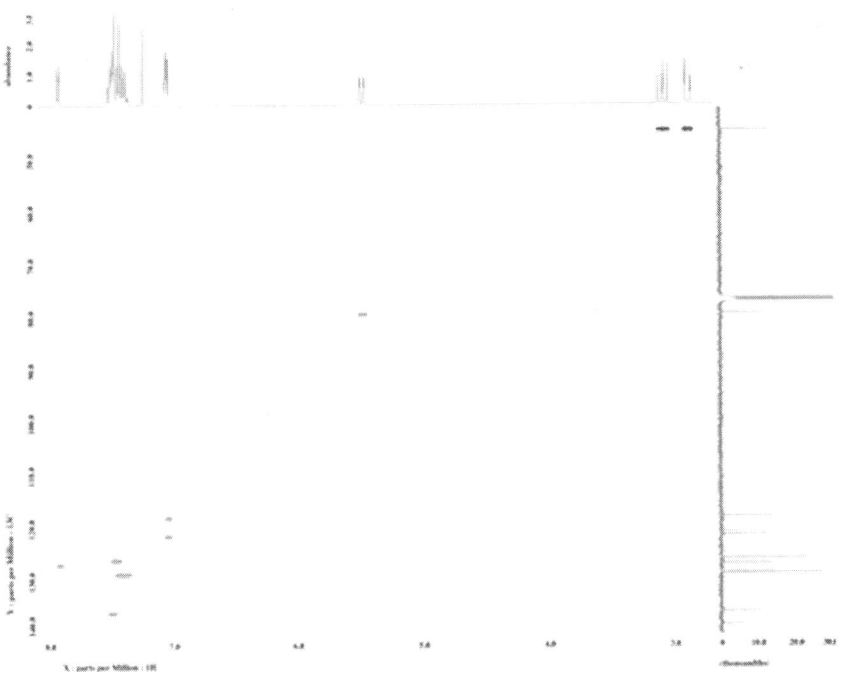

Figure 6. HETCOR of 3-hydroxy-1-(2'-hydroxyphenyl)-3-phenyl-1-propenone.

When a COSY experiment is performed, coupling between the proton on the β carbon to each of the diastereotopic α-hydrogen atoms is clearly evident (Figure 5).

The ^{13}C NMR spectrum further supports the presence of two aliphatic carbons in the product and a HETCOR experiment can be performed to demonstrate that the β-carbon is coupled to two protons that are chemically (and magnetically) distinct from one another (Figure 6).

Tandem Aldol-Michael Reactions of 2-Pyridinecarboxaldehyde with Acetophenone

An even more interesting outcome is obtained from the base-promoted reaction of acetophenone with 2-pyridinecarboxaldehyde. In this reaction, the product isolated is also not the anticipated α,β-unsaturated ketone. The reaction mechanism clearly involves the initial formation of a β-hydroxyketone that *does* dehydrate to produce an α,β-unsaturated ketone; however, the unsaturated intermediate rapidly undergoes 1,4-conjugate (Michael) addition with a second equivalent of the enolate as depicted earlier in scheme 3. The 2-pyridine ring of the aldolate is believed to facilitate conjugate addition by coordinating with the sodium-enolate complex; in essence, acting as a Lewis base catalyst for the reaction. Lewis bases have been shown to catalyze aldol reactions by altering the aggregation state of the metal enolate (*20, 21*). In fact, if pyridine is added to the reaction the yield of diketone is dramatically reduced suggesting that pyridine is competing with the 2-pyridinyl chalcone formed for coordinating the sodium enolate.

It is perhaps more difficult to recognize the anisochrony of the methylene protons in the symmetric 1,5-diketone obtained in this reaction. Clearly, the product lacks a chiral center. However, closer inspection reveals that the geminal protons are not interchangeable with one another by a symmetry operation. Replacing one of the methylene protons with another substituent generates *two* stereogenic carbons (Figure 7).

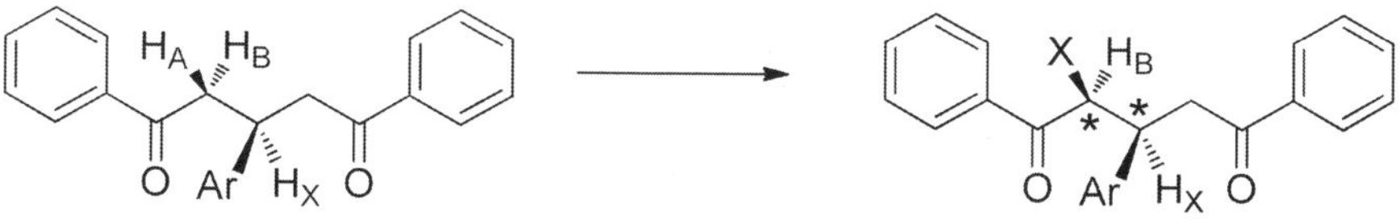

Figure 7. Two new stereocenters form when one of the methylene protons is replaced by a substituent demonstrating that the geminal hydrogen atoms are diastereotopic.

Proton NMR clearly demonstrates the non-equivalency of each set of methylene protons in the large geminal couplings observed (^{2}J = -17 Hz) which can be seen in Figure 8. Splitting diagrams similar to those depicted in Figure 4 can be constructed to illustrate proton-proton coupling and the resulting multiplicities of the signals of the aliphatic hydrogen atoms.

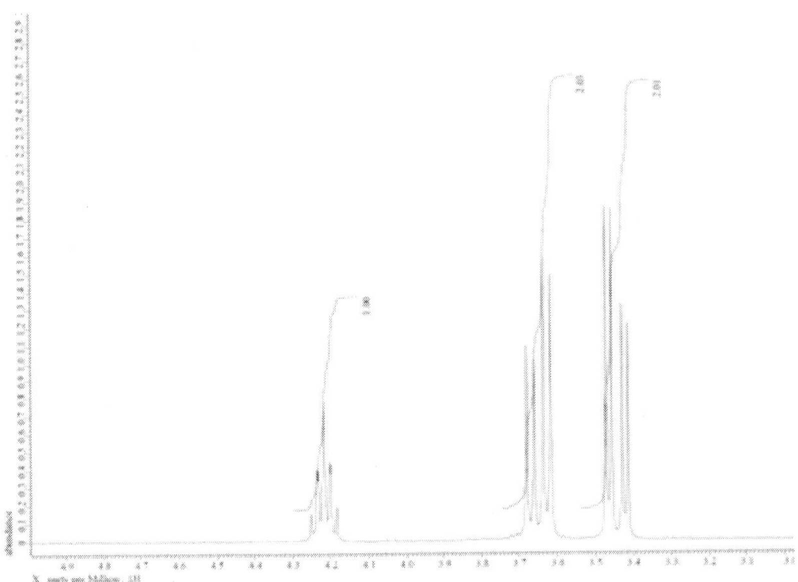

Figure 8. Aliphatic region of the H NMR spectrum of 3-(2-pyridinyl)-1,5-diphenyl-1,5-pentanedione.

Two-dimensional experiments, such as COSY, can also be used to confirm that the methylene protons are not equivalent (Figure 9).

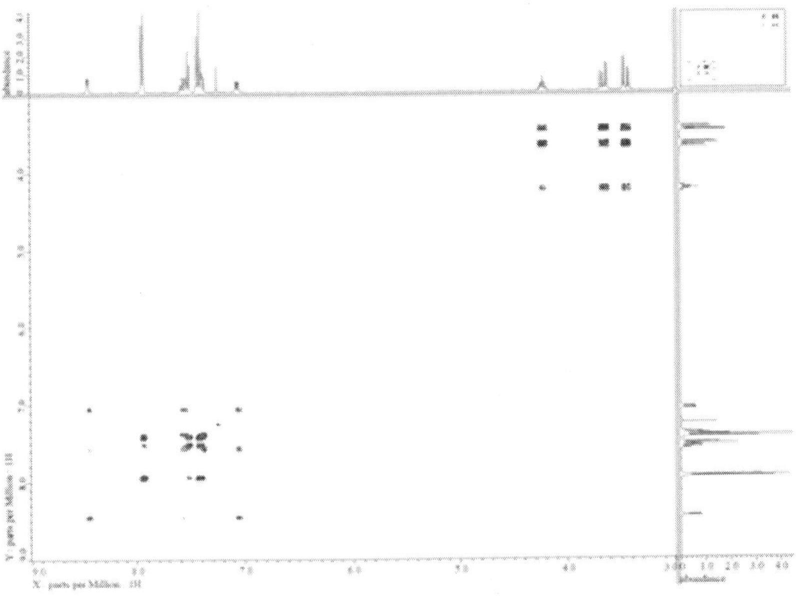

Figure 9. COSY of 3-(2-pyridinyl)-1,5-diphenyl-1,5-pentanedione.

Conclusions

Each of the experiments presented in this chapter highlight concepts discussed in the undergraduate organic chemistry curriculum and the importance of NMR spectroscopy for structural analysis. Since mixed aldol reactions between aromatic aldehydes and ketones typically produce α,β-unsaturated ketones, these experiments are especially useful as guided-inquiry laboratories. Moreover, the pervasiveness of high field instruments in many undergraduate institutions facilitates demonstrations of anisochrony (distinct chemical shifts) of diastereotopic methylene protons. Additionally, theses labs can be used to introduce students to two-dimensional NMR analysis by performing COSY and HETCOR experiments on the products.

References

1. Nielsen, A. T.; Houlihan, W. J. *Org. React.* **1968**, *16*, 1–438.
2. Hajos, Z. G. In *Carbon-Carbon Bond Formation*; Augustine, R. L., Ed.; Marcel Dekker: New York, 1979; Vol. 1, Chapter 1.
3. Mukaiyama, T. *Org. React.* **1982**, *28*, 203–331.
4. Heathcock, C. H. In *Comprehensive Carbanion Chemistry*; Durst, T., Buncel, E., Eds.; Elsevier: Amsterdam, 1983; Vol. II, Part B.
5. Gilbert, J. C.; Martin, S. F. *Experimental Organic Chemistry, A Miniscale and Microscale Approach*, 4th ed; Thomson Brooks/Cole: Belmont, CA, 2006; pp 603–5.
6. Mayo, D. W.; Pike, R. M.; Forbes, D. C. *Microscale Organic Laboratory*, 5th ed.; John Wiley & Sons: Hoboken, NJ, 2011; pp 309–16.
7. Mohrig, J. R.; Hammond, C. N.; Schatz, P. F.; Morrill, T. C. *Modern Projects and Experiments in Organic Chemistry: Miniscale and Standard Taper Microscale*, 2nd ed.; W. H. Freeman: New York, 2003; pp 353–8.
8. Pavia, D. L.; Lampman, G. M.; Kriz, G. S.; Engel, R. G. *Introduction to Organic Laboratory Techniques, a Microscale Approach*, 4th ed.; Thomson Brooks/Cole: Belmont, CA, 2007; pp 339–42.
9. Smith, M. B. *Organic Synthesis*; McGraw-Hill, Inc.: New York, 1994; pp 885–6.
10. Wachter-Jurcsak, N.; Zamani, H. *J. Chem. Educ.* **1999**, *76*, 653–4.
11. Carey, F. A.; Sundberg, R. J. *Advanced Organic Chemistry; Part A: Structure and Mechanisms*, 5th ed.; Springer Science: New York, 2007; pp 682–7.
12. Wachter, N. M.; Finn, M.; Mitchell, J.; Schmelkin, C. Manuscript in preparation.
13. Lambert, J. B.; Shurvell, H. F.; Lightner, D. A.; Cooks, R. G. *Organic Structural Spectroscopy*; Prentice-Hall, Inc.: Upper Saddle River, NJ, 1998; pp 65–72.
14. Wachter-Jurcsak, N. M.; Radu, C.; Redin, K. *Tetrahedron Lett.* **1998**, *39*, 3903–6.
15. Wachter-Jurcsak, N.; Redin, K. *J. Chem. Educ.* **2001**, *78*, 1264–5.
16. Brisbois, R. G.; Batterman, W. G.; Kragerud, S. R. *J. Chem. Educ.* **1997**, *74*, 834–5.

17. Magner, J. T.; Selke, M.; Russell, A. A.; Chapman, O. L. *J. Chem. Educ.* **1996**, *73*, 854–856.
18. Almy, J.; Alvarez, R. M.; Fernández, A. H.; Vázquez, A. S. *J. Chem. Educ.* **1997**, *74*, 1479–1482.
19. Eliel, E. L.; Wilen, S. H. *Stereochemistry of Organic Compounds*; John Wiley & Sons, Inc.: New York, 1994; pp 477–80.
20. Hall, P. L.; Harrison, A. T.; Fuller, D. J.; Collum, D. B. *J. Am. Chem. Soc.* **1991**, *113*, 9575–85.
21. Willard, P. G.; Hintze, M. J. *J. Am. Chem. Soc.* **1990**, *112*, 8604–14.

Chapter 8

Use of HSQC, HMBC, and COSY in Sophomore Organic Chemistry Lab

V. R Miller*

Roanoke College, 221 College Lane, Salem, Virginia 24153
***E-mail: miller@roanoke.edu**

NMR spectroscopy is a very useful technique for experimentally determining the structures of organic compounds. This chapter describes the use of the HSQC, HMBC, and COSY experiments in sophomore Organic Chemistry laboratory for determining the structures of reaction products resulting from Friedel-Crafts acylation and Fischer esterification reactions. These spectra were recorded using the settings that came with the instrument so that the students could concentrate on interpreting the spectra and not be concerned with how the spectra were obtained.

Introduction

The introduction and use of modern instrumentation is a very important aspect of the chemistry curriculum at Roanoke College. In first year General Chemistry laboratory the students use ^{1}H and ^{13}C NMR spectroscopy in the first term to determine an unknown. In the second term they use ^{1}H NMR spectroscopy to characterize aspirin that has been synthesized. In the second year Organic Chemistry laboratory, ^{1}H and ^{13}C NMR spectroscopy are used in the first few weeks of laboratory experiments, with HSQC (Heteronuclear Single Quantum Correlation) being introduced a few weeks later. Our goals for using NMR in organic laboratories are (i) to get the students to think more and understand more about structures, (ii) to allow the students to see for themselves that an experiment or a separation does what it is supposed to do, (iii) to give the students experience with a powerful instrument, and (iv) to improve problem solving skills. Most introductory organic chemistry courses discuss ^{1}H and ^{13}C NMR spectroscopy, but few do much with 2D NMR techniques. For the past five years Roanoke College has been working to incorporate HSQC (Heteronuclear Single Bond Quantum Correlation), HMBC (Heteronuclear Multiple Bond Correlation), and COSY (^{1}H-^{1}H Correlation Spectroscopy) into the first year of Organic Chemistry.

A ^{1}H NMR spectrum gives the most information about the structure of a compound, while a ^{13}C NMR spectrum gives other useful information. However, it is easier to interpret a ^{13}C NMR spectrum than a ^{1}H NMR spectrum. An HSQC spectrum gives little direct structural information, but it is easy to interpret. A signal (cross peak) in an HSQC spectrum shows a correlation between a ^{1}H signal and a ^{13}C signal, thus showing that the hydrogen atom(s) and carbon atom responsible for the ^{1}H and ^{13}C signals is (are) directly bonded to one another. For many chemical reactions in typical organic chemistry laboratory experiments, only one product is expected and its structure can be predicted. Then ^{1}H, ^{13}C, and HSQC NMR spectra can be used to verify the structure of the product and correlate specific signals and their chemical shift values with most of the atoms in the compound. For other reactions however, more than one structure is reasonable and more information is necessary to determine the structure of the product. In these reactions, HMBC and COSY spectra can be very important. This chapter discusses several specific reactions for which ^{1}H, ^{13}C, and HSQC spectra do not provide sufficient information to make a structural determination. In these cases HMBC and COSY spectra can provide additional information to make a definitive structural determination.

Background

While the acquisition of ^{1}H and ^{13}C NMR spectra has been routine for a number of years, recent advances in NMR instrumentation has allowed the acquisition of various 2-D NMR spectra to become routine. NMR instruments now have "push button" acquisition of different 2-D NMR experiments, and this has allowed these techniques to be used in introductory undergraduate organic chemistry courses.

The technology behind the acquisition of some 2-D spectra is quite demanding, particularly for HMBC. The pulse sequence is particularly complicated and needs to be set correctly and implemented exactly (*1*). However, routine acquisition of NMR data has now reached the point that the methods in this chapter have been used without the need for understanding the instrumental complexities involved, and the spectra in this chapter were obtained with the parameters that came with the instrument. In our case, when a student submits his/her sample he/she only has to click on the appropriate buttons (^{1}H, ^{13}C, HSQC, HMBC, and/or COSY) to select the appropriate experiment. Similarly, data retrieval and printing used to be complicated, particularly with 2-D spectra since they required the addition of high-resolution 1-D spectra. Fortunately, software has advanced to the point that a single click on the appropriate button will print the 2-D spectra, and if the appropriate high-resolution 1-D spectra are available, these will be added. Examples of these types of spectra are included in this chapter.

While it is not necessary to understand the math and theory underlying the HSQC and HMBC experiments to use them in determining structures, it is important to be able to interpret the spectra. Figure 1 shows the HSQC spectrum (center box) of butyl butanoate, with the 1-D ^{1}H NMR spectrum on the top and the 1-D ^{13}C NMR spectrum on the left. An HSQC spectrum shows which hydrogen atom(s) is (are) directly attached (one bond) to which carbon atom. A cross peak (circle or series of concentric circles) shows that the hydrogen atom(s) responsible for the signal directly above the cross peak is (are) directly coupled to the carbon atom responsible for the signal directly to the left. The sign of the HSQC cross peaks (often shown in color if phase sensitive HSQC is used) distinguishes between an even versus an odd number of hydrogen atoms on a given carbon atom. Lines were drawn on the spectrum to show which hydrogen atoms were coupled to which carbon atoms. Note that this HSQC spectrum clearly shows the locations of the two hydrogen atoms with chemical shift values near 1.6 ppm. If both the ^{1}H and ^{13}C chemical shifts are very similar, the HSQC cross peaks will overlap and will not be able to be resolved at lower field strengths. This is the case for the two methyl groups with ^{1}H signals near 0.9 ppm in the ^{1}H spectrum with the associated carbons near 14 ppm in the ^{13}C NMR spectrum.

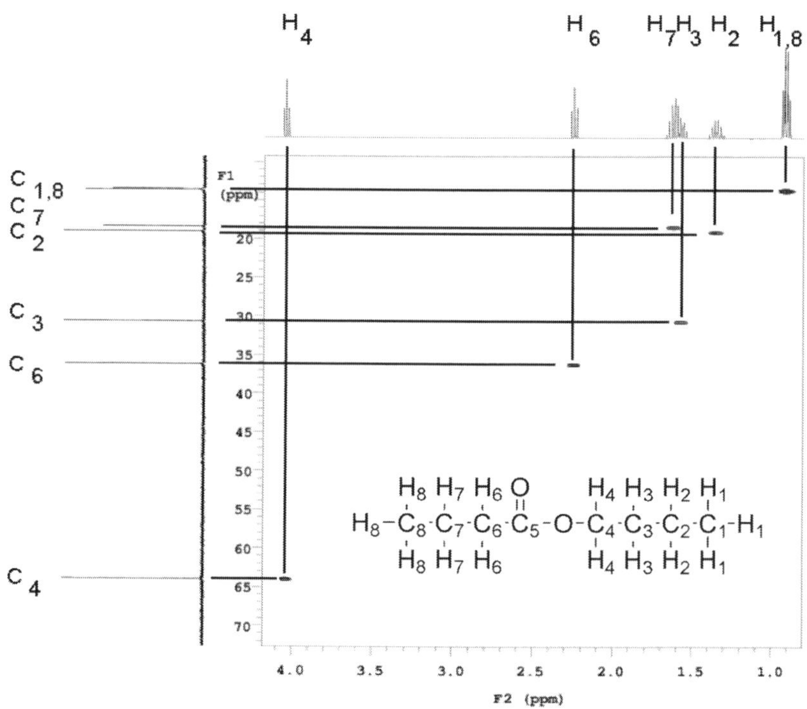

Figure 1. HSQC spectrum of butyl butanoate, with lines drawn in to show which ^{1}H signal is directly coupled to which ^{13}C signal.

While the HSQC experiment gives information as to directly attached hydrogen and carbon atoms, the HMBC experiment gives information as to which carbon atoms are two or three bonds away from a particular hydrogen atom. Unfortunately, only some of the carbon atoms that are two or three bonds away will show cross peaks for a given hydrogen atom. Occasionally, a carbon atom that is one or four bonds away will also give a cross peak.

An HMBC spectrum looks very similar to an HSQC spectrum, except there are typically 1-4 cross peaks for a given hydrogen signal. Figure 2 shows the HMBC spectrum of butyl butanoate. Drawing a line straight down from the hydrogen signal at 4.05 ppm to the first cross peak, and then straight left to the ^{13}C signal at 19 ppm, shows that the hydrogen atom whose signal is at 4.05 ppm is two or three bonds from the carbon whose signal is at 19 ppm. Continuing on down to the next cross peak shows that this same hydrogen is two or three bonds from the carbon whose signal is at 31 ppm. The last cross peak shows that this same hydrogen is two or three bonds from the carbon whose signal is at 174 ppm. The intensities of cross peaks can be quite different from each other so it is sometimes necessary to examine several different intensities (or scaling) of spectra to identify all of them.

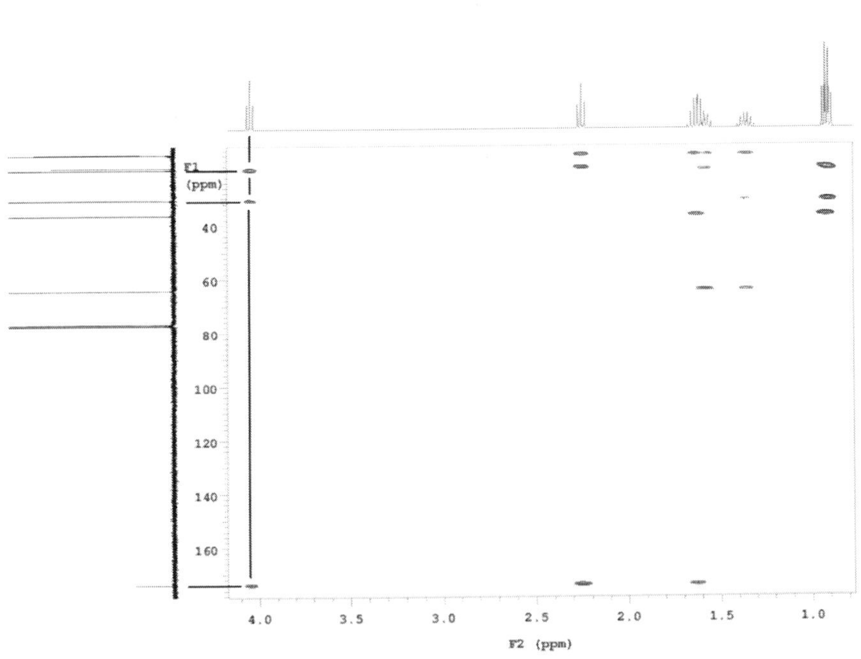

Figure 2. HMBC spectrum of butyl butanoate, with lines drawn in to show which ^{13}C signals are two or three bonds from the hydrogen at 4.05 ppm.

Neither the HSQC nor the HMBC experiments actually count bonds, but instead these experiments are responsive to the magnitude of the spin-spin coupling constant between the atoms. The HSQC experiment is optimized to give a signal for a hydrogen atom and a carbon atom that have coupling constants on the order of 150 Hz. The HMBC experiment is optimized to show a cross peak for a coupling constant in the vicinity of 8 Hz, and is optimized to suppress cross peaks for a coupling constants on the order of 150 Hz. Many good references are available that give more information on the theory and applications of 2D NMR (*2–4*).

Experimental

NMR spectra were recorded with an Agilent 400 MR with about 20 mg of sample in $CDCl_3$. The total experiment time per sample to run ^1H, ^{13}C, HSQC, HMBC, and COSY experiments was about an hour. The experimental parameters used for each experiment were unchanged from the parameters that are preloaded into the Agilent VnmrJ software, Version 3.2, revision A. While the spectra in this chapter are referred to as HSQC, HMBC, and COSY, the spectra that actually were recorded were HSQCAD, gHMBCAD, and gCOSY, respectively, where the g indicates a gradient version of the experiment and AD indicates an adiabatic version of the experiment. ^1H NMR spectra were recorded with an acquisition time of 2.6 s, a relaxation delay of 1.0 s, a 45° observe pulse, and 8 scans, for a total acquisition time of 29 sec. While these settings might not have allowed for complete relaxation, the relative integral values were sufficient for analysis. ^{13}C NMR spectra were recorded with an acquisition time of 1.3 s, a relaxation delay of 1.0 s, a 45° observe pulse, and 256 scans, for a total acquisition time of 9 min, 52 s. For HSQC spectra, the acquisition in F2 (hydrogen) had an acquisition time of 0.15 s and a relaxation delay of 1.0 s with a 45° pulse. The F1 (carbon) acquisition had 96 t1 increments with two scans per t1 increment, for a total acquisition time of 10 min, 37 s. ^1H-^{13}C multiplicity was enabled. The parameters were optimized for a one-bond ^1J X-H coupling constant of 146 Hz. For HMBC spectra the acquisition in F2 (hydrogen) had an acquisition time of 0.15 s and a relaxation delay of 1.0 s with a 45° pulse. F1 (carbon) acquisition had 200 t1 increments with four scans per t1 increment, for a total acquisition time of 33 min, 31 s. The parameters were optimized to detect multiple bond coupling constants of ^nJ X-H of 8 Hz. A two-step ^1J X-H filter was applied. For COSY spectra the acquisition in F2 had an acquisition time of 0.15 s and a relaxation delay of 1.0 s with a 45° pulse. F1 acquisition had 128 t1 increments with one scan per t1 increment, for a total acquisition time of 3 min, 10 s.

The Friedel-Crafts acylation reaction was adapted from the Friedel-Crafts experiment by Pavia et.al. (*5*). In our case, the separatory funnels, round bottom flasks, condensers, and Claisen adaptors were dried overnight in an oven at 100-120 °C. During the reaction the system was protected from moisture by the addition of two drying tubes of anhydrous calcium chloride. For a typical experiment 35 mmol of acetyl chloride in 10 mL of dichloromethane was added over 10 min to

41 mmol of anhydrous aluminum chloride in 10 mL of dichloromethane cooled in ice. Next, 26 mmol of cymene in 10 mL of dichloromethane was added over 10 minutes. The reaction mixture then was removed from the ice water bath and allowed to warm to room temperature over 30 minutes. This mixture was added to 8 mL of concentrated hydrochloric acid and 20 g of ice and then mixed for 10 minutes. The resulting mixture was separated and the aqueous layer extracted with 10 mL of dichloromethane. The combined dichloromethane layers were washed with 2-20 mL portions of saturated aqueous sodium bicarbonate, dried with anhydrous sodium sulfate, and the solvent removed under vacuum. The crude product was then analyzed without further purification.

Friedel-Crafts Acylation Reactions

The Friedel-Crafts acylation experiment is implemented in our second term Organic Chemistry laboratory curriculum. By this time students have had 8-10 experiments in which they have interpreted ^{1}H and ^{13}C NMR spectra, and three to four experiments in which they interpreted HSQC spectra. This prior use of HSQC spectra was concerned with pairing ^{13}C NMR signals with ^{1}H NMR signals. At this stage in their education, students have heard of the HMBC experiment and have used information obtained from a HMBC spectrum, but they have yet to process their own data.

Friedel-Crafts Reaction of *p*-Methyl Anisole with Propionyl Chloride

Most products arising from a Friedel-Crafts acylation reaction can be readily identified using ^{1}H NMR spectroscopy. However, one starting substituted benzene that requires the use of HMBC to completely identify the product in this type of reaction is *p*-methyl anisole. Figure 3 shows the reaction of *p*-methyl anisole with propionyl chloride, with the two potential products that could be formed.

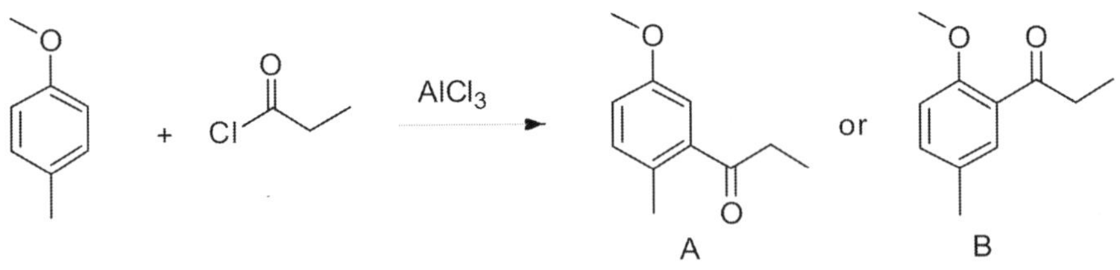

Figure 3. Friedel-Crafts reaction of p-methyl anisole with propionyl chloride.

For ease in identifying the atoms, the numbering system shown in Figure 4 has been used for the two possible products. While this is not a standard numbering scheme, it has been used for this discussion.

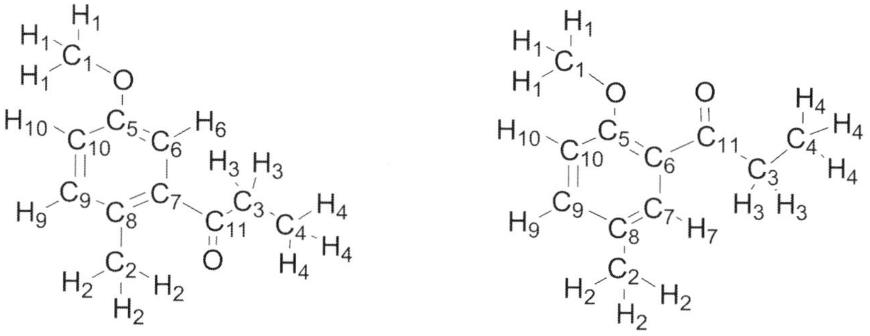

Figure 4. The potential products resulting from the Friedel-Crafts acylation reaction of p-methyl anisole with propionyl chloride.

There are various different paths to take to develop the structure of a compound from NMR data. Experienced users often will start with the ^{1}H NMR spectrum and then go to the ^{13}C and HSQC NMR spectra. If this does not provide enough information, then they will gain additional information from HMBC and /or COSY spectra. One path we are using with our students starts with the ^{13}C NMR spectrum, and then uses the HSQC spectrum to get approximate shift values of the directly attached hydrogen atoms. We then use the ^{1}H NMR spectrum to obtain the accurate chemical shift values, integral values, and appearances of the hydrogen signals.

Discussed below is this latter path using information from a student-prepared sample from the Friedel-Crafts acylation reaction of *p*-methyl anisole with propionyl chloride. Figure 5 shows the ^{13}C NMR spectrum of this product. From this spectrum are obtained the chemical shift values of each of the carbon atoms. These are placed in a table similar to Table 1 below. It is beneficial to record the chemical shift values to the tenth of a ppm because occasionally it is necessary to differentiate signals that are very close to each other.

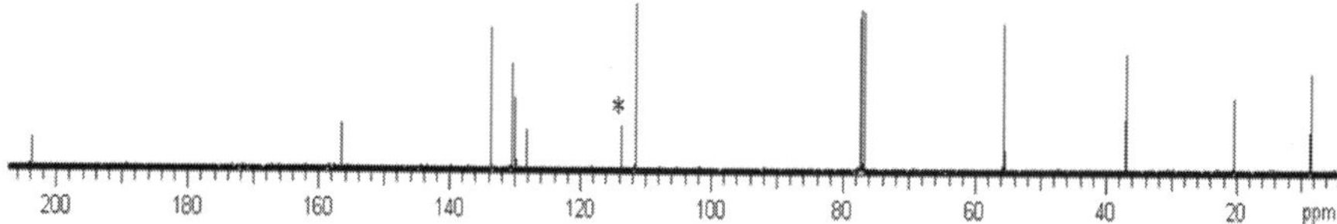

Figure 5. ^{13}C NMR spectrum of the reaction product of p-methyl anisole with propionyl chloride. The peak indicated with a star(*) is p-methyl anisole.

Figure 6 shows the HSQC spectrum, with the cross peaks indicating correlations between carbon atoms and directly bonded hydrogen atoms. These approximate hydrogen chemical shifts are placed in the same row in the table as is the directly attached carbon.

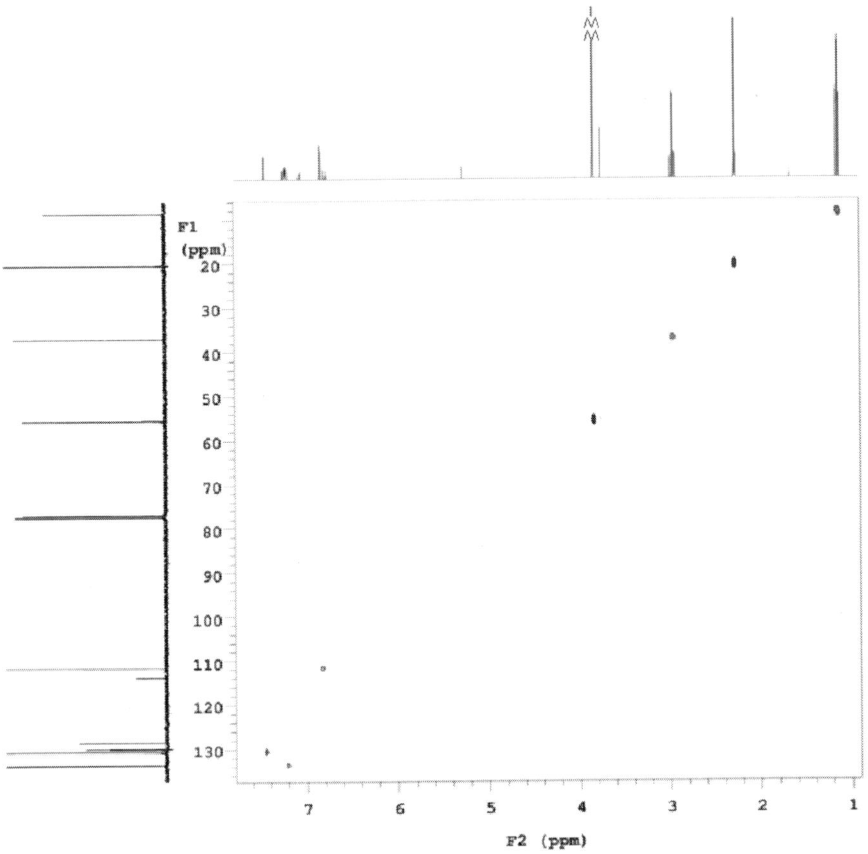

Figure 6. HSQC spectrum of the reaction product of p-methyl anisole with propionyl chloride.

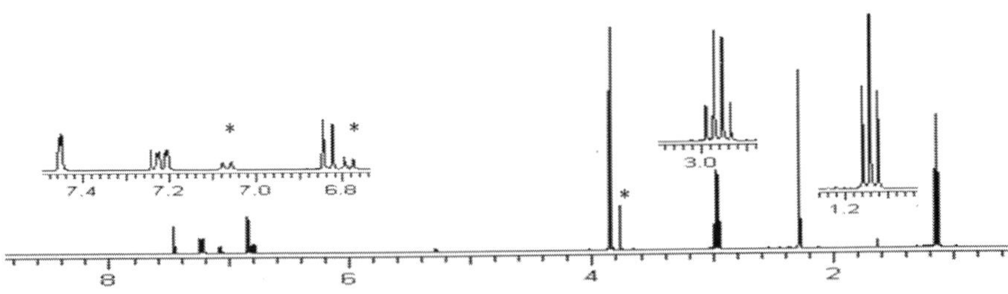

*Figure 7. ^{1}H NMR spectrum of the reaction product of p-methyl anisole with propionyl chloride. The peaks with a star(*) indicate p-methyl anisole.*

The ^{1}H NMR spectrum, Figure 7, is then used to find the exact chemical shift values of each of the hydrogen atoms, with their integration and splitting patterns. This information is also entered into the table. The chemical shift values of the hydrogen atoms are entered to 0.001 ppm. Note that while the carbon chemical shifts are recorded to 0.1 ppm and the hydrogen values are recorded to 0.001 ppm, this does not mean that these values are the "correct" chemical shift values. For ^{13}C and ^{1}H NMR spectroscopy, temperature, concentration, pH, and the presence of other compounds can cause the observed chemical shift values to vary.

Table 1. Chemical shift values for the product of the reaction of *p*-methyl anisole with propionyl chloride as determined from the ^{13}C, HSQC, and ^{1}H NMR spectra

H (ppm)	*H Atom*	*Appearance*	*Int.*	*C (ppm)*	*C Atom*	*HMBC Cross Peaks for a Given (ppm)*	*Note*
				203.7	C11		
				156.5			
7.215		doublet of doublets - long range coupling[a]	1	133.5			
7.453		doublet - long range coupling[a]	1	130.4			
				129.8			
				128.2			
6.832		doublet	1	111.5			
3.843	H1	singlet	3	55.5	C1		
2.959	H3	quartet	2	36.9	C3		
2.278	H2	singlet	3	20.2	C2		
1.138	H4	triplet	3	8.4	C4		

[a] long-range coupling of 2 Hz.

At this stage in the process many assignments can be made. The chemical shift of the carbonyl (C11) is obvious, being the only peak near 200 ppm. The chemical shift of the carbon peak at 55.5 ppm, with its attached hydrogen atom at 3.843 ppm, quickly identifies this as belonging to the methoxy group (C1, H1). The other hydrogen singlet integrates to three (2.278 ppm) belongs to the methyl attached to the aromatic ring (H2). The HSQC spectrum shows that the carbon at 20.2 ppm is the carbon (C2) of this methyl group. The triplet that integrates to three and the quartet that integrates to two belong to the methyl (H4) and methylene (H3), respectively, of the ethyl group. The HSQC spectrum also identifies the carbon atoms of these groups. The hydrogen signals that remained unassigned are a doublet that integrates to 1 with coupling constant of 2 Hz, a doublet of doublets that integrates to 1 which includes a smaller coupling constant of 2 Hz, and another doublet that integrates to one but shows no further coupling. The small 2 Hz coupling constant is indicative of four-bond coupling to the meta position and is very useful in determining structure. The 2 Hz doublet would be the aromatic hydrogen between the propionyl group and the methoxy or the methyl, the doublet of doublets would be the aromatic hydrogen that is *meta* to the singlet, and the doublet without any further coupling would be the other aromatic hydrogen.

With all of this information in hand Table 1 can be filled out. For many reactions that are typically carried out in an organic chemistry laboratory this information might be enough to completely identify the compound. However, for the product of the reaction of *p*-methyl anisole with propionyl chloride this is still not enough information to distinguish between the structures in Figure 4. For this determination, an HMBC spectrum must be used.

Figure 8 shows the HMBC spectrum for this product. The display intensity of the HMBC spectrum has been increased to show all the cross peaks, which unfortunately also reveals noise and artifacts. Signals due to small amounts of starting material are also present. Note that the noise and artifacts can be identified by observing that they do not line up with both a ^{13}C signal and a ^{1}H signal. All three types of signals have been identified with an X in figure 8. Also, the two halves of the spectrum were not taken from the same printing, but from two different printings of the same original data. Thus, the intensities of the ^{1}H peaks are not comparable between the two halves.

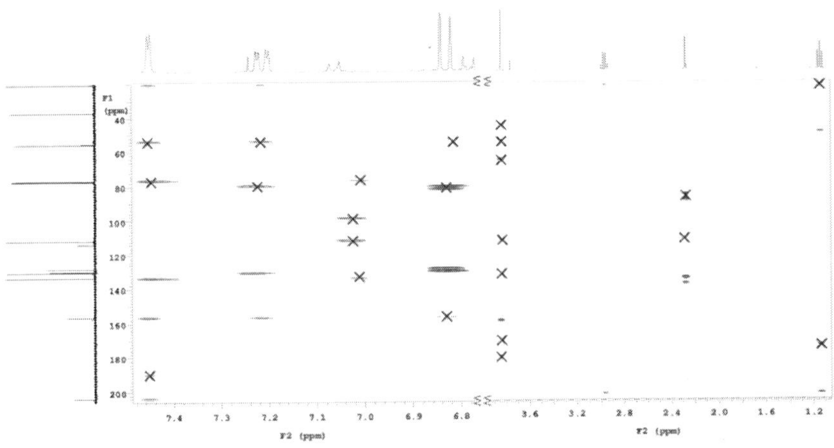

Figure 8. HMBC spectrum of the reaction product of p-methyl anisole with propionyl chloride. See text for explanation of crossed out signals.

Identifying the HMBC signals is not as easy as identifying the HSQC signals. It is necessary to zoom in and out and carefully check that a signal is exactly at the intersection of a ^{1}H signal and a ^{13}C signal. This is because there are noise and artifact peaks that must be identified and ignored. It is also necessary to change the display intensities of the signals because the coupling constants can vary substantially, which in turn causes the actual cross peak intensities to vary.

A key concept in determining whether the acyl group added ortho to the methoxy group or the methyl group is to look at how a carbon attached to the ring is coupled to hydrogen atoms on the ring, and its corollary of how a hydrogen atom on the carbon attached to the ring is coupled to carbon atoms in the ring. This method depends on being able to identify the carbon atom attached to the ring and its attached hydrogen atom. In the acylation of *p*-methyl anisole reaction, the ^{1}H NMR spectrum clearly allows us to identify the methyl group, and the HSQC spectrum shows the hydrogen atoms that pair with the methyl carbon signal.

Table 2. The 2- and 3-bond HMBC cross peaks for the aromatic methyl group of the *p*-methyl anisole product

Atom of aromatic methyl group	*Cross peaks for isomer A*	*Cross peaks for isomer B*
C2	Hydrogen doublet without fine structure (H9)	Hydrogen singlet (H7), hydrogen doublet with fine structure (H9)
H2	Aromatic carbon attached to carbonyl (C7), carbon attached to methyl (C8), carbon attached to hydrogen doublet without fine structure (C9)	Carbon attached to hydrogen singlet (C7), carbon attached to methyl (C8), carbon attached to hydrogen doublet with fine structure (C9)

There would seem to be three different approaches in how to help students use HMBC to determine the structure of the product. These approaches differ in the amount of information given to the students. The most straightforward approach to help the students choose between the two isomers is the approach that requires the least amount of thinking by the students. In this approach it is explained that if the HMBC spectrum has a cross peak between the aromatic methyl hydrogen (H2) and the carbon of the hydrogen doublet, then the acyl group added ortho to the methyl group (isomer A). If the aromatic methyl hydrogen (H2) shows a cross peak to either the aromatic carbon attached to the ^{1}H singlet or to the aromatic carbon of the ^{1}H doublet of doublets, or to both, then the acyl group added ortho to the methoxy group (isomer B). The appearance of a cross peak between H2 (2.278 ppm) and the aromatic carbon (130.4 ppm) attached to the hydrogen doublet (2 Hz coupling) and a cross peak between H2 (2.278 ppm) and the aromatic carbon (133.5 ppm) of the doublet of doublets enables the students to determine that the structure of the product is isomer B. Note that a lack of information, i.e., such a peak is not present, is not useful to consider in HMBC because not all 2- and 3-bond couplings are observed.

Table 3. All 2- and 3-bond HMBC cross peaks for the aromatic methyl group of the *p*-methyl anisole product with propionyl chloride. The shaded signals are not seen

H	*HMBC Cross Peaks for a Given H in Isomer A*	*HMBC Cross Peaks for a Given H in Isomer B*
1	a) C attached to methoxy (C5)	a) C attached to methoxy (C5)
2	a) C attached to methyl (C8) b) C attached to H doublet without fine structure (C9) c) C attached to carbonyl (C7)	a) C attached to methyl (C8) b) C attached to H doublet with fine structure (C9) c) C attached to H singlet (C7)
3	a) C of H triplet (C4) b) C of carbonyl (C11) c) aromatic C attached to carbonyl (C7)	a) C of H triplet (C4) b) C of carbonyl (C11) c) aromatic C attached to carbonyl (C6)
4	a) C of methylene (C3) b) C of carbonyl (C11)	a) C of methylene (C3) b) C of carbonyl (C11)
6	a) C of H doublet with fine structure (C10) b) aromatic C attached to methoxy (C5) c) aromatic C attached to carbonyl (C7) d) C of carbonyl (C11) e) C attached to methyl (C8)	NA
7	NA	a) C of H doublet with fine structure (C9) b) aromatic C attached to methoxy (C5) c) C aromatic C attached to carbonyl (C6) d) C of carbonyl (C11) e) C attached to methyl (C8) f) C of methyl (C2)
9	a) C attached to methyl (C8) b) C attached to methoxy (C5) c) C of methyl (C2) d) C attached to carbonyl (C7) e) C attached to H doublet with fine structure (C10)	a) C attached to methyl (C8) b) C attached to methoxy (C5) c) C of methyl (C2) d) C attached to H singlet (C7) e) C attached to H doublet (C10)
10	a) C attached to methoxy (C5) b) C attached to methyl (C8) c) C attached to H doublet (C9) d) C attached to H singlet (C6)	a) C attached to methoxy (C5) b) C attached to methyl (C8) c) C attached to H doublet with fine structure (C9) d) C attached to carbonyl (C6)

A second approach, and one that requires somewhat more thinking by the student, is to have the students concentrate on the HMBC cross peaks of the methyl group attached to the aromatic ring and not to be concerned about the other HMBC cross peaks. They should examine the structure of one isomer and determine with which aromatic hydrogen atoms the aromatic methyl carbon (C2) should be coupled, and with which aromatic carbons the aromatic methyl hydrogen atoms (H2) should be coupled and put this information into a table. Then have them do the same for the other isomer. Such a table for acylated *p*-methyl anisole is shown in Table 2. Comparing which HMBC cross peaks would be seen and which actually are seen allows for the determination of the structure of the product. The observation of all five of the cross peaks for isomer B is strong evidence that B is the product.

Table 4. Chemical shift values and their assignment to atoms for the product of the reaction of p-methyl anisole with propionyl chloride. The non-shaded assignments were determined by ^{1}H and ^{13}C spectra. The HSQC spectrum was used to put the hydrogen and its directly attached carbon on the same line. The shaded assignments required an HMBC spectrum

H (ppm)	*H Atom*	*Appearance*	*Int.*	*C (ppm)*	*C Atom*	*HMBC Cross Peaks for a Given H, ppm*	*Note*
				203.7	C11		1
				156.5	C5		6
7.215	H9	doublet split into another doublet by long range coupling[7]	1	133.5	C9	156.5, 130.4,20.2	3
7.453	H7	singlet split into a doublet by long range coupling[7]	1	130.4	C7	203.7, 156.5, 133.5, 20.2	3
				129.8	C8		6
				128.2	C6		6
6.832	H10	doublet	1	111.5	C10	129.8, 128.2	5
3.843	H1	singlet	3	55.5	C1	156.5	2
2.959	H3	quartet	2	36.9	C3	203.7, 8.4	2
2.278	H2	singlet	3	20.2	C2	133.5, 130.4[8], 129.8[8]	2,4
1.138	H4	triplet	3	8.4	C4	203.7, 36.9	2

[1] C determined from the ^{13}C NMR spectrum; [2] H and C determined from ^{1}H, ^{13}C, and HSQC spectra; [3] That these H atoms are coupled to the methyl carbon attached to the aromatic ring (20.2 ppm) gives evidence that the structure is B; [4] That this H atom is coupled to the carbon atom of the doublet split into another doublet by long range coupling (133.5 ppm) and to the carbon of the singlet (130.4 ppm) gives evidence that the structure is B; [5] Once the structure is established, these are determined from the ^{1}H, ^{13}C, and HMBC spectra; [6] Determined from HMBC spectrum; [7] Long-range coupling of 2 Hz; [8] This assignment is tenuous and more HMBC resolution would be needed to confirm that there are cross peaks for both of these carbons.

A third approach, and one that requires the most thinking by the student, is to have them first make a table of all the 2- and 3-bond HMBC cross peaks that could be seen in both isomers. Students then make a table of the cross peaks that are present in the HMBC spectrum. Correct comparison of these two tables will lead to the correct structure and allow assignment of all of the NMR signals to atoms in the molecule. Table 3 shows the predicted cross peaks for the two products.

An advantage of this approach is that while not all the HMBC signals are necessary to identify the compound, the consistency of the extra information can lead to the discovery errors in the pairing of atoms, and to the discovery of errors in identifying the compound. A disadvantage of this approach is that it takes considerably more time.

Adding the HMBC cross peaks to Table 1 and making the rest of the assignments provides Table 4.

This table shows the experimental chemical shift values and also shows the assignments of these signals to atoms, i.e., the pairing of observed hydrogen and carbon signals to their corresponding molecular sites. Immediately below the table are notes on the assignments that give the logic behind the assignments. This is a summary of the reasoning discussed above. The assignment of the signal without an attached hydrogen atom at 156.5 ppm is made by noting that it shows cross peaks with H1, H7, and H9. The assignments of C8 to the peak at 129.8 ppm and C6 to the peak at 128.2 ppm are tenuous and are based on interpreting the somewhat broad cross peak between the hydrogen at 2.278 ppm and the carbon(s) at 130 ppm as containing two cross peaks.

Figure 9 shows the structure of the product with the atoms labeled as to their chemical shifts. Some people find such visualization more useful than a table of numbers.

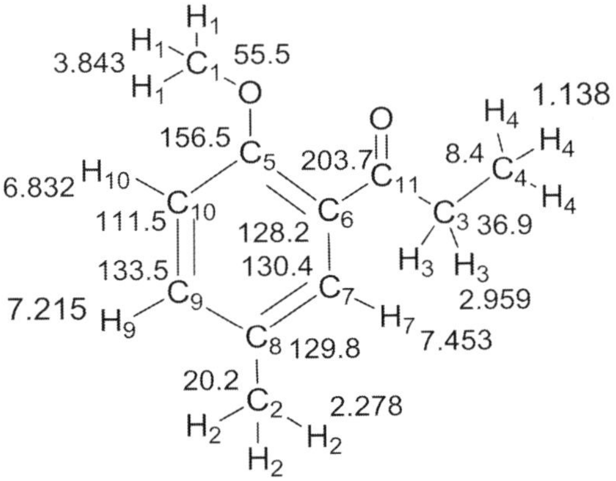

Figure 9. Structure of the product of the reaction of p-methyl anisole with propionyl chloride, with the atoms labeled as to their chemical shift values.

In summarizing these approaches, the first approach explains to the students that if the HMBC spectrum has a particular cross peak (a cross peak between a specific carbon peak and a specific hydrogen peak), then the compound has a specific (stated) structure. The second approach explains to them exactly where to look, but they have to determine what cross peaks are associated with which isomer. The third approach involves the most thinking, requiring them to look at all of the couplings. The best approach to use for a laboratory experiment will depend upon how much prior use of the HMBC experiment they have had before this experiment, on how much time is available for helping students individually or in small groups, and finally on the capabilities of the students.

A fourth approach, that of just telling the students to use HMBC to determine the structure, will leave them quite frustrated. (Even faculty members can be frustrated using this approach.) Unfortunately, this is what we did the first time we made extensive use of HMBC. We have found that it is better to ease them into the HMBC experiment by using it in several lab experiments, and supplying them with fewer hints each time. This year we have gone even one step further. In the sixth experiment of the first term students acquired an HSQC spectrum of their carbonyl-containing product and used it to determine directly attached carbon and hydrogen atoms. In a handout we tell them that by using an HMBC experiment, a technique they would learn in the second term, they would then be able to determine that a particular hydrogen atom was two or three bonds from a carbonyl carbon. With this information they could then assign chemical shift values to all the atoms in the compound.

For the Friedel-Crafts acylation experiment described above it should be noted that a more sophisticated analysis of the ^{1}H chemical shift values can lead to the correct structure of the product of this reaction without the use of the HMBC experiment. Based on the combined upfield ortho substituent effect of the methoxy group and the downfield para effect of the carbonyl, the aromatic hydrogen at 6.832 ppm is ortho to the methoxy group, structure B in Figure 4. At Roanoke College we do not go into the analysis of chemical shifts that much and therefore do not use this approach. In the following example this more sophisticated analysis of the ^{1}H chemical shift values approach does not work and the HMBC experiment must be used.

Products of the Reaction of *p*-Cymene with Acetyl Chloride.

A more challenging product to analyze is from the reaction of *p*-cymene (1-isopropyl-4-methylbenzene) with acetyl chloride, Figure 10. One complication of this reaction is that of determining which of the methyl groups is attached to the aromatic ring and which is attached to the carbonyl group. The HMBC spectrum allows for this determination. Another complication is that both isomers are produced in about a 7:1 ratio.

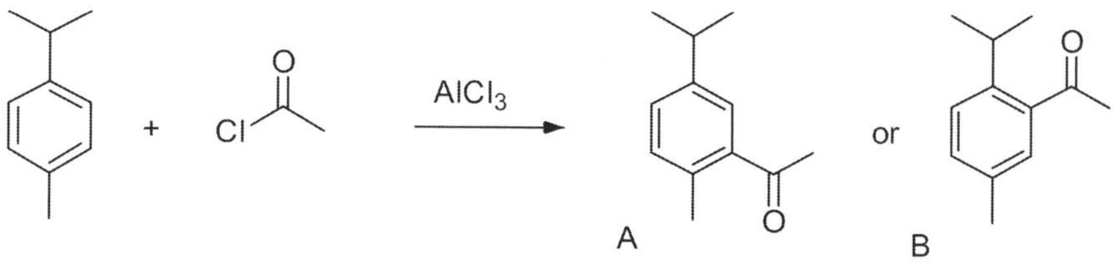

Figure 10. Reaction of p-cymene with acetyl chloride, showing the two possible isomers.

Figure 11 shows the ^{13}C NMR spectrum of the product arising from the reaction of *p*-cymene with acetyl chloride.

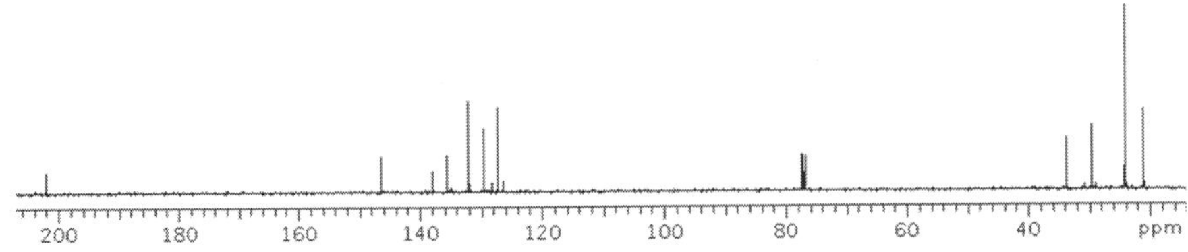

Figure 11. ^{13}C NMR spectrum of the reaction of p-cymene with acetyl chloride, showing a major and a minor product.

While some, but not all of the peaks from the second isomer can be seen, the assumption that the 11 tallest peaks belong to the more abundant isomer is correct. From this spectrum the chemical shift values of the carbons are obtained and placed in a table similar to Table 5. While a careful consideration of the chemical shift values can be used to identity several signals, the only one that we use with our students is the carbonyl at about 206 ppm.

Figure 12 shows the HSQC spectrum for this reaction.

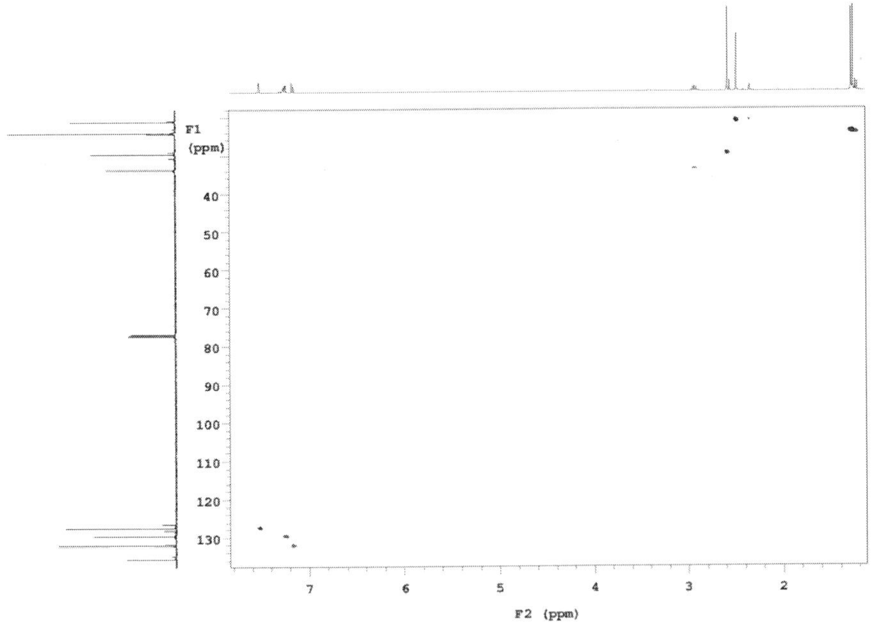

Figure 12. HSQC spectrum of the major product of the reaction of p-cymene with acetyl chloride. The small cross peak at 2.35 ppm belongs to the minor product.

These approximate hydrogen chemical shifts are placed in the same row in the table as is the directly attached carbon.

Then the ^{1}H NMR spectrum, Figure 13, is used to find the exact chemical shift values of each of the hydrogen atoms, and their areas and appearances. This information in entered into the table.

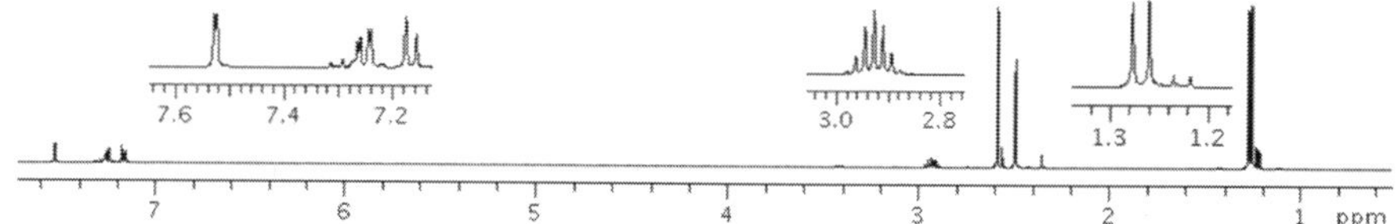

Figure 13. ^{1}H NMR spectrum (with expansions) of the reaction of p-cymene with acetyl chloride, showing the major and a minor product.

Table 5. Spectral data (^{1}H, ^{13}C, and HSQC) and partial peak assignments for the major product of the Friedel-Crafts reaction of *p*-cymene with acetyl chloride

H (ppm)	*H Atom*	*Appearance*	*Int.*	*C (ppm)*	*C Atom*	*HMBC Cross Peaks for a Given H (ppm)*	*Note*
				202.0			
				146.3			
				137.8			
				135.5			
7.165		doublet	1	132.0			
7.252		doublet split into another doublet by long range coupling [1]	1	129.5			
7.525		singlet split into a doublet by long range coupling*	1	127.3			
2.927	H2	septet	1	33.7	C2		
2.584		singlet	3	29.6			
2.490		singlet	3	21.1			
1.267	H1	doublet	6	24.0	C1		

[1] Long range coupling of 2 Hz.

At this stage the table has the entries as shown in Table 5. This information is still not enough to identify the major isomer of this Friedel-Crafts reaction.

Figure 14 shows structures of the two most reasonable products, along with a numbering scheme. Again, this is not a standard numbering scheme but is used in the subsequent discussion.

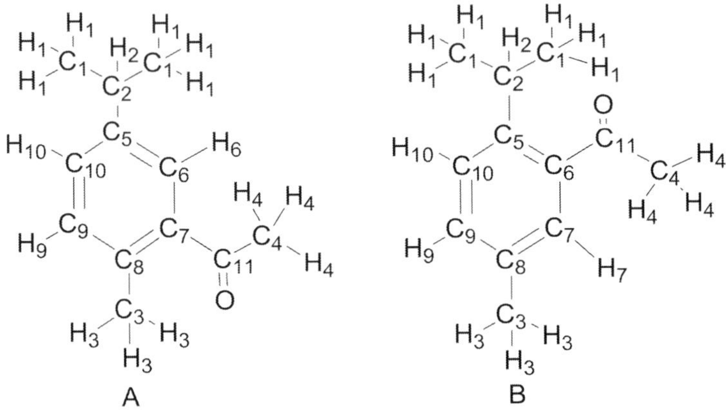

Figure 14. Two isomeric products of the reaction of p-cymene with acetyl chloride.

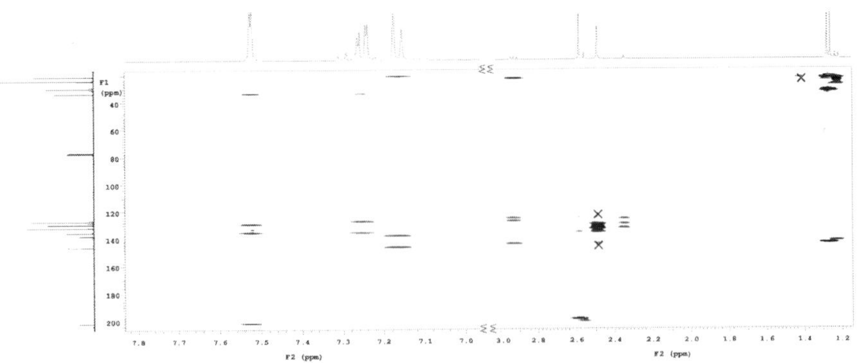

Figure 15. HMBC spectrum of the major product of the reaction of p-cymene with acetyl chloride. The crossed out signals are noise peaks. The small peaks at 2.56 and 1.23 ppm on the ^{1}H axis belong to the minor product.

Table 6. Spectral data (^1H, ^{13}C, HSQC, and HMBC) and complete peak assignments for the major product of the Friedel-Crafts reaction of *p*-cymene with acetyl chloride. The non-shaded assignments were determined by ^1H and ^{13}C spectra. The HSQC spectrum was used to put the hydrogen and its directly attached carbon on the same line. The shaded assignments required HMBC

H (ppm)	*H Atom*	*Appearance*	*Int.*	*C (ppm)*	*C Atom*	*HMBC Cross Peaks for a Given H, ppm*	*Note*
				202.0	C11		1
				146.3	C5		7
				137.8	C7		7
				135.5	C8		7
7.165	H9	doublet	1	132.0	C9	146.3, 137.8, 21.1	5
7.252	H10	doublet split into another doublet by long range coupling [8]	1	129.5	C10	135.5, 127.3, 33.7	6
7.525	H6	singlet split into a doublet by long range coupling[8]	1	127.3	C6	202.0, 135.5, 129.5, 33.7	6
2.927	H2	septet	1	33.7	C2	146.3, 129.5, 127.3	2
2.584	H4	singlet	3	29.6	C4	202.0, 137.8, 127.3	3
2.490	H3	singlet	3	21.1	C3	137.8, 135.5, 132.0	4
1.267	H1	doublet	6	24.0	C1	146.3, 33.7, 24.0[9]	2

[1] C determined from ^{13}C spectrum; [2] H and C determined from ^1H, ^{13}C, and HSQC spectra; [3] H determined from HMBC spectrum, C from ^{13}C and HSQC spectra; [4] H determined by elimination, C from ^{13}C and HSQC spectra; [5] That the aromatic methyl hydrogen atoms (H3) are coupled to the aromatic carbon attached to the doublet without long range coupling is evidence that the structure has the acyl group added ortho to the aromatic methyl group. This structure also is supported by the coupling between the aromatic doublet without long range coupling hydrogen and the carbon of the aromatic methyl group (C3); [6] That the methine hydrogen (H2) is coupled to the carbon of the doublet split into another doublet by long range coupling and to the carbon of the singlet with long range coupling is evidence that the structure has the acyl group added ortho to the aromatic methyl group. This structure also is supported by the coupling between the hydrogen doublet split into another doublet by long range coupling and the hydrogen singlet with long range coupling to the methine carbon (C2); [7] C determined from HMBC; [8] Long range coupling of 2 Hz; [9] This strong cross peak can be interpreted as being to the equivalent C three bonds away.

At this stage, a few assignments can be made. The larger doublet at 1.26 ppm and the larger septet at 2.93 ppm can be assigned to the isopropyl group. The two singlets near 2.5 ppm are associated with the methyl on the aromatic ring and the methyl attached to the carbonyl, but which is which cannot be determined yet. The singlet with long range coupling can be paired with the aromatic hydrogen between the isopropyl and acetyl groups, the doublet with long range coupling can be paired with the aromatic hydrogen meta to the aromatic hydrogen singlet, and the doublet without long range coupling can be paired with the aromatic hydrogen para to the aromatic singlet. However, both isomers in Figure 14 have this arrangement, and more information is needed.

The HMBC spectrum, shown in Figure 15, can provide this additional information. The intensity of the HMBC spectrum has again been increased to show all the cross peaks, but this also means that noise and artifacts become more apparent. Again, noise and artifacts have been identified with an X. The small peaks at 2.56 and 1.23 ppm on the ^{1}H axis belong to the minor product.

The HMBC cross peak between the carbonyl carbon and the singlet at 2.584 ppm with an area of three shows that this methyl belongs to the acetyl group that was added. By elimination, the other methyl group, the singlet at 2.490 ppm with an area of three, must be the one attached to the aromatic ring.

A key point to help the students with this product is to have them concentrate on the aromatic methyl group and/or the methine group. Either one of these two groups can be used with HMBC to identify the adjacent groups. For example, if the hydrogen atoms on the aromatic methyl are coupled with the clean doublet (the doublet not showing long range coupling), then the acetyl group added ortho to the methyl group. This means that the structure of the product can be determined from only one HMBC peak, relating to either the hydrogen or the carbon on either the methyl group or the isopropyl group. Being able to identity several of these peaks, however, gives the student more confidence in their determined structure. The notes below in Table 6 provide summary information on these various avenues of reasoning. The different approaches as to how much information to tell the students with this compound is similar to the approaches discussed above with *p*-methyl anisole.

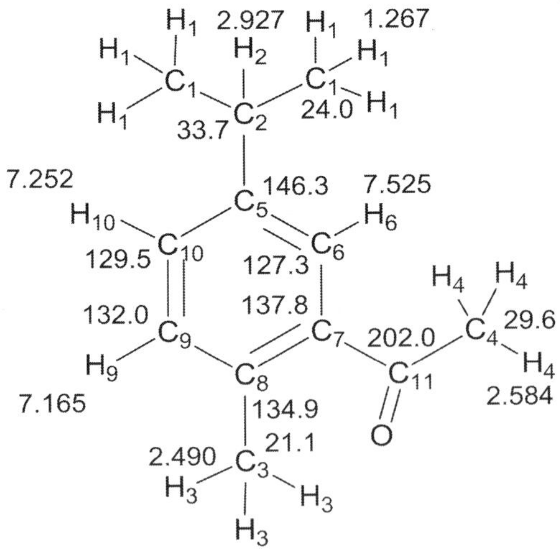

Figure 16. Structure of the major product of the reaction of p-cymene with acetyl chloride, with the chemical shifts of the atoms given.

Table 6 gives the complete assignment of the NMR signals to the individual atoms in the major product of this Friedel-Crafts reaction, product where the acetyl group adds ortho to the methyl of *p*-cymene. Figure 16 gives the chemical shift values of all the atoms in this major isomer.

Fisher Esterification and COSY Spectra

Another 2-D NMR experiment that can be very useful in certain circumstances is a COSY spectrum. This spectrum gives information as to which hydrogen atoms are three bonds away from each other. Again, this 2-D technique does not actually count bonds, but responds to coupling constants. The typical range for three-bond coupling is 6-9 Hz, for which most COSY spectra are optimized

While there are many cases in which analysis by COSY has been very important, most of these have been fairly complicated molecules. This means it can be difficult to find compounds suitable for organic chemistry labs. One compound that is suitable for COSY analysis is butyl butanoate, a compound that can be prepared in a Fisher esterification reaction. While we have used COSY spectral analysis in our Organic Chemistry laboratories, and have done Fisher esterification reactions, we have not prepared the butyl butanoate product. This ester and other esters were prepared in a student independent study project, Figure 17.

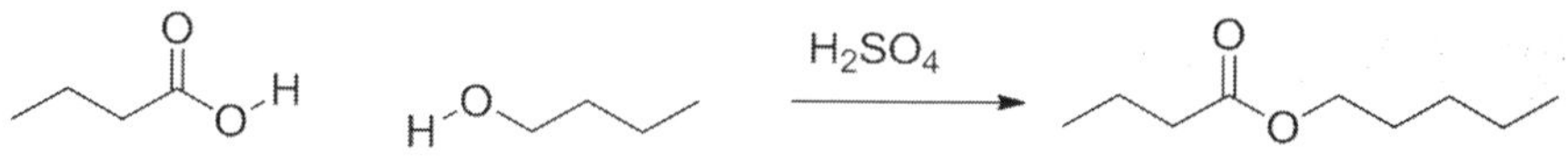

H_2SO_4

Figure 17. Preparation of butyl butanoate.

Use of the COSY Spectrum To Distinguish between Butyl Butanoate and Propyl Pentanoate

One pair of compounds whose structure cannot be unequivocally determined by ^1H, ^{13}C, and HSQC spectra, but can be determined by including a COSY spectrum, is butyl butanoate and propyl pentanoate. While these are not two isomers that can be generated from a single reaction, they are two isomers that are quite similar, differing only in the number of methylene groups on the acid side of the ester and on the alcohol side. Their structures are shown in Figure 18. An Organic Chemistry laboratory experiment in which one or the other might be produced is an esterification reaction in which the student does not know whether the starting materials are 1-butanoic acid and 1-butanol, whose product would be structure A in Figure 18, or 1-pentanoic acid and 1-propanol, whose product would be structure B in Figure 18

```
 H8 H7 H6 O      H4 H3 H2 H1                H8 H7 H6 H5 O      H3 H2 H1
 |  |  |  ||     |  |  |  |                 |  |  |  |  ||     |  |  |
H8-C8-C7-C6-C5-O-C4-C3-C2-C1-H1           H8-C8-C7-C6-C5-C4-O-C3-C2-C1-H1
 |  |  |         |  |  |  |                 |  |  |  |         |  |  |
 H8 H7 H6        H4 H3 H2 H1                H8 H7 H6 H5        H3 H2 H1
            A                                          B
```

Figure 18. Structure and numbering of butyl butanoate (A) and propyl pentanoate (B).

Figure 19 shows the ^{1}H and ^{13}C NMR spectra of a commercial sample of butyl butanoate. In the ^{13}C NMR spectrum the carbonyl carbon and the carbon attached to the oxygen can be identified, but this is of no help in distinguishing between the two isomers in Figure 18. In the ^{1}H NMR spectrum the methylene adjacent to the oxygen and the methylene adjacent can be identified, but again, this is of no use in distinguishing between the isomers. Even in the absence of overlapping signals, the ^{1}H NMR spectrum cannot provide the desired information. The information that is needed is whether there is a two carbon chain or a three carbon chain attached to the methylene attached to the oxygen (or attached to the carbonyl). A COSY spectrum can provide this information.

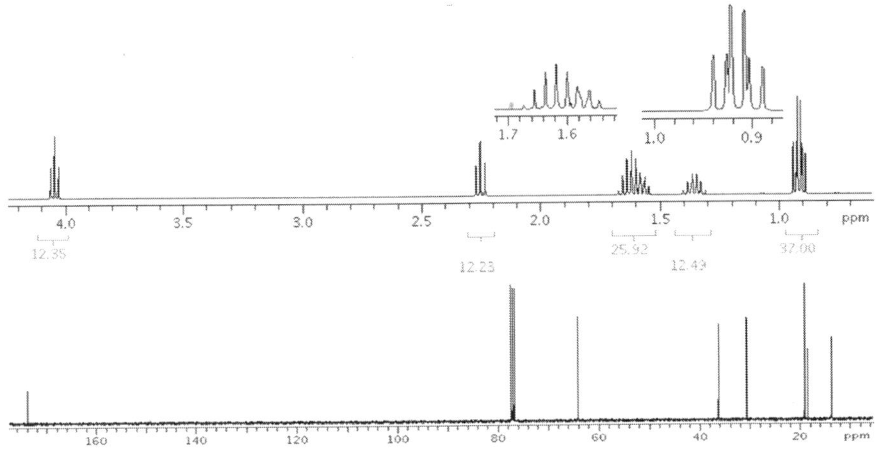

Figure 19. ^{1}H (top) and ^{13}C (bottom) spectra of butyl butanoate, with areas and expansions.

The COSY spectrum of this compound, Figure 20, shows that the three bond coupling sequence starting from the hydrogen adjacent to the oxygen is 4.046, 1.57, 1.355, 0.907 ppm. This shows that there are four carbons with hydrogen atoms in a row. The spectrum also shows that a three bond coupling sequence starting from the hydrogen atom attached to the carbonyl. Adding this sequence of three carbons with their attached hydrogen atoms to the carbonyl carbon gives a second four-carbon chain. Since there are two four-carbon chains and not a three-carbon chain and a five-carbon chain, the structure is A, butyl butanoate and not B, propyl pentanoate.

Butyl butanoate is a good example of the usefulness of a COSY spectrum. Even though there is considerable overlap in some of the signals, the sequence of hydrogen groups on both sides of the ester functional group can be determined with the pre-set experimental parameters on the NMR instrument. Running the experiment at higher resolution (longer acquisition times for ^{1}H, more t1 increments, and better shimming) confirmed the assignments, but did not make reading the spectra significantly easier.

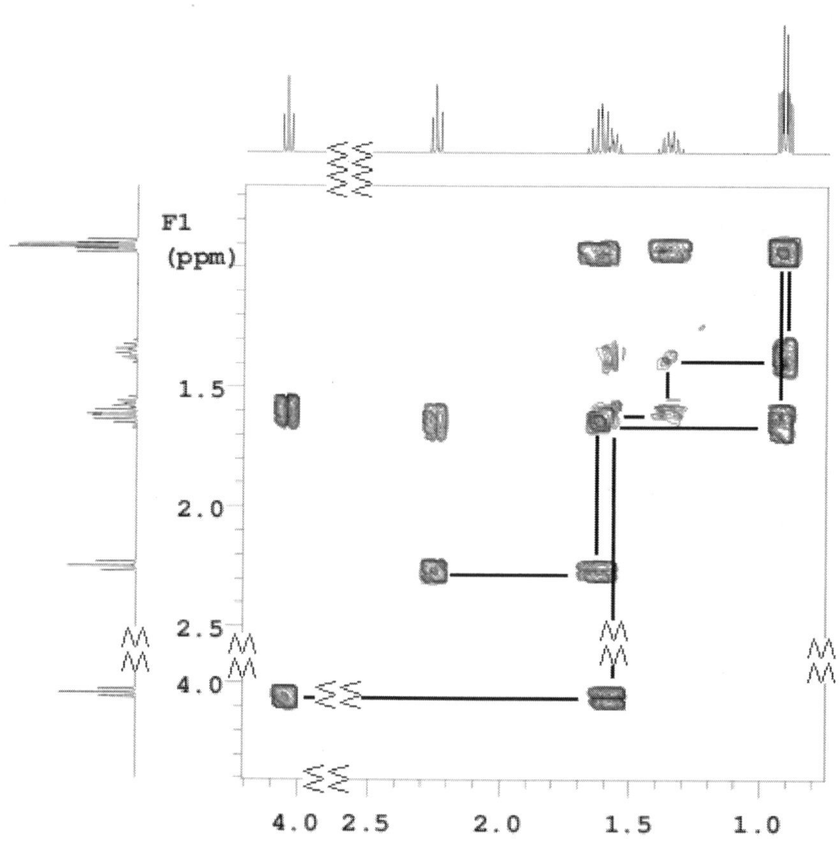

Figure 20. COSY spectrum of butyl butanoate with lines interpreting the three bond coupling.

While butyl butanoate shows the power of COSY, the structure also can be determined using HMBC. Table 7 gives the chemical shift values and assignments to atoms of a commercial sample of butyl butanoate, and Figure 21 show the structure of butyl butanoate with the atoms labeled with chemical shift values.

Table 7. Chemical shift values and their assignment to atoms for butyl butanoate. The non-shaded assignments were determined by ^{1}H and ^{13}C spectra. The HSQC spectrum was used to put the hydrogen and its directly attached carbon on the same line. The shaded assignments required COSY

H (ppm)	*H Atom*	*Appearance*	*Int.*	*C (ppm)*	*C Atom*	*HMBC Cross Peaks for a Given H, ppm*	*Note*
				173.8	C5		1
4.046	H4	triplet	2	64.0	C4	173.8, 30.7, 19.1	2
2.250	H6	triplet	2	36.2	C6	173.8, 18.5, 13.6[5]	2,3
1.57[6]	H3	multiplet	2	30.7	C3	64.0, 19.1	4
1.355	H2	sextet	2	19.1	C2	64.0, 30.7, 13.6[5]	4
1.62[6]	H7	multiplet	2	18.5	C7	173.8, 36.2, 13.6[5]	4
0.921[5]	H8	triplet	3	13.65[5]	C8	36.2, 18.5	4
0.907[5]	H1	triplet	3	13.62[5]	C1	30.7, 19.1	4

[1] C determined from ^{13}C spectrum; [2] C and H determined from ^{1}H, ^{13}C, and HSQC spectra; [3] Since HMBC shows that the triplet adjacent to the carbonyl is coupled to a methyl provides evidence that the structure is A and not B. That the structure is A not B can also be determined using the COSY spectrum; [4] Once the structure is known, C and H are determined using the COSY spectrum; [5] Even though the peaks can be distinguished, it cannot be determined which hydrogen is attached to which carbon; [6] The exact shift cannot be determined because the signals appear to have considerable second order character.

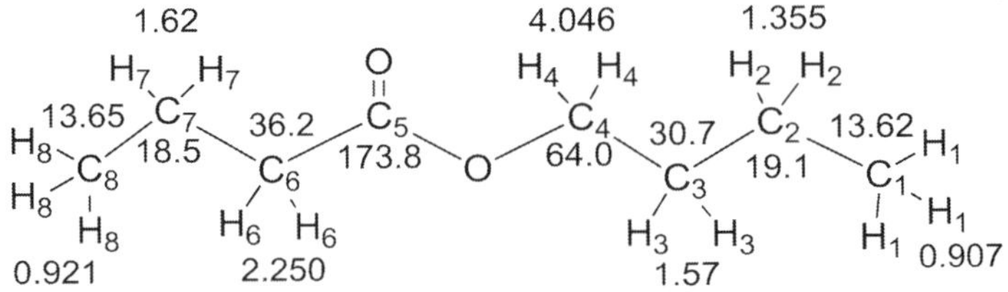

Figure 21. Structure of butyl butanoate with atoms labeled with chemical shifts, in ppm.

Conclusions

^{1}H and ^{13}C NMR spectroscopy have been important in the Organic Chemistry curriculum at Roanoke College for many years. With the acquisition of an automated sample changer, we have tried different ways to introduce 1-D and 2-D NMR spectroscopy early into the introductory organic laboratory sequence, with various degrees of success. At first we introduced 2-D NMR spectroscopy in the middle of second semester, but found that students needed more experience with it. Then we tried starting off with 1-D and 2-D NMR spectroscopy in the first three weeks of first semester laboratory. Here we found that the students did not have the needed vocabulary or familiarity with structures. Then we tried introducing 1-D and 2-D NMR spectroscopy, along with vocabulary and structures during the first four weeks. This worked better, but students seemed overwhelmed. During this time NMR spectroscopy was also presented in lecture after the middle of the first semester, creating a disconnect between laboratory and lecture content. Currently, what seems to work best for our curriculum is to introduce NMR integration and splitting pattern matching in the first laboratory, with reinforcement of these concepts over the next several experiments. NMR spectroscopy prediction and interpretation are then introduced in class and laboratory in the middle of the term.

In our first laboratory students work in teams of two to learn to (i) draw organic structures using a commercial drawing program; (ii) retrieve, expand, integrate, and print proton NMR spectra; and (iii) use integration and splitting pattern matching with spectra of known compounds to identify spectra of pure compounds and of mixtures.

Over the next several weeks, students run (prepare samples, place them in the queue of the sample changer, and choose the spectra to run) and process their own spectra using splitting pattern matching to confirm the success of the experiment. Several weeks later, NMR theory and interpretation is presented in lecture and laboratory. After that, most of the subsequent experiments include interpretation of ^{1}H, ^{13}C, and HSQC spectra as part of the product analysis. HMBC and COSY analysis are introduced in the second semester and incorporated into experiments such as the Friedel-Crafts acylation and Fischer esterification reactions.

A very important concept in organic chemistry is molecular structure, and NMR spectroscopy is a very good tool for determining structure and emphasizing its importance. While much information can be learned from just ^{1}H and ^{13}C NMR, we feel 2-D NMR experience is also important. HSQC spectra are very useful for relating ^{13}C signals to specific atoms, but not very useful directly in structure determination. HMBC spectra are more useful in structure determination, but only after a thorough ^{1}H, ^{13}C, and HSQC analysis. We have not used NOESY yet, but would like to in the future.

Molecular structure is a key concept in organic chemistry and NMR spectroscopy is an important tool in the determination of molecular structure. While significant structural information can be obtained from routine ^{1}H and ^{13}C NMR analysis, we believe that exposure to 2-D NMR analysis techniques provides a unique experience for undergraduate students. Regardless of a student's future in the field of chemistry and NMR interpretation, teaching students to use different pieces of information, including 2-D NMR will better prepare them to think logically and draw reasonable conclusions based on data interpretation.

References

1. Alvarado, E. Two-dimensional experiments with VnmrJ 2.2 – University of Michigan, URL www.umich.edu/~chemnmr/docs/2D_experiments-v2.pdf (accessed January 2013).
2. Simpson, J. H. *Organic Structure Determination Using 2-D NMR Spectroscopy: a Problem-Based Approach*; Elsevier: Boston, MA, 2008.
3. Richards, S. A.; Hollerton, J. C. *Essential Practical NMR for Organic Chemistry*; Wiley: West Sussex, United Kingdom, 2011.
4. Reynolds, W. F. *Heteronuclear Multiple Bond Correlation (HMBC) Spectra*; Encyclopedia of Magnetic Resonance; Wiley Online Library, published Online 15 SEP 2010.
5. Pavia, D. L.; Lampman, G. M.; Kriz, G. S.; Engel, R. G. *Introduction to Organic Laboratory Techniques: A Small Scale Approach*, 2nd ed.; Brooks/Cole: Belmont, CA, 2005.

Inorganic/Heteronuclear Chemistry

Chapter 9

^{31}P NMR Spectroscopy in an Undergraduate Inorganic Curriculum

Chip Nataro,[*,1] Chelsea L. Mandell,[2] and Margaret A. Tiedemann[1]

[1]Department of Chemistry, Lafayette College, Easton, Pennsylvania 18042
[2]Department of Chemistry, Texas A&M University, College Station, Texas 77842
[*]E-mail: nataroc@lafayette.edu

^{31}P NMR spectroscopy is a useful means of characterizing inorganic compounds and as such should be part of an undergraduate inorganic chemistry curriculum. This continuously evolving curriculum finds its roots in the research laboratory with results from the research laboratory translating into examples for the classroom and exercises in the teaching laboratory. Two extensions of previously reported laboratory exercises are reported herein. In addition to outlining the ^{31}P NMR spectroscopy portion of a two-course inorganic curriculum, the insights of two students at varying points in their progress through this curriculum are presented.

Introduction

To discuss the role of NMR spectroscopy in an inorganic curriculum, a brief introduction to the current curriculum is warranted. A chemistry major at Lafayette College would traditionally take General Chemistry I and II during their first year. They will first be introduced to NMR spectroscopy in Organic Chemistry I and II during sophomore year. There are two inorganic courses offered which also discuss NMR spectroscopy, Inorganic Chemistry I and Inorganic Chemistry II. Inorganic Chemistry I is offered in the spring semester and is typically taken in the sophomore or junior year. However, the course is open to any student, providing the prerequisites are met. Incoming students can place out of General Chemistry I and instead can take General Chemistry II in their first semester. If they wish to take a chemistry course in the second semester, Inorganic Chemistry I is the only course for which they will have the prerequisites. In addition, the course is not required for biochemistry majors, but some take it as an elective during senior year. This makes a class with diverse levels of experience; the majority of the students have learned about NMR spectroscopy in Organic Chemistry, but some have never been exposed to the technique at all. Inorganic Chemistry II is offered in the fall and is almost exclusively taken by seniors.

In order to provide additional insights into this curriculum, two students that are at different stages in their careers are serving as co-authors. Chelsea Mandell [CM] offers the perspective of a recent graduate. She took Inorganic Chemistry I in the spring of her sophomore year having completed one semester of Organic Chemistry. She then took Inorganic Chemistry II in the fall of her senior year. Margaret Tiedemann [MT] took Inorganic Chemistry I during her second semester at Lafayette, after completing General Chemistry II the previous semester. Both students worked extensively in the Nataro [CN] research lab. The reflections of the students on their experience will be presented throughout this manuscript.

Student Insights

[CN]: Prior to taking Inorganic Chemistry I, what experience did you have with NMR spectroscopy?

[CM]: I was first exposed to NMR spectroscopy while doing research, using mainly $^{31}P\{^{1}H\}$ and ^{1}H. I understood $^{31}P\{^{1}H\}$ well enough to monitor my reactions and match peaks to specific compounds. I worked with symmetric phosphine compounds so one peak was starting material while the other peak would be product. If two peaks were present, it meant that my starting materials had not reacted completely. Proton NMR spectroscopy on the other hand was just something I knew I had to do after obtaining a product. I did not truly grasp the concept of analyzing ^{1}H NMR spectra until I took Organic Chemistry I. In Organic Chemistry I, I learned what ^{13}C NMR spectra looked like and how to interpret ^{1}H NMR spectra through integration, simple coupling and chemical shift. Research continued to enhance my understanding to include checking for purity within my ^{1}H spectra and being able to pick out some of the solvent impurities. The following semester I took Inorganic Chemistry I.

[MT]: Before Inorganic Chemistry I, I only had limited experience with NMR spectroscopy through high school chemistry classes. A mention was given in AP Chemistry and a few spectra were shown in an extracurricular reading assignment. The main thing I gained is that the actual instrument is very expensive! During a trip to Lafayette College as a prospective new student, I was shown the college's NMR instrument on a tour of the chemistry department and reminded of the instruments usage and the super-cooled magnet inside. A few spectra and the instrument itself were all that I had seen of NMR spectroscopy until I reached Inorganic I.

Inorganic Chemistry I: Introduction

While a detailed examination of the NMR instrument is a bit too advanced for Inorganic Chemistry I, it is important for the students to have some appreciation for the technique. To accomplish this, some sort of common ground is necessary. Since all of the students in this course have completed General Chemistry, our discussion begins with Bohr line spectra, a topic that is reviewed in Inorganic Chemistry I a few weeks prior to the NMR spectroscopy discussion. In particular it is important for students to remember that energy is either gained or lost for there to be a transition of an electron from one energy level to another. This translates nicely to NMR spectroscopy where the spin of the nucleus has two different energy states when subjected to an external magnetic field. The transition between these two energy states requires energy, which can be measured. NMR spectroscopy is the technique that measures these energy differences.

Student Insights

[CN]: How did your experience with NMR spectroscopy in Organic Chemistry help prepare you for Inorganic Chemistry? Did your experience in Inorganic Chemistry help clarify any concepts from Organic Chemistry?

[CM]: Organic Chemistry helped me thoroughly understand the basics of interpreting proton spectra. With Inorganic Chemistry as the next step, it was just applying these same basic concepts but to different nuclei. Since most of the students in Inorganic Chemistry had already taken Organic Chemistry, the material for interpreting multinuclear spectra was covered rather quickly. I was glad that I had previously taken Organic Chemistry before Inorganic Chemistry because of these more rapid explanations. Since Organic Chemistry mainly focused on ^{1}H NMR spectroscopy I initially had difficulty understanding the splitting patterns of other NMR active nuclei, especially when they were combined within the same molecule. Inorganic Chemistry helped explain the satellite peaks surrounding the chloroform peak in ^{1}H NMR spectra that I'd sometimes see. It also explained why in ^{13}C NMR spectroscopy there was no need to consider coupling between adjacent carbon atoms. In Organic Chemistry, a description of how NMR spectroscopy worked was given, and how, at that time, I felt a lack of need to understand how the experiment worked. However, when it was explained to me a second time, in a separate class, I became aware of the importance of knowing these concepts. And knowing them actually ended up helping me understand splitting patterns better.

[CN]: How did your experience with NMR spectroscopy in Inorganic Chemistry help prepare you for Organic Chemistry? Did your experience in Organic Chemistry help clarify any concepts from Inorganic Chemistry?

[MT]: When I reached Organic Chemistry, I was familiar with the theory behind the instrument and the splitting patterns. I also regarded interpreting spectra differently that my classmates, because the spectra looked familiar and were not as intimidating. I had a better handle on interpreting simple patterns such as doublets and triplets and viewed the peaks as part of a puzzle that I had to fit together. This was a skill I learned from Inorganic I, as well as a general

confidence about the content. However, in the beginning of Organic Chemistry, I made solving spectra more difficult than I really needed to because I tried to find more complex splitting patterns than was necessary. I knew patterns such as doublets of doublets existed based on spectra from Inorganic, but those didn't ever seem to appear in Organic. There was a slight disconnect in my mind of the spectra I had seen in Inorganic and the spectra I was trying to solve in Organic. Overall, the familiarity I had with the instrument from Inorganic helped me to understand concepts re-introduced in Organic.

In our curriculum, the introduction to ^{1}H NMR spectroscopy takes place in Organic Chemistry I. The introduction is fairly quick and shortly thereafter students can examine complex spectra using chemical shift, coupling and integration as guides. Because of the relatively small chemical shift range (0-12 ppm is typical) and the vast numbers of organic compounds, students are taught about the ranges to expect protons near certain functionalities. Coupling is introduced, beginning with the simple quartet and triplet of an isolated ethyl group. More complex patterns are then introduced. However, coupling to multiple protons in different environments, for example the methylene protons on carbons 4 and 5 in 2-hexanone, can significantly complicate the observed patterns. Additional complications, such as long-range coupling, are briefly presented as the students become more adept at examining spectra. These components allow the students to learn quite a bit about this technique, but there are some additional aspects that can be covered in the Inorganic Chemistry curriculum.

Student Insights

[CN]: Both of you were involved in inorganic research prior to taking Organic Chemistry. How did your research experience help in your understanding of NMR spectroscopy?

[CM]: Organic Chemistry never stressed the idea that different nuclei can be beneficial in determining a product's structure as well as for monitoring a reaction for completeness. NMR spectroscopy in research helped me realize its extended value of more than just two useful nuclei and that it is actually helpful in research and not just used in teaching laboratories. As I was discussing earlier, proton NMR spectroscopy was completely confusing prior to Organic Chemistry. Even after having several sessions on trying to learn and interpret spectra, I was not able to after the summer. Once NMR spectroscopy appeared in Organic Chemistry I easily picked up interpreting proton spectra, everything that I had been taught over the summer made complete sense. Due to the summer research I feel that I picked up interpreting spectra much more quickly than I would have without it.

[MT]: Having experience using the NMR instrument during research helped me understand when to utilize NMR spectroscopy and the factors that affect spectra. I realized the importance of chemical shift in phosphorus NMR spectra, and the vast examples of molecules I used provided a lot of opportunities to examine various spectra. Research applied many concepts that were introduced in Inorganic but become more important in interpreting spectra in Organic labs. Already being able to identify residual starting material, impurities or background noise on spectra when I reached Organic lab was a very useful skill.

However, hydrogen NMR spectra were extremely confusing when I encountered them during research. I understood that hydrogen is NMR active, but the spectra with many peaks and indistinguishable coupling patterns were basically foreign. When I studied NMR spectroscopy in Organic Chemistry I, I realized that the key to interpreting the spectra from my research was really looking at the shifts of the peaks to confirm the types of hydrogen present. Essentially, I used what I knew about interpreting phosphorus spectra to interpret hydrogen spectra as well.

I consider research to be an incredibly important part of undergraduate education. The use of $^{31}P\{^{1}H\}$ NMR spectroscopy, in particular for monitoring reactions, is critical in my research. As my research program grew, it became more apparent that multinuclear NMR spectroscopy needed to be included in the Inorganic curriculum. The close tie between research and the classroom means that many of the examples come directly from my research. These can certainly be adapted to meet individual needs and preferences.

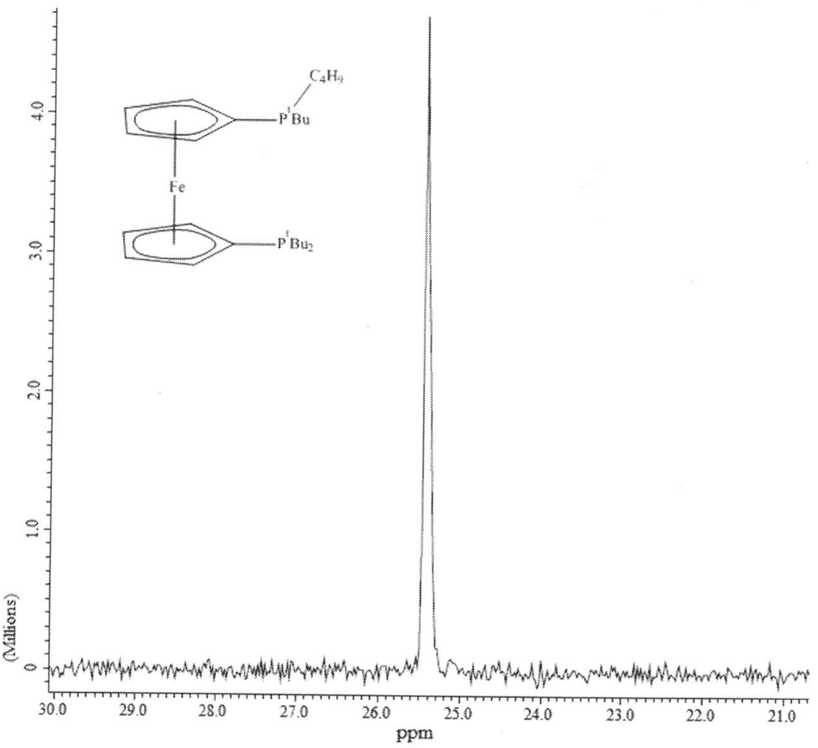

Figure 1. $^{31}P\{^{1}H\}$ NMR spectrum of dtbpf (one of the tert-butyl groups is given as a formula for students that are unfamiliar with functional groups) in CH_2Cl_2.

Inorganic Chemistry I: Introduction to ^{31}P NMR Spectroscopy

In terms of natural abundance and spin, ^{31}P and ^{1}H are very similar which readily allows for building on the foundation laid in the Organic Chemistry course. However, there are two aspects that require immediate attention: proton decoupling and chemical shift. ^{13}C NMR spectroscopy is only briefly discussed in our organic curriculum, so students typically do not immediately recognize what is meant by ^{31}P{^{1}H} NMR spectra. Usually, a brief discussion is sufficient for students to recall that the {^{1}H} means the spectrum is proton decoupled. The best way for students to fully appreciate the difference between proton coupled and proton decoupled spectra is an example such as that offered by 1,1′-bis(di-tertbutylphosphino)ferrocene (dtbpf). While this molecule (Figure 1) may be a bit intimidating at first (ferrocene is a topic in Inorganic Chemistry II), I will poll the class about how many phosphorus atoms are in the molecule and then how many different phosphorus environments are present. Students generally are able to conclude that the phosphorus atoms are equivalent.

The ^{31}P{^{1}H} NMR spectrum of dtbpf (Figure 1) displays a singlet at 25.5 ppm. At this point the students are not informed that this spectrum was taken in methylene chloride (CH_2Cl_2 not CD_2Cl_2). Students that have taken Organic Chemistry are typically taught to consider chemical shifts of 0-12 ppm for proton NMR spectra. When asked about ^{13}C NMR spectrum typically at least one student will remember that the range of chemical shifts is larger than that observed in a ^{1}H spectrum. The take home message for students at this point is that the range of chemical shifts can be quite large and is very dependent on the nucleus that is being observed. While discussing the range of chemical shifts it is also important to discuss what affects the chemical shift of a nucleus such as nearby functional groups, which is a topic thoroughly covered in Organic Chemistry. As my course has students that have not taken Organic Chemistry, it is important to slightly modify this statement to the atoms or groups surrounding the atom being considered. In this particular example it would be the two *tert*-butyl groups and the ferrocenyl group.

Next, the effect of having coupling to protons is examined. In order to ensure that all students are on equal footing, this discussion starts with an examination of chloroethane. The students are asked to determine the number of different types of protons in the molecule. From this point, the basis for coupling to the methyl group is examined. There are two protons on the neighboring carbon atom of the methylene group. Students are asked to determine what possible orientations the spins of these two protons can have in relation to the external field. The typical first answer is either both aligned with or both aligned against the external field. Once those two are proposed, the combination of opposing spins is finally suggested. These three different possibilities give the triplet splitting pattern that can be fit by the equation, $2nI+1$ where n is the number of coupled atoms and I is the spin of the coupled nucleus (this equation is typically simplified in our organic curriculum to $n+1$). In addition the 1:2:1 intensity pattern can be developed by considering the possible combinations to give the various alignments. There is only one possible arrangement of the spins that will give both spins aligned with the external field (symbolized as ↑/↑). Similarly, there is only one combination in which the two spins will be aligned against the external field (symbolized as ↓/↓). However, there are two possible arrangements (↑/↓ and ↓/↑) in which the spins are opposite. The intensity pattern is given by the probability of forming the different neighboring states. A similar analysis of the methylene protons gives the combination of all aligned (↑/↑/↑) two aligned (↑/↑/↓, ↑/↓/↑ and ↓/↑/↑), one aligned (↑/↓/↓, ↓/↑/↓ and ↓/↓/↑) and none aligned (↓/↓/↓) which corresponds to the 1:3:3:1 pattern of a quartet. Pascal's triangle can then be introduced as a way to simplify this analysis.

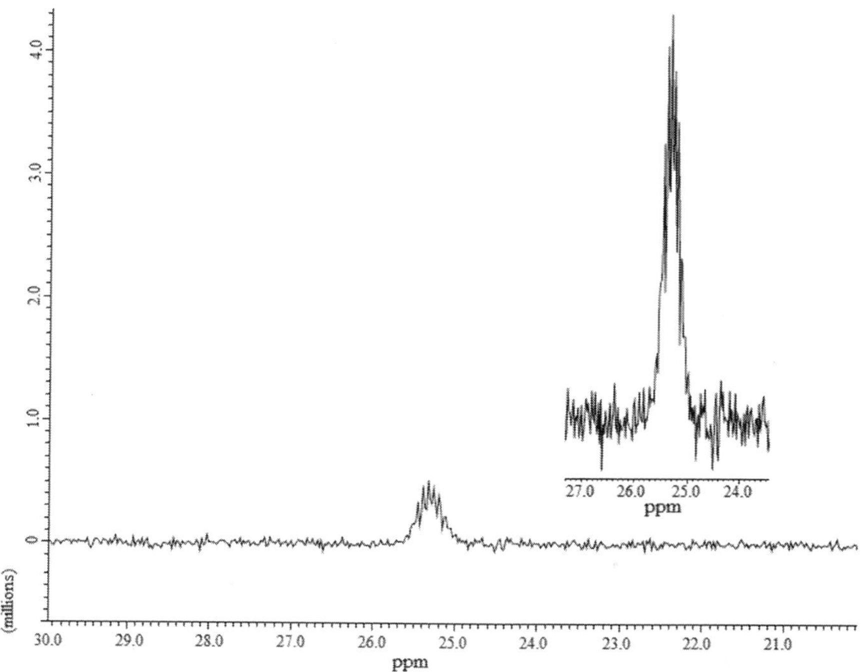

Figure 2. ^{31}P NMR spectrum of dtbpf in CH_2Cl_2 inset with a zoomed in picture of the spectrum.

At this point, our analysis of dtbpf can continue. The class is polled about how many different types of protons there are in the molecule. With 44 total hydrogen atoms, students tend to be a bit hesitant in saying that there are three different types. We will then discuss which protons might couple to the phosphorus atom. Generally, the protons of the *tert*-butyl groups are the first mentioned. The ring protons are mentioned as also being possible. If a student does not ask directly, I will ask them to consider if there can be coupling through the iron. They typically do not think it would happen. With the diverse nature of the class, it is not worth going into significant detail other than briefly mentioning that the distances and angles between the atoms do not allow for coupling through the iron center. This leaves a possible 22 protons that can couple to phosphorus. Assuming the coupling constant to all of the protons is the same (which is unlikely), the equation n+1 suggests a 23 line pattern. In our sophomore organic course, students are taught to just consider neighbors for the n+1 calculation. For example, in 1-chloropropane, the protons of the central methylene group would be expected to give rise to a six line pattern due to the five neighboring proton atoms. Given the complexity of this particular example, the introduction of 'tree diagrams' seems premature at this point in the discussion. The spectrum (Figure 2) does not appear as a clean 23 line pattern, but students can see there is coupling. It is worth noting that the vertical axis for the ^{31}P NMR spectrum of dtbpf in Figure 2 is the same as that in Figure 1. Both spectra were acquired using the same sample and the same number of scans. In summary, while proton coupling can provide useful information, it is typically more useful to obtain the proton decoupled spectrum.

Inorganic Chemistry I: Monitoring Reactions

The next topic to be introduced is the use of $^{31}P\{^1H\}$ NMR spectra to monitor a reaction. This topic is not covered in our organic curriculum because students do not perform microscale syntheses in a deuterated solvent. Additional options that are not employed include performing a larger scale reaction in a deuterated solvent (which would be quite expensive) or in a solvent without protons (CS_2 and CCl_4 being possible, albeit extremely undesirable options). However, monitoring reactions by NMR spectroscopy is critical to my research, and inspired the inclusion of multinuclear NMR spectroscopy as part of the Inorganic Chemistry curriculum.

At this point, students are informed that the $^{31}P\{^1H\}$ NMR spectrum of dtbpf was obtained in CH_2Cl_2. When the class is polled as to the significance of this information, a student will typically recall that they obtained their ^1H NMR spectra in $CDCl_3$ for Organic Chemistry laboratory. Even if they do not recall the specific solvent, they generally remember that the solvent was deuterated. I then ask the students why they use this solvent. This question usually takes some time to reach the conclusion that if the solvent was not deuterated, the only thing we would see in the spectrum would be the solvent; for example, a 0.020 g sample of a compound with a molecular mass of 100 g/mol would contain $2.0x10^{-4}$ moles of the analyte while the 0.50 mL of $CHCl_3$ in which the sample was dissolved contains $6.2x10^{-3}$ moles of $CHCl_3$. This makes the mole fraction of analyte to solvent approximately 0.03. After pointing out this concern, the students are asked why CH_2Cl_2 can be used for the $^{31}P\{^1H\}$ spectrum of dtbpf. Although it often requires significant prodding, the class is able to reach the conclusion that phosphorus and not proton is being observed, therefore the solvent is 'invisible'. There are two potential complications worth pointing out. First, the solvent could contain phosphorus, although this is fairly unlikely. Second, this does mean that shimming and locking cannot be performed on the sample. Given the diverse experience of the students, it is not worth spending a significant amount of time going over the details of shimming and locking. I will mention that shimming and locking is performed by looking at the signal for the deuterium in the solvent. The instrument is then fine tuned to give the best spectrum possible. Since we do not use a deuterated solvent, we cannot shim and lock, which can lead to lower quality spectra. However, the spectra are sufficient for determining if a reaction is complete, and the benefit of being able to take a sample directly from a reaction mixture is more than enough to outweigh the lower quality of the spectra. It is possible to put a sealed capillary containing the deuterated version of the solvent into the NMR tube and locking on that signal, but that is likely too much detail for this discussion.

The reaction of dtbpf with $[Pd(MeCN)_2Cl_2]$ in CH_2Cl_2 can be monitored by $^{31}P\{^1H\}$ NMR spectrum (Figure 3). In this particular example, the reaction occurs fairly quickly, so stoichiometric control was used for pedagogical purposes. In a research setting, various factors such as time, reaction stoichiometry and temperature, will play important roles in monitoring a reaction. Inevitably, at least one student that has taken Organic Chemistry will question how they can determine the chemical shift of the product. Those students are used to the tables that appear in most organic textbooks that list expected chemical shifts for various functional groups. Students are informed that the tables are not as well developed for other nuclei such as phosphorus. They are also not expected to predict chemical shifts, but rather draw logical conclusions based on the information available. In this example, the important point is if the students can conclude whether or not a reaction is taking place based upon the spectra presented.

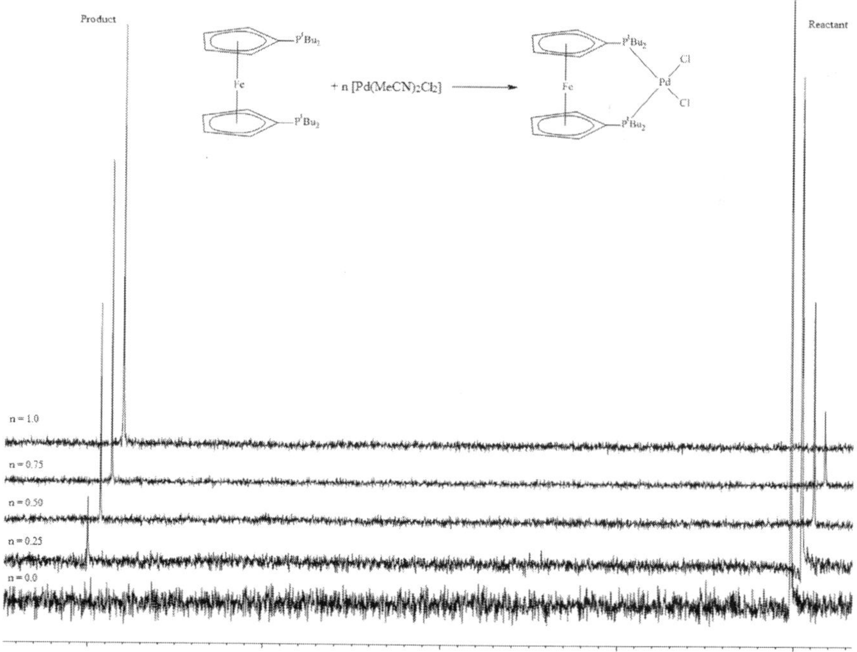

Figure 3. $^{31}P\{^1H\}$ spectra for the reaction of dtbpf with varying amounts of $[Pd(MeCN)_2Cl_2]$.

Inorganic Chemistry I: Coupling Example 1

Thus far, the discussion has only investigated molecules with one type of phosphorus. The next example continues to build on the foundation, but includes several new concepts. Another tool in structure elucidation learned in Organic Chemistry is the integrated area of the peaks. The anion, 2,5-di-t-butyl-1,3,4-triphosphacyclopentadienyl (*1*) (Figure 4), is a good example to discuss integration and it also has coupling between phosphorus atoms. Upon showing the structure of the anion, students are asked how many different phosphorus environments are present and by this point can readily identify the two different environments. The question of whether or not there will be coupling between the two types of phosphorus is often met with some hesitation. Some students seem to think that proton decoupling removes all coupling, while others seem to think that coupling can only occur to protons. It is important to emphasize that coupling can occur between any NMR active nuclei unless they are specifically decoupled like the protons in this example. The $^{31}P\{^{1}H\}$ NMR spectrum displays a doublet at 245 ppm and a triplet at 252 ppm (Figure 4) (*2*). In reporting the chemical shifts for these two signals, the center peak of the triplet is chosen while for the doublet, the average of the two peaks is reported.

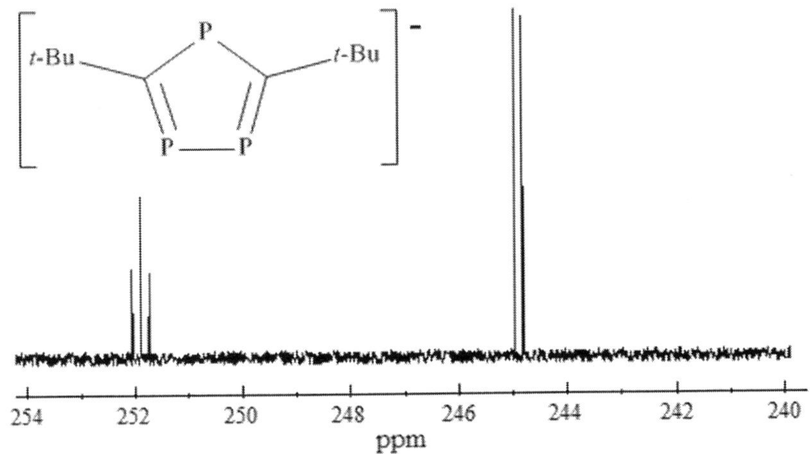

Figure 4. Simulated $^{31}P\{^{1}H\}$ NMR spectrum of 2,5-di-t-butyl-1,3,4-triphosphacyclopentadienyl anion (3).

The coupling constant ($^{2}J_{P\text{-}P}$) is 47 Hz, and it should be pointed out that this value is seen both in the doublet and triplet. This coupling constant is significantly larger than what students would have encountered in Organic Chemistry. The effects of distance between coupling atoms, mass of coupling atoms, and angle between coupling atoms can be considered. In this case, the mass of the atoms and the distance between them are the most relevant points. Generally, the heavier the atom the larger the coupling constant, therefore a larger coupling constant to phosphorus is not unusual. Also in this example, there are only two bonds between the phosphorus atoms. In organic systems, there are typically at least three bonds unless the protons are diastereotopic. In that case the coupling can be quite large (*4*), although our students seem to have minimal recollection of this. Finally, the relative integration of the two signals can be considered. It is important to note that the peaks for a given signal are relative to each other; for example, the 1:2:1 ratio of the triplet does not mean that the smaller peaks in the triplet have the same intensity as the peaks in the doublet which are 1:1 relative to each other. However, when the entire signal is considered, the integrated area of the doublet is twice that of triplet, as there are two equivalent phosphorus atoms comprising that signal.

Inorganic Chemistry I: Coupling Example 2

The next example was selected as an introduction to using tree diagrams for analyzing peaks and because it has coupling between two different nuclei. The compound, $[CpRu(dppe)(\eta^1\text{-}P_4)][PF_6]$ (*5*) (Figure 5) presents four different signals that must be considered. Students tend to ignore the PF_6^- at first, but once it is pointed out they readily pick out the four different phosphorus environments.

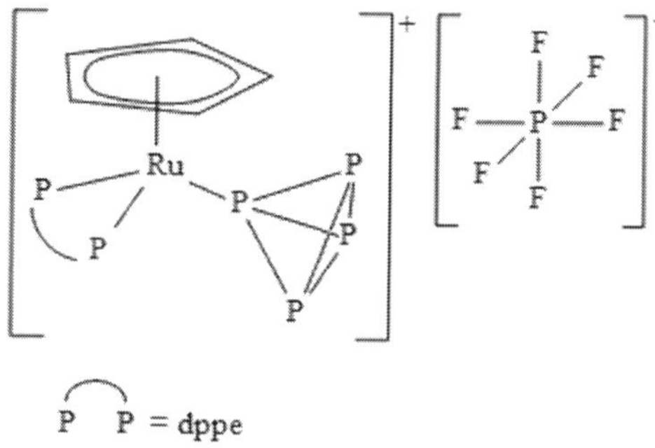

Figure 5. $[CpRu(dppe)(\eta^1\text{-}P_4)][PF_6]$.

At first glance the $^{31}P\{^1H\}$ NMR spectrum (Figure 6) appears quite complicated. Students are informed that coupling need only be considered to NMR active nuclei that are separated by two bonds or less, that ^{19}F is a spin = ½ nucleus and that it has a natural abundance of 100%. After briefly discussing whether or not phosphorus can couple with fluorine and the structure of PF_6^-, the septet at -143.1 ppm, which is observed in the chemical shift window from -120 to -170 ppm can be assigned to the anion. The large coupling constant, 709.5 Hz, can be noted in particular as it is a one bond coupling ($^1J_{P\text{-}F}$), which is unlikely to be seen in Organic Chemistry.

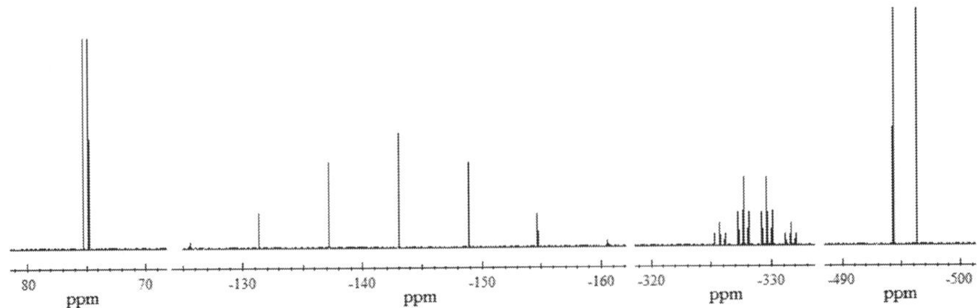

Figure 6. Simulated $^{31}P\{^{1}H\}$ spectrum of $[CpRu(dppe)(\eta^1\text{-}P_4)][PF_6]$ (3).

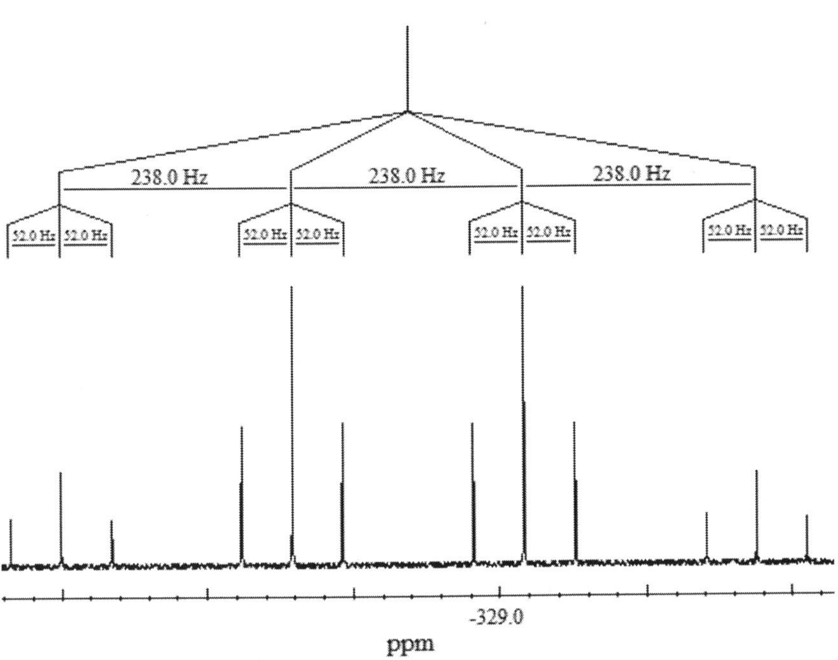

Figure 7. Tree diagram for the q of t in $[CpRu(dppe)(\eta^1\text{-}P_4)][PF_6]$ (3).

In looking at the phosphorus environments in the cation, the phosphorus atoms of the dppe ligand will give rise to a doublet as will the phosphorus atoms of the P_4 ligand that are not coordinated to the ruthenium. Based on integration, magnitude of the coupling constant and the chemical shift, the peak at -495.5 ppm ($^{1}J_{P\text{-}P}$ = 238.0 Hz) can be assigned to the phosphorus atoms of the P_4 ligand that are not coordinated to ruthenium. The remaining doublet at 75.1 ppm ($^{2}J_{P\text{-}P}$ = 52.0 Hz) can be assigned to the phosphorus atoms of the dppe ligand. At this point, the remaining signal must be for the phosphorus of the P_4 ligand that is coordinated to ruthenium. In order to fully analyze this signal, tree diagrams are introduced (Figure 7). The initial line represents the atom of interest if there was no coupling. It should be centered at -328.8 ppm. This phosphorus is coupled to the three other phosphorus atoms of the P_4 ligand with a coupling constant of 238.0 Hz and the two phosphorus atoms of the dppe ligand with a 52.0 Hz coupling constant. Starting with the largest coupling, the initial signal is split into a quartet. The two inner peaks are symmetrically split from the initial peak by 119.0 Hz. The two outer peaks are separated by 238.0 Hz from the neighboring inner peak. Based on Pascal's triangle, the intensities of the peaks at this point are 1:3:3:1 (or ⅛, ⅜, ⅜, ⅛) (*6*). Each line in the resulting quartet is then split into a triplet. The center line of each triplet remains unmoved from the line in the quartet and the other two lines are separated by ±52.0 Hz from the center. A triplet has a 1:2:1 intensity pattern which gives the final pattern, a quartet of triplets, a 1/32:2/32:1/32; 3/32:6/32:3/32; 3/32:6/32:3/32; 1/32:2/32:1/32 intensity pattern.

Inorganic Chemistry I: Magnetic Inequivalence

The next NMR spectroscopy topic that is introduced in the Inorganic Chemistry I course is magnetic inequivalence. This is a topic that is only briefly mentioned in this course. It is something of which the students should be aware, but not necessarily be expected to decipher. The ^{31}P NMR spectrum of hexafluorotricyclophosphazene, $N_3P_3F_6$ (Figure 8), is an excellent example of this phenomenon (*7*). When the structure is shown to the students and they are asked to predict what the ^{31}P spectrum would look like, they are typically confident that it should be a triplet due to coupling to the two ^{19}F nuclei bonded to a specific phosphorus atom. The spectrum is significantly more complicated. The students are asked to choose one of the phosphorus atoms in the molecule. Upon making the selection, the $^{1}J_{P\text{-}F}$ for that particular phosphorus is evident. However, this selection now makes the four remaining fluorine atoms inequivalent as they would have $^{3}J_{P\text{-}F}$ coupling to the chosen phosphorus atom. Once the fluorine atoms are no longer equivalent, the phosphorus atoms become inequivalent, and therefore, $^{3}J_{P\text{-}F}$ coupling is observed.

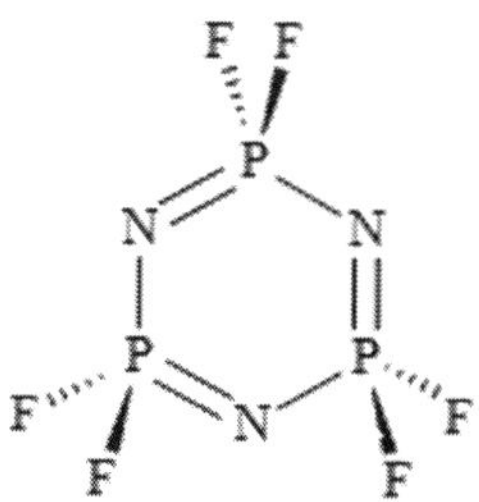

Figure 8. Structure of $N_3P_3F_6$.

Inorganic Chemistry I: Nuclei with I ≠ ½

The final NMR spectroscopy topic covered in the Inorganic Chemistry I course is a brief introduction to nuclei that have spins other than ½. For this course, careful examination of the ^{13}C NMR spectrum of a deuterated solvent is a sufficient introduction. The spectrum of acetone-d_6 displays a singlet for the carbonyl carbon at 206.3 ppm and a 1:3:6:7:6:3:1 septet (*8*) at 29.9 ppm for the methyl carbons (Figure 9).

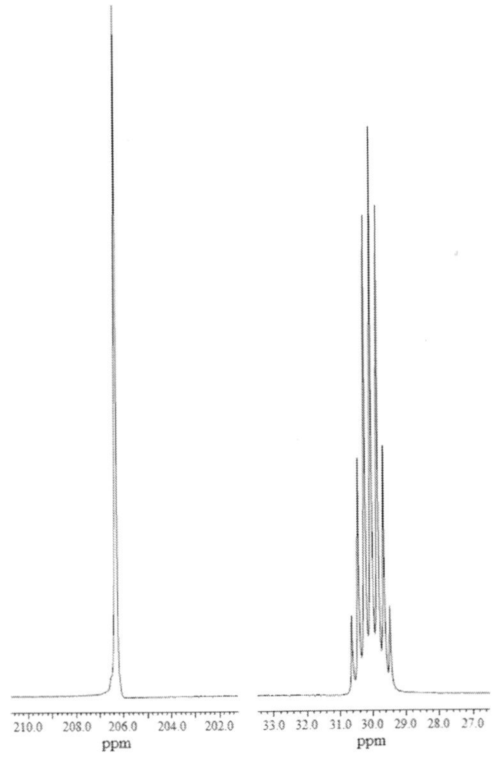

Figure 9. $^{13}C\{^1H\}$ NMR spectrum of acetone-d_6.

Student Insights

[CN]: How did these laboratory exercises aid in your understanding of the material?

[CM]: The lab portion of Inorganic Chemistry helped reinforce lab techniques and characterization methods that I used in research, especially in regards to the borohydride and copper lab. Conducting the second lab was the first time I had encountered a paramagnetic species. I never quite understood why it was so hard to get a clean NMR spectrum of one before, until that day. Even though I had seen the Pt satellite signals during my research, now with the background on NMR active nuclei they made much more sense and were predictable.

[MT]: The laboratory exercises provided my first experience with NMR spectra not from a book or lecture. The imperfect spectra were a challenge amidst the unfamiliar laboratory techniques and the upper-class classmates. Because most laboratory techniques are introduced in Organic Chemistry, I relied heavily on the knowledge of other students to get through the lab period. However, during the analysis of the laboratory exercises, I was able to step up and use what I had learned in class to make sense of the NMR spectroscopy results.

Laboratory Extension

Inorganic Chemistry I has the option to be taken with or without the laboratory. Students that take the laboratory receive hands on experience with ^{31}P NMR spectroscopy. Two of the laboratory exercises employ ^{31}P NMR spectroscopy as a means of characterization. The first is an adaptation of a lab involving the coordination of borohydride to a copper center (*9*). IR spectroscopy is the only means of characterization reported in this lab in particular because it is used to determine the coordination mode of the borohydride ligand. The incorporation of ^{31}P NMR spectroscopy has proven to be a very useful addition to this exercise. Students obtain the ^{31}P NMR spectra of PPh_3, the borohydride compound ($(Ph_3P)_2CuBH_4$) and the two products of the thermal decomposition of $(Ph_3P)_2CuBH_4$. Continuing with the theme from lecture, the students are not asked to focus on the fine details of the chemical shift for each compound. However, the spectrum of PPh_3 gives a peak at approximately -5 ppm in the $^{31}P\{^1H\}$ NMR spectrum. Coordination of the PPh_3 to copper gives a peak at -1 ppm for $(Ph_3P)_2CuBH_4$. The thermal decomposition of $(Ph_3P)_2CuBH_4$ gives two products. One is soluble in methylated spirits and gives a peak at -5 ppm; this can be identified as PPh_3 by comparison to the starting material. The other decomposition product, H_3BPPh_3, is insoluble in methylated spirits and gives rise to a peak at 21 ppm.

The second laboratory exercise incorporating ^{31}P NMR spectroscopy involves the coordination of 1,1′-bis(diphenylphosphino)ferrocene (dppf) to a variety of transition metal centers (*10*). This laboratory experiment provides the only discussion of the NMR spectrum of paramagnetic compounds in the curriculum and allows students to perform Evan's method on these paramagnetic samples (*11*). In addition, the platinum compound allows for a brief introduction to NMR active nuclei that are not 100% abundant. This topic is covered in greater detail in Inorganic Chemistry II.

Student Insights

[CN]: After completing Inorganic Chemistry I, were there any topics that you were still not entirely comfortable with?

[CM]: Multiple NMR active nuclei within the same compound were still a bit confusing for me and would tend to slow me down, especially when forced to predict a spectrum. Also, my understanding of what exactly was happening

within the instrument was limited. However, by senior year with the additional practice from my research and taking CHEM 440 Structure Determination ([CN] this class is a senior level elective with a focus on NMR spectral interpretation), I no longer had these problems or at least could work through similar problems much faster and had a better grasp on understanding the fundamentals behind NMR spectroscopy.

[MT]: As Inorganic Chemistry I provided my first actual instruction for NMR spectroscopy, I left the class with limited confidence when solving spectra. As I began to use the instrument during research, I could understand the phosphorus spectra and how the difference in shift of the appropriate peaks signified a reaction. However, the hydrogen spectra were essentially still unrecognizable. The spectra of aromatic and alkyl groups did not provide clear splitting patterns, so I became confused about the similarities and differences between phosphorus and hydrogen spectra.

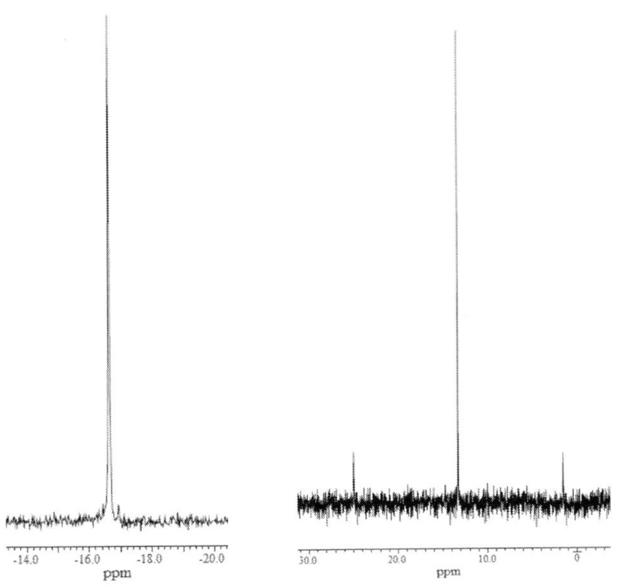

Figure 10. $^{31}P\{^1H\}$ *NMR spectrum of dppf in* CH_2Cl_2 *and* $^{31}P\{^1H\}$ *NMR spectrum of* $[PtCl_2(dppf)]$ *in* CH_2Cl_2.

Inorganic Chemistry II: NMR Active Nuclei Less than 100% Abundant

The Inorganic Chemistry II course briefly reviews the concepts from Inorganic Chemistry I and then introduces NMR spectroscopy with nuclei that have a natural abundance of less than 100%. Like dtbpf, dppf displays a singlet in the $^{31}P\{^1H\}$ NMR spectrum. Upon reaction with $[PtCl_2(MeCN)_2]$, the compound $[PtCl_2(dppf)]$ is formed as indicated by the downfield shift in the ^{31}P signal to 13.3 ppm (Figure

10). The new peak is still referred to as a singlet, but it has platinum-195 satellites with a $^{1}J_{P\text{-}Pt}$ of 3780 Hz. Platinum-195 is spin ½ and has a natural abundance of 33.832% (*12*). In examining the spectrum, the central singlet comprises 66.168% of the signal as that is the natural abundance of platinum isotopes that are not NMR active. The remaining 33.832% of the signal is divided equally among the satellite peaks; in other words, each satellite comprises 16.916% of the signal. Therefore, the peaks have a relative ratio of 1:3.9116:1 which is different from the 1:2:1 ratio of a binomial triplet. It is also important to note that the $^{1}J_{P\text{-}Pt}$ coupling constant is measured between the two satellite peaks, not between the central peak and the satellite peaks.

Figure 11. $^{31}P\{^{1}H\}$ NMR spectrum of $[Pt(dppf)(\mu\text{-}Cl)]_2^{2+}$ in $CDCl_3$.

Laboratory Extension

A possible extension of the dppf laboratory exercise is the reaction of $[PtCl_2(dppf)]$ with NaBArF (BArF = $[B(3,5\text{-}C_6H_3(CF_3)_2)_4]^-$) (*13*). A 0.0038 g (0.0046 mmol) sample of $[PtCl_2(dppf)]$ was dissolved in 1 mL of CH_2Cl_2 in an NMR tube. To this solution, 0.0043 g (0.0050 mmol) of NaBArF was added. The solution immediately changed color from yellow to orange. The product displays a singlet in the $^{31}P\{^{1}H\}$ NMR spectrum at 19.0 ppm in CH_2Cl_2 with a $^{31}J_{P\text{-}Pt}$ of 3840 Hz (Figure 11). Determining the product of this reaction can provide an interesting intellectual challenge for the students. With just the $^{31}P\{^{1}H\}$ NMR spectrum, it would appear that there is not a lot of information available. However, there are very important conclusions that can be drawn from the spectrum. First, the spectrum is different than that of the starting material indicating that some form of a reaction has taken place. However, there is not a large difference in the chemical shift which suggests the product is not significantly different (i.e., a different Pt valence, coordination number or geometry) from the starting material. Second, a single signal indicates that the product only has one phosphorus environment regardless of how many phosphorus atoms are actually in the product. Finally, the presence of the platinum-195 satellites indicates that the phosphorus atom(s) in the product are still coordinated to platinum.

Students can generally deduce that the NaBArF removes a chloride ligand to form NaCl which leaves an electron deficient, divalent platinum (MLX_2, 14 electron) (*14*). The product is assumed to be a sixteen electron, divalent platinum (ML_2X_2) species. Students might initially suggest coordination of the solvent or the $BArF^-$. While good ideas, when asked why neither of those answers is possible, students can reach the conclusion that the single phosphorus environment eliminates the possibility of having both a chloride and another ligand in addition to the dppf (*15*). This leaves dimerization as the only possible solution. In order for the product to be symmetric, dimerization must occur via two bridging chloride ligands (Figure 12) (*16*).

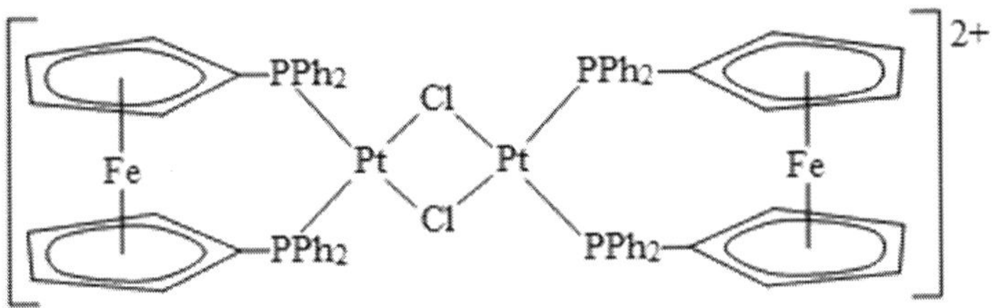

Figure 12. $[Pt(dppf)(\mu\text{-}Cl)]_2^{2+}$.

A further extension of the presence of NMR active nuclei that have less than a 100% natural abundance is that any coupling to other nuclei will also be seen in satellites. The cation $[Pt(dippf)(PPh_3)]^{2+}$ (dippf = 1,1′-bis(di-*iso*propylphosphino)ferrocene) (Figure 13) has two different phosphorus environments.

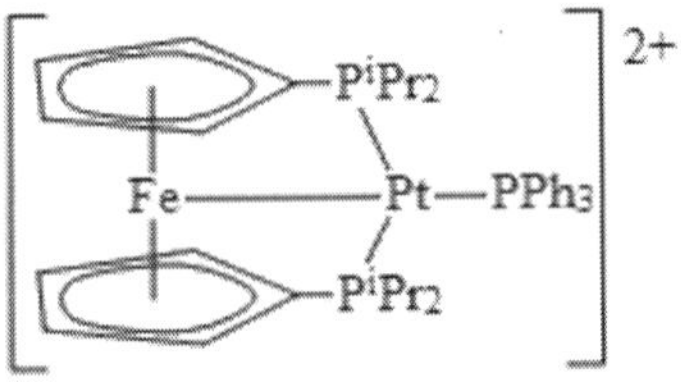

Figure 13. $[Pt(dippf)PPh_3)]^{2+}$.

The $^{31}P\{^1H\}$ NMR spectrum displays a triplet ($^2J_{P\text{-}P}$ = 17.3 Hz) at 16.9 ppm for the PPh_3 ligand and a doublet at 5.52 ppm for the dippf ligand (Figure 14) (*12*). The platinum-195 satellites display the same coupling patterns and constants as the central peak; the satellites for the doublet are doublets with $^2J_{P\text{-}P}$ = 17.3 Hz and the satellites for the triplet are triplets with the same coupling constant. The $^1J_{P\text{-}Pt}$ for the doublet is 2200 Hz and for the triplet it is 4170 Hz.

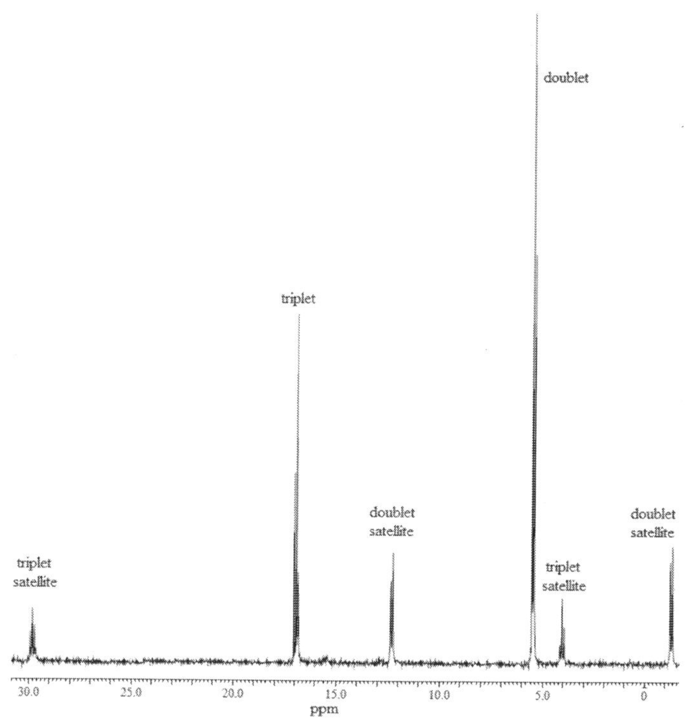

Figure 14. $^{31}P\{^1H\}$ NMR spectrum of $[Pt(dippf)PPh_3]^{2+}$ in acetone-d_6.

Inorganic Chemistry II: Magnetic Inequivalence with NMR Active Nuclei Less than 100% Abundant

The final topic of interest is magnetic inequivalence caused by an NMR active nucleus that is less than 100% abundant. This is illustrated nicely by 1,2-bis(dicyclohexyl)phosphine selenide ($dcpeSe_2$) (*17*). While the phosphorus atoms appear equivalent, the presence of selenium is a complication. Selenium-77 is a spin ½ nuclei that has a natural abundance of 7.63% (*12*). Therefore, 85.3% of the $dcpeSe_2$ will not contain an NMR active selenium, 14.1% will have one NMR active selenium atom and 0.582% will have two ^{77}Se atoms (Figure 15).

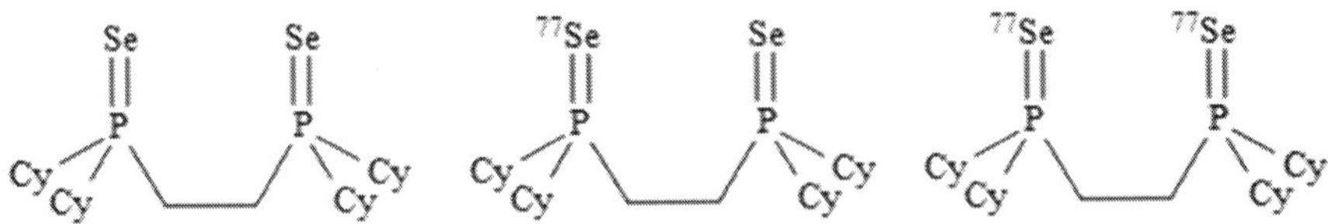

Figure 15. Isotopomers of dcpeSe$_2$.

The $^{31}P\{^1H\}$ spectrum (Figure 16) of dcpeSe$_2$ shows a singlet at 57.6 ppm which can be attributed to the dcpeSe$_2$ molecules that do not contain an NMR active selenium atom.

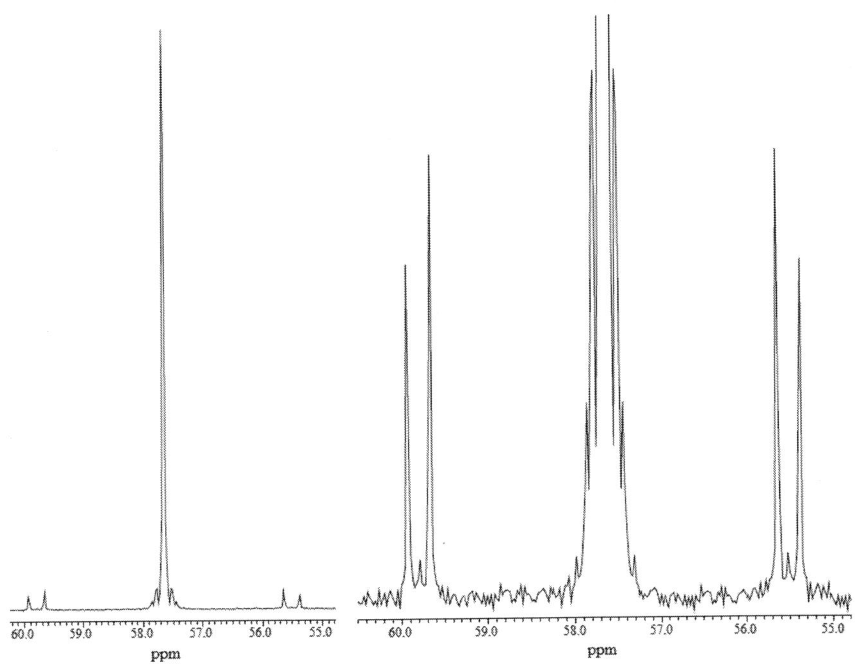

Figure 16. $^{31}P\{^1H\}$ NMR spectrum of dcpeSe$_2$ (the spectrum on the right is zoomed in on the satellite peaks).

In the dcpeSe^{77}Se molecules, the phosphorus atoms are no longer equivalent. This gives rise to a doublet of doublets ($^3J_{P\text{-}P}$ = 45.0 Hz and $^4J_{P\text{-}Se}$ = 17.3 Hz) centered at 57.6 ppm for the phosphorus that is not bonded to the ^{77}Se. The phosphorus bonded to ^{77}Se appears as the satellites (a doublet of doublets) centered at 59.8 and 55.5 ppm ($^1J_{P\text{-}Se}$ = 714 Hz and $^3J_{P\text{-}P}$ = 45.0 Hz). Finally, the phosphorus atoms are equivalent in the relatively small number of dcpe^{77}Se$_2$ molecules. These appear as a doublet ($^1J_{P\text{-}Se}$ = 714 Hz) centered at 57.6 ppm that can be described as the small peaks in the center of the major satellite peaks.

Student Insights

[CN]: Upon completing the curriculum and starting graduate studies, what is your assessment of your comprehension of multinuclear NMR spectroscopy?

[CM]: Even with the combination of Organic Chemistry, Inorganic Chemistry, Structure Determination and research, I was still nervous that my NMR spectral interpretation skills would not be up to expectation for graduate school. Instead, they tended to surpass expectations. I came to graduate school only having used ^{1}H, ^{13}C *and* ^{31}P *NMR spectroscopy. Within the first few weeks of working in lab I easily added* ^{19}F *and* ^{11}B *NMR spectroscopy to my repertoire. Because of my background I had no trouble easily converting between different NMR active nuclei and interpreting spectra. Due to all the hands-on time I had with the instrument itself, my NMR spectroscopy training through the school went by faster and smoother than the normal route. Currently, I am en route on synthesizing a paramagnetic species. Thanks to the Inorganic lab dealing with paramagnetic species, that at the time I thought I'd never see again, I can continue to build on the foundation from my undergraduate education.*

Table 1. Comparison of NMR spectroscopy topics in Inorganic and Organic curricula

	Organic	*Inorganic*
Coupling equation	n + 1	2nI + 1
Spin values	½	½ and more
NMR active nuclei	^{1}H and ^{13}C	Many
Emphasis	^{1}H	^{31}P
Coupling to	other protons	any NMR active nuclei
Chemical shift ranges	0-12 ppm	depends on nucleus
Coupling constants	>20 Hz	depends on nuclei
Coupling patters	important	important
Chemical shift	defined by functionality	emphasis on changes
Solvent	$CDCl_3$	any
$\{^{1}H\}H$	briefly mentioned	covered in depth
Integration	important	important

Conclusions

Presenting multinuclear NMR spectra in an inorganic chemistry curriculum is important in order for students to gain a better appreciation for the power of this technique. While there is some overlap with the topics in NMR spectroscopy covered in our organic chemistry curriculum, the differences provide a more complete perspective to our students (Table 1). However, the challenge of having classes in which some students have taken Organic Chemistry while others have not is greatest when covering NMR spectroscopy. This curriculum continues to develop primarily with the inclusion of new research results. The general outline is adaptable to other nuclei or research interests. In particular, paramagnetism and nuclei that are not spin ½ are not significant components of my research at this time and as a result are not major components of this outline. As students have pointed out, the fundamentals taught in this and other courses are useful. But it is when they can apply these concepts to 'real world' samples in the research lab that they fully develop their understanding of this material.

Acknowledgments

We thank the Kresge Foundation for funding the acquisition of our Jeol 400 MHz NMR spectrometer.

References

1. Cowley, A. H.; Hall, S. W. *Polyhedron* **1989**, *8*, 849.
2. With limited emphasis on chemical shifts, it is unlikely that a student will question the similarity of the chemical shifts for these two apparently very different environments for phosphorus. However, if the question is raised or if the instructor wishes, the aromaticity of this anion can be considered.
3. Spectra were simulated using WINDNMR. Reich, H. J. *J. Chem. Educ.* **1995**, *72*, 1086.
4. The range of coupling constants for diastereotopic protons is 0-30 Hz, the largest H-H coupling constants in typical organic compounds. Silverstein, R. M., Webster, F. X., Kiemle, D. J. *Spectroscopic Identification of Organic Compounds*, 7th ed.; John Wiley & Sons, Inc.: Hoboken, NJ, 2005; p 198.
5. Di Vaira, M.; Peruzzini, M.; Costantini, S. S.; Stoppioni, P. *J. Organomet. Chem.* **2006**, *691*, 3931.
6. Nataro, C.; McNamara, W. R.; Maddox, A. F. In *Modern NMR Spectroscopy in Education*; Rovnyak, D.; Stockland, R. A., Eds.; ACS Symposium Series 969; American Chemical Society: Washington, DC, 2007; p 246.
7. Kapicka, L.; Dastych, D.; Richterova, V.; Alberti, M.; Kubacek, P. *Magn. Res. Chem.* **2005**, *43*, 294.
8. For a system in which I = ½ (such a ^{1}H) the intensities for a septet are 1:6:15:20:15:6:1. Since ^{2}H is I = 1, the intensities are different (see ref 1).
9. Platt, A. W. G. In *Inorganic Experiments*, 3rd ed.; Woollins, J. D., ed.; Wiley-VCH: Weinheim, 2010; p 58.
10. Nataro, C.; Fosbenner, S. M. *J. Chem. Educ.* **2009**, *86*, 1412.
11. Girolami, G. S., Rauchfuss, T. B., Angelici, R. J. In *Synthesis and Technique in Inorganic Chemistry: A Laboratory Manual*, 3rd ed.; University Science Books: Sausalito, CA, 1999; p 117.
12. Rosman, K. J. R.; Taylor, P. D. P. *Pure Appl. Chem.* **1999**, *71*, 1593.

13. Mandell, C. L.; Tiedemann, M. A.; Gramigna, K. M.; Diaconescu, P. L., Dougherty, W. G.; Kassel, W. S.; Nataro, C. Manuscript in preparation.
14. The Covalent Bond Classification system for electron counting in transition metal systems is used in the Inorganic Chemistry II course (Parkin, G. in *Comprehensive Organometallic Chemistry III*; Crabtree, R. H., Mingos, D. M. P., Eds.; Elsevier: Oxford, 2006; Volume 1, Chapter 1).
15. Students may get creative and suggest any number of other ideas (*e.g.,* a fluxional process or a geometry change at the Pt center). It is best to apply Occam's razor.
16. A possible monomeric structure with a Fe-Pt dative bond is a second possibility but is not observed with dppf, see ref 13.
17. Mandell, C. L.; Tiedemann, M. A.; O'Connor, A. R.; Chan, B.; Nataro, C. Manuscript in preparation.

Chapter 10

Using ^{195}Pt and ^{31}P NMR To Characterize Organometallic Complexes: Heteronuclear Coupling in the Presence of Geometric Isomers

Daron E. Janzen,* Mainong Hang, and Hannah M. Kaup

Department of Chemistry and Biochemistry, St. Catherine University, St. Paul, Minnesota 55105
***E-mail: dejanzen@stkate.edu**

A laboratory experiment has been developed for the Advanced Inorganic Chemistry laboratory involving the synthesis and characterization of air-stable organometallic platinum(II) cyclometallated complexes. Using one-dimensional ^{195}Pt, ^{31}P, and ^{1}H NMR data, students are able to assign isomeric composition and identify the nature of the product isomers obtained. This experiment introduces students to heteronuclear coupling effects, nonroutine NMR nuclei, and the complexity of NMR spectra introduced by nuclei whose isotopic abundance is neither very high (^{1}H) or very low (^{13}C).

Introduction

Use of one-dimensional ^{1}H and ^{13}C NMR spectroscopy has become routine in the undergraduate chemistry laboratory curriculum, particularly in organic chemistry. This is not surprising since NMR spectroscopy is typically introduced in introductory organic chemistry courses as the most important tool for structural determination. Increasingly, one-dimensional DEPT (Distortionless Enhancement by Polarization Transfer) spectra and two-dimensional COSY (COrrelation SpectroscopY), HMQC (Heteronuclear Multiple Quantum Coherence), and HMBC (Heteronuclear Multiple Bond Correlation) spectra have been introduced in organic laboratory experiments. Use of heteronuclear two-dimensional experiments involving ^{1}H and ^{13}C such as HMQC and HMBC begin to develop student expertise with coupling between different types of nuclei for purposes of structural assignment, but spin-spin coupling between these nuclei is not directly observed as the ^{13}C domain is typically ^{1}H decoupled. In fact, most undergraduate students have never seen or acquired a proton coupled ^{13}C NMR spectrum! Besides a typical lack of experience with heteronuclear coupling, undergraduate laboratory experiments using NMR of nuclei other than ^{13}C and ^{1}H are uncommon. Despite being largely relegated to inorganic lab courses, a growing number of laboratory experiments that observe ^{19}F (*1–6*) and ^{31}P (*7–22*) nuclei have been developed. Less common examples include $^{10/11}B$ (*23*), $^{14/15}N$ (*24–27*), ^{17}O (*26, 28*), ^{23}Na (*29*), ^{27}Al (*30*), ^{29}Si (*26*), ^{59}Co (*31*), ^{77}Se (*13*), ^{95}Mo (*32*), and ^{195}Pt (*14, 33, 34*). The experiment described here not only provides students experience with heteronuclear coupling, but also with other NMR nuclei that have vastly different natural abundances, chemical shifts, and coupling constants.

Table 1. Selected properties of NMR nuclides[a]

Isotope	*I*	*γ (x10⁷ rad T⁻¹s⁻¹)*	*Natural Abundance*	*Receptivity Relative to ¹H*	*Frequency (MHz)*
^{1}H	1/2	26.75	99.99	1	400
^{13}C	1/2	6.73	1.07	$1.8x10^{-4}$	100
^{19}F	1/2	-25.18	100.00	$8.3x10^{-1}$	376
^{31}P	1/2	10.84	100.00	$6.6x10^{-2}$	162
^{10}B	3	2.87	19.58	$3.9x10^{-3}$	42.8
^{11}B	3/2	8.58	8.42	$1.3x10^{-1}$	128
^{14}N	1	1.93	99.63	$1.0x10^{-3}$	28.8
^{15}N	1/2	-2.71	0.37	$3.9x10^{-6}$	40.4
^{17}O	5/2	3.63	0.04	$1.1x10^{-5}$	54.4
^{27}Al	5/2	6.98	100.00	$2.1x10^{-1}$	104
^{29}Si	1/2	-5.32	4.70	$3.7x10^{-4}$	79.5
^{59}Co	7/2	6.30	100.00	$2.8x10^{-1}$	94.8
^{77}Se	1/2	5.12	7.58	$5.3x10^{-4}$	76.4
^{95}Mo	5/2	-1.75	15.92	$5.2x10^{-4}$	26.1
^{195}Pt	1/2	5.84	33.80	$3.4x10^{-3}$	86.0

[a] Values obtained from reference (*35*).

Laboratory experiments that utilize NMR nuclei beyond ^{1}H, ^{13}C, ^{19}F and ^{31}P need to be assessed in terms of ease of acquisition of spectra. As instrument time allocated per student (or lab group) is often limited, ease of data collection plays a large role in the feasibility of utilizing less common NMR nuclei. Important considerations include nuclear spin, relative natural abundance, range and type of probe available, and relative sensitivity. Attention must also be paid to temperature and solvent sensitivity of chemical shifts as well as ease of use of chemical shift standards. The availability of an autosampler and the ability to queue experiments for acquisition outside class time may also play an important role in the practicality of experiments employing less common NMR nuclei. Table 1 (*35*) summarizes the properties of selected NMR nuclides on a 400 MHz spectrometer.

Availability of a tunable broadband probe is required for acquisition of a variety of NMR spectra since each nucleus resonates at a different frequency.

Some probes may also have auto-tuning capabilities, allowing for fast switching between acquisitions of different nuclei. Many tunable probes have access to a large range of frequencies for direct detection (often ^{31}P to ^{15}N), but may not be able to access some low frequency nuclei (such as ^{95}Mo). Low frequency tunable probes are available to perform routine low frequency nuclei measurements, but this necessitates switching the probe, which is time consuming and inconvenient to other users who need routine ^{1}H and ^{13}C NMR data. These probes are uncommon and an added expense. For educational purposes, there are likely other experiments (such as the one outlined in this paper) that will pedagogically achieve the same results, that allow for the observation of heteronuclear coupling information to nuclei outside the probes detection range.

The nuclear spin must also be taken into account when performing experiments on atypical NMR nuclei. Quadrupolar nuclei (nuclear spin > ½) are very sensitive to local symmetry, with significant line broadening in low symmetry environments. Nuclei with small chemical shift ranges in low symmetry environments may yield little useful information since broad lines may limit discrimination of chemical shift changes or multiple nuclei in unique environments may be too similar. Line broadening of spin- ½ nuclei are not subject to the same symmetry sensitivity as quadrupolar nuclei, so signals are less broad. Nuclear spin also affects the spin-spin splitting patterns observed with nuclei of the same type as well as with coupled heteronuclei. These patterns are subject to the 2nI + 1 rule.

Relative receptivity is a measure of detection response of a nuclide compared to ^{1}H. The receptivity is dependent on the gyromagnetic ratio, the natural abundance, and the spin of the nucleus. The gyromagnetic ratio, γ, of each nucleus strongly affects the relative sensitivity as receptivity is proportional to γ^3. Low γ nuclei are less sensitive, requiring more transients and thus more time to collect a spectrum with an acceptable signal:noise ratio (e.g. ^{14}N, ^{15}N, and ^{95}Mo). Isotopic abundance varies widely amongst the nuclides of interest. High isotopic abundance can offset some of the effects of low γ values for some nuclides. Also, use of an inverse probe can enhance signal intensity for low natural abundance and/or low γ nuclei. High spin values also enhance receptivity, although quadrupolar nuclei introduce additional linewidth issues.

The properties of ^{195}Pt make this nuclide ideal for the undergraduate laboratory. While the gyromagnetic ratio of ^{195}Pt is slightly smaller than ^{13}C, the relative receptivity is almost 20 times that of ^{13}C. The most common chemical shift reference, $[PtCl_6]^{2-}$ in D_2O, is stable and easy to use. The chemical shift range for ^{195}Pt is very large (-6000 to +7500 ppm), though for complexes in the most studied oxidation state (Pt^{II}) this range is roughly -6000 to -1000 ppm (*36*).

Coupling constants involving ^{195}Pt are usually observable and $^{(1\text{-}3)}J$ couplings with ^{195}Pt have been observed with ^{1}H, ^{13}C, ^{15}N, ^{31}P, ^{19}F, and other nuclei (*37*). Coupling of ^{195}Pt appears both in ^{195}Pt NMR spectra as well as in spectra of coupled nuclei. For example, coupling of ^{1}H nucleus with ^{195}Pt appears in a ^{1}H NMR spectrum as satellites with a total integration of 33.8% of a given peak, indicating the 33.8% natural abundance of ^{195}Pt compared with the 66.2% of NMR inactive nuclides of platinum.

In recent years, a few laboratory experiments utilizing ^{195}Pt NMR have appeared. Brittingham et al. have developed an experiment involving binuclear phosphine-bridged platinum complexes (*14*). An experiment published by Berry also utilizes platinum phosphine complexes (*33*). Cis and trans isomers of diamminedichloroplatinum(II) have been studied using ^{195}Pt NMR in an experiment described by Arvanitis et al. (*34*). Each of these experiments take advantage of the heteronuclear coupling of ^{195}Pt with other nuclei in the respective systems studied.

The laboratory experiment described here presents a series of complexes easily synthesized and characterized by numerous one-dimensional NMR spectra of nuclides that include ^{1}H, ^{13}C, ^{31}P, and ^{195}Pt. Students synthesize a platinum (II) coordination complex and a cyclometallated organometallic phosphine complex. Heteronucelar coupling between ^{195}Pt, ^{1}H, ^{13}C, and ^{31}P are observed in various one-dimensional spectra, with some of the same coupling constants measured from the perspective of each nucleus. In addition, geometric cis and trans isomers are present in both the starting material and cyclometallated complex. Students must sort out coupling constants as well as isomeric composition from spectra of modest complexity. Conveniently, NMR spectra of starting materials and products are all published for comparison with student results if desired.

Synthesis

The synthetic scheme for this experiment is shown in Scheme 1. Both steps in this reaction scheme afford reasonable yields and neither products nor reagents are air-sensitive. Both reactions are carried out in air in standard laboratory glassware and require no special drying of solvents. Benzonitrile, $PtCl_2$, and $P(mes)_3$ (mes = 2,4,6-trimethylphenyl) are commercially available and can be used as purchased.

Synthesis of $[Pt(PhCN)_2Cl_2]$

This compound was prepared by a modification of the method used by Anderson et al (*38*). Benzonitrile (28 mL) was heated in a 100-mL round bottom flask to 100°C with stirring. Platinum (II) chloride (1.00 g, 3.76 mmol) was added slowly to the hot benzonitrile solution. After all the $PtCl_2$ was added, heating of the reaction continued for 10 min, after which time the hot solution was filtered. To the filtrate, 190 mL of petroleum ether was added, causing precipitation of a yellow solid. After allowing the solution to cool to room temperature, the solid was collected on a medium porosity glass fritted funnel, rinsed with additional petroleum ether (3 x 15 mL) to remove unreacted $PtCl_2$, and dried *in vacuo* to yield 1.54 g (3.26 mmol, 87%) as a mixture of cis/trans isomers, with the trans isomer typically predominating. This reaction can be scaled down for individual student preparation, or a single batch on this scale will suffice as starting material for a class of 10 students.

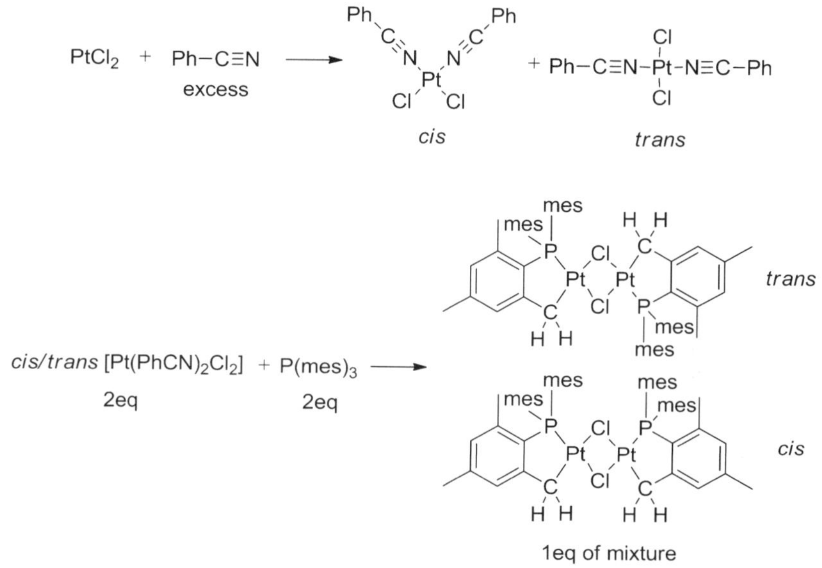

$PtCl_2$ + Ph−C≡N (excess) ⟶ *cis*-$[Pt(PhCN)_2Cl_2]$ + Ph−C≡N−Pt(Cl)$_2$−N≡C−Ph (*trans*)

cis/trans $[Pt(PhCN)_2Cl_2]$ (2eq) + $P(mes)_3$ (2eq) ⟶ *trans* + *cis* (1eq of mixture)

Scheme 1. Synthesis of cis- and trans-$[Pt(PhCN)_2Cl_2]$*and cis- and trans-*$[Pt\{CH_2C_6H_2(CH_3)_2Pmes_2\text{-}C,P\}(\mu\text{-}Cl)]_2$

Several synthetic procedures have been developed to yield single isomers of $[Pt(PhCN)_2Cl_2]$, but this compound readily isomerizes in various solvents. (*39*). The use of $[Pt(PhCN)_2Cl_2]$ in the subsequent reaction is insensitive to the isomeric composition. Optionally, this starting material may be purchased if less time is available.

Synthesis of $[Pt\{CH_2C_6H_2(CH_3)_2Pmes_2\text{-}C,P\}(\mu\text{-}Cl)]_2$

This compound was prepared by a modification of the method used by Fornies et al (*41*). Solid $[Pt(PhCN)_2Cl_2]$ (0.150 g, 0.317 mmol) and $P(mes)_3$ (0.123 g, 0.317 mmol) were mixed in 2-methoxyethanol (5 mL) in a 25 mL round bottom flask. The reaction was heated at reflux with stirring for 1 hr. The solids dissolve into solution during heating, and the product precipitates out of the solution as the reaction proceeds. The reaction mixture was allowed to cool to room temperature and the gray/white precipitate was collected using a medium porosity glass fritted funnel, washed with diethyl ether (15 mL), and dried *in vacuo* to yield 0.098 g of product (0.079 mmol, 50% yield). This synthesis produces a mixture of cis/trans isomers, with the trans product as the major isomer.

This complex was previously reportedly synthesized employing a different procedure (*40*). The procedure reported here is similar to that of Fornies et al. who utilized this reaction to synthesize the related *o*-tolyl phosphine analog (*41*). The synthetic scheme used requires less time (1 h vs. 6 h), no added base, and does not require recrystallization or column chromatography for purification.

Data Acquisition

The products and starting materials were characterized using ^{1}H, ^{13}C, ^{31}P, and ^{195}Pt NMR spectroscopy. All NMR spectra were collected on a JEOL ECS-400 NMR spectrometer and processed using Delta NMR software (*42*) or ACD/NMR Processor Academic Edition (*43*). Spectra were acquired in $CDCl_3$ at 20°C at concentrations of approximately 10 mM for $[Pt\{CH_2C_6H_2(CH_3)_2Pmes_2\text{-}C,P\}(\mu\text{-}Cl)]_2$ or 20 mM for $[Pt(PhCN)_2Cl_2]$ and $P(mes)_3$ since larger quantities of these were available. Residual solvent peaks were used to reference ^{1}H (7.24 ppm) and $^{13}C\{^{1}H\}$ (77.16 ppm) NMR spectra. $^{31}P\{^{1}H\}$ NMR spectra were externally referenced to phosphoric acid (0 ppm) and $^{195}Pt\{^{1}H\}$ NMR spectra were externally referenced to $[PtCl_6]^{2-}$ (0 ppm). ^{1}H, ^{13}C, and ^{31}P NMR spectra were acquired with routine parameters. ^{195}Pt NMR spectra were acquired at 85.94 MHz. A pulse width of 4.83 μs was used with a relaxation delay of 0.1 s. This corresponds to a 30° tip angle (14.5 μs, 90° pulse). An offset of -3000 ppm was used with a sweep width of 3000 ppm. Each scan contained 32768 data points, providing a resolution of 6.6 Hz. ^{195}Pt NMR spectra with acceptable signal/noise were acquired with 2500 scans in 10 min under these conditions. Processing parameters used for the ^{195}Pt NMR spectra were identical to ^{13}C with the exception of single exponential function applied before the Fourier Transform. Some systems call this the line broadening parameter. The width of the single exponential applied to the ^{195}Pt FIDs was 50 Hz. As the relaxation delay was short (0.1 s), the decoupler was turned on during the pulse and acquisition. The NOE resulting from the decoupling scheme precluded quantitation of isomers from the ^{195}Pt NMR spectra. The same is true of ^{13}C and ^{31}P NMR spectra collected in the method described.

Results

The NMR spectra of the starting materials show a variety of features, including the heteronuclear coupling and the presence of isomers. Table 2 summarizes the NMR data collected for starting materials and products.

Trimesityl phosphine, $P(mes)_3$, shows interesting heteronuclear coupling in its NMR spectra. The ^{1}H NMR spectrum is easily interpreted since only one aromatic proton resonance is present along with two unique methyl signals in a 2:1 ortho:para ratio. Since no ^{3}J coupling is possible, no ^{1}H-^{1}H spin-spin splitting is observed. However, close examination of the one aromatic resonance shows a small ^{4}J ^{1}H-^{31}P coupling. The ^{13}C NMR spectrum of $P(mes)_3$ displays ^{1}J, ^{2}J, and ^{3}J ^{13}C-^{31}P coupling (Figure 1).

Table 2. Selected NMR data summary

Chemical Shifts (ppm) and Coupling Constants

	Compound 1[a]	*Compound 2*[a]	*Compound 3*[a]
^{1}H	6.77 (^{4}J ^{31}P-^{1}H = 3.1 Hz) 2.25, 2.04	b	(*cis*) 3.14; (*trans*) 3.15 ^{2}J ^{1}H-^{195}Pt = 104 Hz
^{13}C	142.82 (^{2}J ^{31}P-^{13}C = 17.2 Hz); 137.71, 131.70 (^{1}J ^{31}P-^{13}C = 19.1 Hz) 129.87, 22.96 (^{3}J ^{31}P-^{13}C = 16.2 Hz) 21.15	(*trans*) 135.14, 133.56, 129.49, 115.20, 109.41 (*cis*) 135.31 133.72, 129.45, 116.76, 108.97	c
^{31}P	-5.8	d	(cis) 9.13 (^{1}J ^{31}P-^{195}Pt = 5194 Hz) (*trans*) 8.30 (^{1}J ^{31}P-^{195}Pt = 5116 Hz)
^{195}Pt	d	(*cis*) -2272 (*trans*) 2333	(*cis*) -3212 (^{1}J ^{195}Pt- ^{31}P = 5216 Hz) (*trans*) -3289 ^{1}J ^{195}Pt-^{31}P = 5100 Hz

[a] Compound 1 = $P(mes)_3$, Compound 2 = $[Pt(PhCN)_2Cl_2]$, Compound 3 = $[Pt\{CH_2C_6H_2(CH_3)_2Pmes_2\text{-}C,P\}(\mu\text{-}Cl)]_2$; [b] not interpreted; [c] not acquired; [d] not applicable.

Aromatic carbons 1 and 2, as well as the ortho methyl groups (carbon 5), appear as doublets with similar coupling constants. Assignment of the ^{13}C NMR spectrum is aided by the differences in numbers of equivalent aromatic carbons (1:2:2:1), the number of unique quaternary carbons (three), and their proximity to the ^{31}P nucleus. The ^{31}P NMR spectrum exhibits a singlet with a typical aromatic phosphine chemical shift. No ^{31}P-^{13}C coupling is observed under the data collection conditions of this experiment since the ^{31}P linewidth is of a similar magnitude as the size of the coupling constant determined from the ^{13}C NMR spectrum.

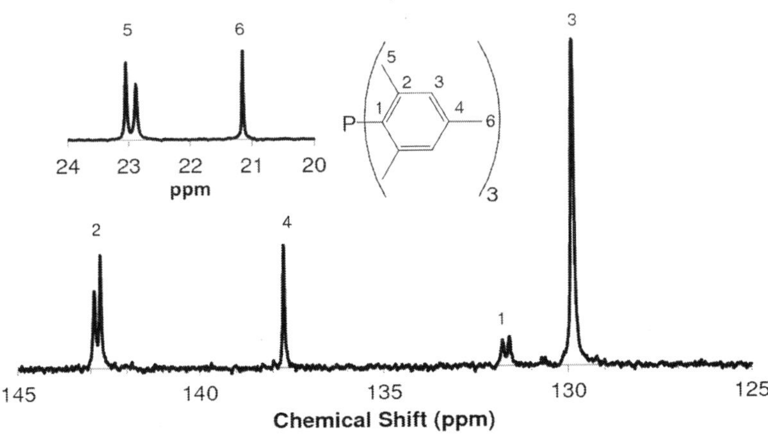

Figure 1. ^{13}C NMR spectrum of $P(mes)_3$. The aliphatic region inset intensity scale is approximately half that of the aromatic region.

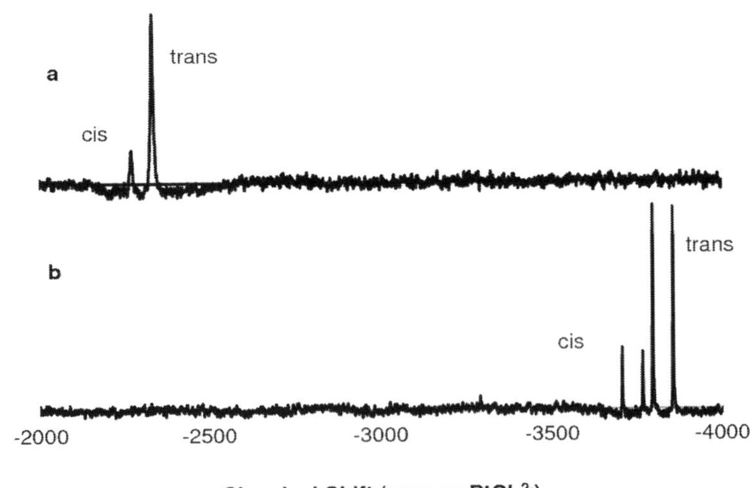

Figure 2. ^{195}Pt NMR spectra of (a) $[Pt(PhCN)_2Cl_2]$ and (b) $[Pt\{CH_2C_6H_2(CH_3)_2Pmes_2\text{-}C,P\}(\mu\text{-}Cl)]_2$.

While the ^{1}H NMR spectrum of $[Pt(PhCN)_2Cl_2]$ is simple, two isomers are present in solution. As the protons on the phenyl rings are relatively insensitive to the cis/trans isomeric arrangements, little is gleaned from this spectrum. The ^{195}Pt NMR spectrum, however, provides clear evidence that two unique Pt environments are present in solution (Figure 2a).

Two singlet peaks are observed at -2334 ppm and -2272 ppm. Assuming each resonance is the result of a mononuclear Pt species, this spectrum can be rationalized as a mixture of the cis and trans isomers.

The ratio of the isomers may vary according to the preparation of the $[Pt(PhCN)_2Cl_2]$. Using the procedure reported here, the major isomer is the trans isomer. The ^{195}Pt NMR resonances are assigned based on comparison with reported chemical shifts (*39*). The trans isomer ^{195}Pt is more shielded than the cis isomer as the π back donation from the metal is less effective than in the cis isomer (*39*). Conversely, the electron density at the platinum center is slightly lower for the cis isomer, giving rise to a more downfield ^{195}Pt chemical shift.

The ^{13}C NMR spectrum of $[Pt(PhCN)_2Cl_2]$ also shows the presence of two similar species, as two sets of peaks with similar ratios between the major and minor species. Initially, students may interpret the complexity as coupling of ^{195}Pt with ^{13}C. The large number of bonds between many of the carbons and platinum as well as the absence of symmetrical ^{195}Pt satellites quickly rules this out, however. Using the data collection parameters of this experiment, ^{2}J ^{13}C-^{195}Pt coupling is not observed. While these coupling constants have been reported, long acquisition times make their observation prohibitive. These signals are also difficult to observe because the nitrile carbons are quaternary and the close quadrupolar ^{14}N atom both contribute to the long relaxation of these carbons (*39*). No ^{13}C-^{195}Pt coupling is observed in the ^{195}Pt spectrum as the approximate 250 Hz ^{2}J ^{13}C-^{195}Pt coupling in the cis and trans isomers is too small to observe compared to the resolution and linewidth of the ^{195}Pt spectra. While the ^{13}C NMR data is reported for $[Pt(PhCN)_2Cl_2]$ in Table 2, 20,000 scans were necessary to obtain a spectrum with an acceptable signal to noise ratio. As collection of this data requires overnight acquisition (17 h), it is likely impractical for each student to collect this data.

The final product in this experiment, the dimer $[Pt\{CH_2C_6H_2(CH_3)_2Pmes_2\text{-}C,P\}(\mu\text{-}Cl)]_2$, shows rich heteronuclear coupling information in the presence of geometric cis/trans isomers. While ^{1}H, ^{31}P, and ^{195}Pt NMR spectra were acquired for this dimer, a ^{13}C NMR spectrum was not collected because sufficient scans to observe heteronuclear coupling were impractical due to the long acquisition time required. The ^{1}H NMR spectrum is complicated as the cyclometallation has reduced the symmetry of the ligated $P(mes)_3$ considerably.

The most important feature of this spectrum lies in the 2.95-3.35 ppm region (Figure 3). This region shows the hydrogen atoms attached to the cyclometallated carbon. Two unequal integration values are discernible, with flanking ^{195}Pt satellites. The similar chemical shifts, but unequal integrations, indicate the presence of cis/trans isomers. While unique ^{2}J ^{1}H-^{195}Pt coupling constants for each isomer are not resolved, an average coupling constant of 105 Hz can be measured. This value is typical for cyclometallated platinum(II) phosphine complexes (*37, 44*). The coupling of the hydrogen to ^{195}Pt is confirmed by the relative integration ratio of ^{195}Pt satellites to the uncoupled proton of 17:66:17. The aromatic region shows the broken symmetry of the cyclometallation of one of the mesityl groups. The non-cyclometallated methyl region is complicated to assign, but the presence of isomers is confirmed by two sets of similar chemical shift resonances with consistent integration ratios (about 3:1 trans/cis). Complete assignment of this spectrum is unnecessary.

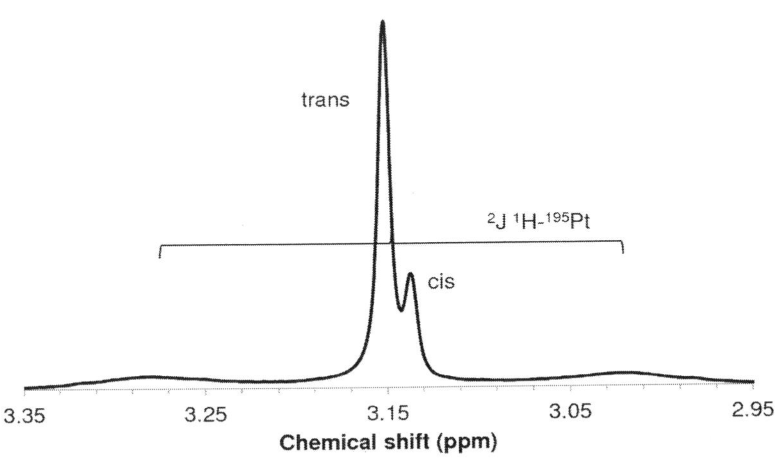

Figure 3. Partial ^{1}H NMR spectrum of [Pt{$CH_2C_6H_2(CH_3)_2$Pmes$_2$-C,P}(μ-Cl)]$_2$.

The ^{195}Pt NMR spectrum of [Pt{$CH_2C_6H_2(CH_3)_2$Pmes$_2$-*C,P*}(μ-Cl)]$_2$ also shows heteronuclear ^{195}Pt-^{31}P coupling and the presence of two doublets as the cis/trans isomers (Figure 2b). While the ^{195}Pt NMR linewidth is large, the chemical shift differences of these isomers as well as the magnitude of the ^{195}Pt-^{31}P coupling constants are sufficient for clear separation of these isomers in the ^{195}Pt NMR spectrum. The single ^{31}P environment for each isomer coupled to the respective ^{195}Pt nucleus gives rise to the appearance of two doublets. The difference in the coupling constants, though only about 2%, is large enough to distinguish between isomers as the magnitude of these is so large(^{1}J ^{195}Pt-^{31}P = 5100 Hz *trans*, 5216 Hz *cis*). The upfield shift of these ^{195}Pt resonances of [Pt{$CH_2C_6H_2(CH_3)_2$Pmes$_2$-*C,P*}(μ-Cl)]$_2$, relative to the starting material [Pt(PhCN)$_2$Cl$_2$], is consistent with increased electron density at the ^{195}Pt nucleus in the cyclometallated complex due in large part to the coordination of the carbon.

Heteronuclear ^{31}P-^{195}Pt coupling and the presence of cis/trans isomers are also observed in the ^{31}P NMR spectrum (Figure 4). As ^{195}Pt is not 100% abundant, satellites are present for each of the two unique isomeric ^{31}P resonances. Essentially identical ^{31}P-^{195}Pt coupling constants are measured from the ^{195}Pt satellite doublets in the ^{31}P NMR spectrum as in the ^{195}Pt NMR spectrum. While distribution of the isomers cannot be determined quantitatively from these spectra, a rough integration ratio of 3:1 trans/cis is found in the ^{31}P NMR spectrum. While the chemical shift differences of the cis and trans isomers is small, the narrow linewidth of the ^{31}P NMR spectrum and large coupling constants make observation and interpretation of these spectra possible. The major isomer is assigned to the trans isomer based on reported values (*40*).

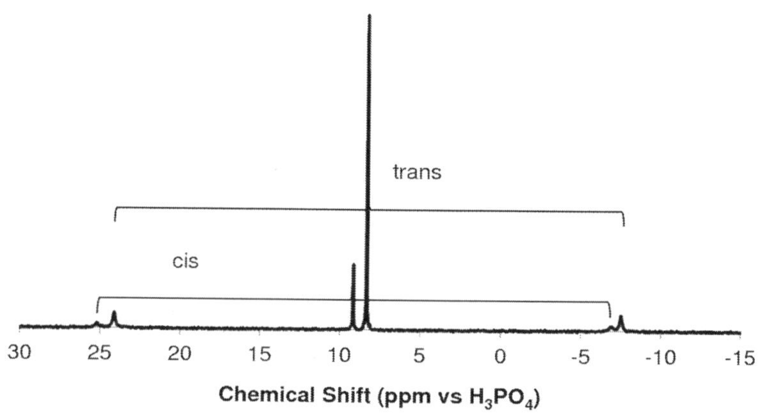

Figure 4. ^{31}P NMR spectrum of $[Pt\{CH_2C_6H_2(CH_3)_2Pmes_2\text{-}C,P\}(\mu\text{-}Cl)]_2$. ^{195}Pt satellites for each isomer are indicated by brackets.

Students are required to interpret the NMR spectrum of the starting materials and products, compiling tables of chemical shifts and coupling constants as well as assigning the most relevant peaks and calculating rough percentages of isomers based on integration (where applicable). Students are reminded that solid quantitative information cannot be gained through integration of the ^{195}Pt, ^{31}P, and ^{13}C NMR spectra obtained in the manner described for this experiment. Students are asked to discuss how heteronuclear coupling effects observed in the final product $[Pt\{CH_2C_6H_2(CH_3)_2Pmes_2\text{-}C,P\}(\mu\text{-}Cl)]_2$ appear differently in the ^{31}P and ^{195}Pt NMR spectra, including a semi-quantitative measure of the integration of peaks (^{195}Pt satellites and decoupled ^{31}P) for a given isomer in the ^{31}P spectrum. The most relevant literature references may be provided alongside the experimental instructions, or students can be directed to perform a literature search using SciFinder to help in the analysis of the results they obtain.

Further Experiments

This laboratory exercise can easily be expanded if the instructor desires a larger project with more characterization and further comparison with the literature. Additional ^{1}H and ^{13}C NMR spectra of benzonitrile can be acquired to observe effects on these ligand chemical shifts upon coordination to platinum(II).

Though the platinum compound used to synthesize $[Pt(PhCN)_2Cl_2]$ is $PtCl_2$, an insoluble polymeric material, a related compound K_2PtCl_4 is readily soluble in water. A ^{195}Pt NMR spectrum can be obtained of K_2PtCl_4 to measure ^{195}Pt chemical shift changes upon replacing two chloride ligands with benzonitrile ligands. A combination of two-dimensional ^{1}H-^{13}C NMR spectra (HMQC and HMBC) can be collected to make complete unambiguous assignments for samples of $P(mes)_3$ and $[Pt(PhCN)_2Cl_2]$, with the long distance coupling information gained by HMBC experiments being crucial for assigning the quaternary carbons in each of these samples. A two-dimensional ^{1}H-^{195}Pt HMQC spectrum of $[Pt\{CH_2C_6H_2(CH_3)_2Pmes_2\text{-}C,P\}(\mu\text{-}Cl)]_2$ can be obtained. This experiment would demonstrate the two unique ^{2}J ^{1}H-^{195}Pt couplings for the two isomers present. The infrared spectra of benzonitrile and $[Pt(PhCN)_2Cl_2]$ can also be collected. Coordination induced shifts in the nitrile stretching frequencies can be rationalized, as well as the identification of cis and trans isomer stretches of $[Pt(PhCN)_2Cl_2]$, based on a group theory analysis. Idealized C_{2v} symmetry of the cis isomer yields two IR-active nitrile stretches, while the D_{2h} trans isomer only has one IR-active nitrile stretch. The intensity of the cis and trans $[Pt(PhCN)_2Cl_2]$ nitrile IR stretches can also be compared with the major isomer identification made by ^{195}Pt NMR. Additional verification of the ^{195}Pt and ^{31}P NMR assignments of the major and minor isomers of $[Pt\{CH_2C_6H_2(CH_3)_2Pmes_2\text{-}C,P\}(\mu\text{-}Cl)]_2$ can be performed by acquiring NMR spectra in more or less polar solvents. More polar solvents will decrease the trans/cis isomer ratio as the trans isomer has an expected negligible dipole moment (*41*). The cis isomer with its significant dipole moment will be stabilized in a more polar solvent relative to the trans isomer. Structural comparison of the major solution isomer of $[Pt\{CH_2C_6H_2(CH_3)_2Pmes_2\text{-}C,P\}(\mu\text{-}Cl)]_2$ with the published X-ray crystal structure (trans isomer) can also be added (*45*). The crystallographic information file (.cif) of *trans*-$[Pt\{CH_2C_6H_2(CH_3)_2Pmes_2\text{-}C,P\}(\mu\text{-}Cl)]_2$ can be found in the Cambridge Structural Database (*46*) and analyzed using the free software Mercury (*47*), both available from the Cambridge Crystallographic Data Centre.

Conclusions

This undergraduate laboratory experiment develops student expertise with multinuclear NMR methods. Students encounter a variety of different chemical shift ranges, coupling constants, and the effects of natural abundance on spectral appearance. Using a sequence of straightforward reactions that do not require inert conditions, students access a system with moderate complexity (heteronuclear coupling in the presence of geometric isomers) that has been previously fully characterized. NMR active nuclei, such as ^{195}Pt, with reasonable natural abundance, can provide simple spectra with ease of acquisition commensurate with the undergraduate laboratory. The intent of our work is to encourage others to incorporate nonroutine NMR nuclei into the laboratory curriculum. With experiments such as this, students begin to think of NMR spectroscopy in a broader and richer context.

Acknowledgments

The authors would like to acknowledge the National Science Foundation – Course Curriculum and Laboratory Instruction grant program (Award # 0836842) and St. Catherine University for providing funds for the purchase of a JEOL ECS 400 MHz NMR with autotune unit and tunable broadband probe. The authors also acknowledge Dr. Letitia Yao, NMR Research Associate, NMR Lab, Department of Chemistry, University of Minnesota for her aid in editing this manuscript.

References

1. Peterman, K. E.; Lentz, K.; Duncan, J. *J. Chem. Educ.* **1998**, *75*, 1283–1284.
2. Pohl, N.; Schwarz, K. *J. Chem. Educ.* **2008**, *85*, 834–835.
3. eMendonca, D. J.; Digits, C. A.; Navin, E. W.; Sander, T. C.; Hammond, G. B. *J. Chem. Educ.* **1995**, *72*, 736–739.
4. Dobbie, R. C. *J. Chem. Educ.* **1976**, *53*, 129.
5. Jensen, A. W.; O'Brien, B. A. *J. Chem. Educ.* **2001**, *78*, 954–955.
6. Hawrelak, E. J. In *Modern NMR Spectroscopy in Education*; Rovnyak, D., Stockland, R., Eds.; ACS Symposium Series 969; American Chemical Society: Washington, DC, 2007; pp 288–299.
7. Berry, D. E.; Carrie, P.; Fawkes, K. L.; Rebner, B.; Xing, Y. *J. Chem. Educ.* **2010**, *87*, 533–534.
8. Li, W.; Kagan, G.; Hopson, R.; Williard, P. G. *J. Chem. Educ.* **2011**, *88*, 1331–1335.
9. Fenton, O. S.; Sculimbren, B. R. *J. Chem. Educ.* **2011**, *88*, 662–664.
10. Glidewell, C.; Pogorzelec, P. J. *J. Chem. Educ.* **1980**, *57*, 740–741.
11. Faust, K. E.; Storhoff, B. N. *J. Chem. Educ.* **1989**, *66*, 688–689.
12. Cano, M.; Campo, J. A. *J. Chem. Educ.* **1993**, *70*, 948–950.
13. Dakternieks, D.; Dyson, G. A.; O'Connell, J. L.; Schiesser, C. H. *J. Chem. Educ.* **1994**, *71*, 168–169.
14. Brittingham, K. A.; Schreiner, S.; Gallaher, T. N. *J. Chem. Educ.* **1995**, *72*, 941–944.
15. Lee, M. *J. Chem. Educ.* **1996**, *73*, 184.
16. Linn, D. E. *J. Chem. Educ.* **1999**, *76*, 70–72.
17. Queiroz, S. L.; de Araujo, M. P.; Batista, A. A.; MacFarlane, K. S.; James, B. R. *J. Chem. Educ.* **2001**, *78*, 87–88.
18. Al-Ajlouni, A. M.; Bose, R. M.; Volckova, E. *J. Chem. Educ.* **2001**, *78*, 83–87.
19. Cerrada, E.; Laguna, M. *J. Chem. Educ.* **2005**, *82*, 630–633.
20. Pappenfus, T. M.; Hermanson, D. L.; Ekerholm, D. P.; Lilliquist, S. L.; Mekoli, M. L. *J. Chem. Educ.* **2007**, *84*, 1998–2000.
21. Nataro, C.; Fosbenner, S. M. *J. Chem. Educ.* **2009**, *86*, 1412–1415.
22. Zee, B.; Howard, K. In *Modern NMR Spectroscopy in Education*; Rovnyak, D., Stockland, R., Eds.; ACS Symposium Series 969; American Chemical Society: Washington, DC, 2007; pp 234–244.
23. Zanger, M.; Moyna, G. *J. Chem. Educ.* **2005**, *82*, 1390–1392.

24. Schaeffer, C. D.; Myers, L. K.; Coley, S. M.; Otter, J. C.; Yoder, C. H. *J. Chem. Educ.* **1990**, *67*, 347–349.
25. Grehn, L.; Ragnarsson, U.; Welch, C. J. *J. Chem. Educ.* **1997**, *74*, 1477–1479.
26. Samples, M. S.; Yoder, C. H.; Schaeffor, C. D. *J. Chem. Educ.* **1987**, *64*, 177–178.
27. Rovnyak, D.; Thompson, L. E.; Selzler, K. J. In *Modern NMR Spectroscopy in Education*; Rovnyak, D., Stockland, R., Eds.; ACS Symposium Series 969; American Chemical Society: Washington, DC, 2007;pp 219–23.
28. Zielinski, T. J.; Grushow, A. *J. Chem. Educ.* **2002**, *79*, 707–714.
29. Peters, S. J.; Stevenson, C. D. *J. Chem. Educ.* **2004**, *81*, 715–717.
30. Davis, C. M.; Dixon, B. M. *J. Chem. Educ.* **2011**, *88*, 309–310.
31. Borer, L. L.; Russell, J. G.; Settlage, R. E.; Bryant, R. G. *J. Chem. Educ.* **2002**, *79*, 494–497.
32. Minelli, M. In *Modern NMR Spectroscopy in Education*; Rovnyak, D., Stockland, R., Eds.; ACS Symposium Series 969; American Chemical Society: Washington, DC, 2007; pp 276–287.
33. Berry, D. E. *J. Chem. Educ.* **1994**, *71*, 899–902.
34. Arvanitis, G. M.; Wilk, K. L. *Chem. Educ.* **1997**, *2*, 1–10.
35. Grushow, A. *NMR Periodic Table.* http://arrhenius.rider.edu:16080/nmr/NMR_tutor/periodic_table/nmr_pt_frameset.html (accessed Aug 10, 2012).
36. Still, B. M.; Kumar, P. G. A.; Aldrich-Wright, J. R; Price, W. S. *Chem. Soc. Rev.* **2007**, *36*, 665–686.
37. Priqueler, J. R.; Butler, I. S.; Rochon, F. D. *Appl. Spectrosc. Rev.* **2006**, *41*, 185–226.
38. Anderson, G. K.; Lin, M.; Sen, A.; Gretz, E. *From Inorganic Syntheses: Reagents for Transition Metal Complex and Organometallic Syntheses*; Angelici, R. J., Ed.; Inorganic Syntheses; John Wiley & Sons, Inc.: Hoboken, NJ, 2007; Vol. 28, pp 60−63.
39. Rochon, F. D.; Melanson, R.; Thouin, E.; Beauchamp, A. L.; Bensimon, C. *Can. J. Chem.* **1996**, *74*, 144–152.
40. Alyea, E. C.; Malito, J. *J. Organomet. Chem.* **1988**, *340*, 199–126.
41. Fornies, J.; Martin, A.; Navarro, R.; Sicilia, V.; Villaroya, P. *Organometallics* **1996**, *15*, 1826–1833.
42. *Delta NMR Processing and Control Software v 4.3.6*; JEOL USA, Inc.: Peabody, MA, 2006.
43. *ACD/NMR Processor Academic Edition v 12.01*; Advanced Chemistry Development, Inc.: Toronto, Canada, 2012.
44. Clark, H. C; Goel, A. B.; Goel, R. G.; Goel, S. *Inorg. Chem.* **1980**, *19*, 3220–3225.
45. Alyea, E. C.; Ferguson, G.; Malito, J.; Ruhl, B. L. *Organometallics* **1989**, *8*, 1188–1191.
46. *Cambridge Structural Database v 5.33*; Cambridge Crystallographic Data Centre: Cambridge, U.K., 2012.
47. *Mercury v 3.0*; Cambridge Crystallographic Data Centre: Cambridge, U.K., 2012.

Chapter 11

Beyond Ordinary Undergraduate Experiences: Routine Measurements with Heteronuclear, Heterogeneous, and Paramagnetic Samples

Patrick J. Desrochers*

Department of Chemistry, University of Central Arkansas, Conway, Arkansas 72035
***E-mail: patrickd@uca.edu**

High field instrumentation has made pulsed ^{1}H and ^{13}C NMR spectroscopy routine in undergraduate curricula. Less common in undergraduate experiences is NMR spectroscopy of heteronuclear centers (^{2}H, ^{11}B, ^{19}F, ^{27}Al, ^{31}P, or ^{77}Se). Similarly, paramagnetic samples are typically avoided because of their sometimes unpredictable effect on spectra. This summary describes NMR measurements that were performed by undergraduates in teaching and undergraduate research laboratories. An array of heteronuclear, heterogeneous (resin supported), and paramagnetic samples were studied. A primary product of these efforts is the broader training students received in practical NMR instrumentation and applications, complementary to their traditional organic (^{1}H, ^{13}C) NMR spectroscopy experience. A secondary result is the emergent, dynamic NMR spectroscopy environment in the department, leading to yet more novel experiments for both teaching and research applications.

Introduction

Modern FT-NMR instrumentation is increasingly common in undergraduate chemistry departments nationwide thanks in part to ACS CPT mandates (*1*) and financial commitments like the NSF's CCLI programs. Accordingly, ^{1}H and

^{13}C NMR measurements are now routine components of second-year organic offerings. These experiments also include what used to be more complex research experiments like HETCOR, HMQC, COSY and DEPT 135.

Undergraduate research is an engine for educational innovation. M. J. Molina emphasized the research/teaching synergy when he stated,

> "...explain[ing] my views to students with critical and open minds, I find myself continually challenged to go back and rethink ideas. I know teaching and research as complementary, mutually reinforcing activities (*2*)."

It is natural therefore, that the challenges of research add to the growing array of NMR experiments and applications. Coupled with adequate instrumentation, this makes it possible for current undergraduate chemistry students to earn experience with NMR as a widely applicable tool to solve chemical problems.

Table 1. Various heteronuclei accessible to modern FT-NMR instruments

Nucleus (% nat.ab.)	*Frequency[a] (MHz)*	*Representative Application*	*Ref.*
^{11}B (80.4%)	96.2	Boranes, metal scorpionates	(*3*)
^{19}F (100%)	282.2	-CF_3 appendages in compounds	(*4*)
^{27}Al (100%)	78.2	Solution phase equilibria	(*5, 6*)
^{31}P (100%)	121.4	Metal coordination spheres	(*7*)
^{51}V (99.8%)	78.9	Catalysts and bioactive compounds	(*8, 9*)
^{59}Co (100%)	71.2	Metal coordination spheres	(*10*)
^{77}Se (7.6%)[b]	57.2	Selenates, radical precursors	(*11*)
^{195}Pt (33.8%)	64.5	Platinum nucleic acid complexes	(*12*)

[a] Frequency for a 300 MHz (7.05 Tesla) FT-NMR. [b] While not highly abundant, ^{77}Se has a high relative receptivity (three times ^{13}C).

Table 1 summarizes some of the nuclei that have been studied using magnetic resonance in undergraduate research and/or teaching settings in our institution. While not exhaustive, this list does illustrate the breadth of the periodic table that is accessible to modern NMR instrumentation. The practical familiarity that results from studying such an array of nuclei means that students and instructors earn confidence in NMR spectroscopy as a precise element-specific tool for experiments beyond simple characterizations, e.g, ones that probe kinetic and thermodynamic properties of reactions and molecules that participate in them.

Undergraduates are not commonly introduced to NMR spectroscopy of nuclei beyond ^1H and ^{13}C, or measurements on heterogeneous or paramagnetic samples. The two-semester immersion into ^1H and ^{13}C NMR, that is typical of organic offerings (including organic spectroscopy), can leave students with a kind of spectroscopic myopia where every chemical problem is seen only through these two ubiquitous nuclei. Maslow's axiom: "It is tempting, if the only tool you have is a hammer, to treat everything as if it were a nail, " succinctly expresses this pitfall (*13*). For example, students find hyperfine splittings usually limited to ^{1}H spectra, and so they may become conditioned that this is the only place to expect it. After all, ^{13}C NMR spectra are invariably proton decoupled, conditioning students not to consider hyperfine issues in these or other spectra. This can lead students to forget that hyperfine coupling is a broad and general quality of magnetic resonance and not just limited to proton spectra.

Other perceived limitations abound. Students are typically taught to prepare samples as concentrated solutions and to use deuterated solvents. While this works for most samples, it doesn't work for *all* systems. Few undergraduate departments possess the capability to record NMR spectra of solid samples (magic angle probes), so that experiments involving heterogeneous samples might not even be considered. The advent of cost effective high field superconducting magnets has made the need of paramagnetic-shift reagents obsolete in most NMR experiments. In fact it is more common in undergraduate experiences that paramagnetic samples and impurities are *avoided* due to their influence on line-widths and sometimes to the unpredictable effect on chemical shifts.

Individual challenges encountered by some of the undergraduate research projects in this department motivated the experiments described here. This experience in turn motivated new experiments for advanced laboratories and encourages the development of new experiments for traditional organic courses. A major goal of this work is to convince undergraduates that NMR spectroscopy is a broadly applicable tool to solve a myriad of chemical problems and that modern instrumentation and novel chemical systems are breaking down perceived limitations. This reinforces the knowledge that NMR spectroscopy is a workhorse tool for precise characterizations of established and emerging chemical systems.

Experimental Methods

FT-NMR spectra were recorded using a JEOL ECX-300 spectrometer. This particular model requires manual probe tuning using two sets of potentiostatic knobs at the base of the magnet. One set serves the high frequency probe (^{1}H and ^{19}F) and the other set tunes all other low-frequency nuclei accessible by this model instrument, ranging from $\nu(^{14}N) = 21.7$ MHz to $\nu(^{31}P) = 121.4$ MHz.

Samples were typically prepared using borosilicate-glass NMR tubes (Norel). Airfree samples are readily prepared by loading degassed NMR tubes using cannula or thin syringe needles through rubber septa fitted to the top of the tube. An even better system is the style of tube with a threaded cap and PTFE/rubber septum (purchased through Kontes, K897635-0800). Boron-free tubes (Norel) were used to record ^{11}B spectra reported here. Conventional borosilicate glass tube can be used for these spectra, provided allowance is made for the broad ^{11}B resonance centered near 0 ppm in all spectra (*14*, *15*). The use of post-acquisition software can help alleviate this issue. In general, fast relaxation of quadrupolar nuclei like ^{11}B meant shorter relaxations delays could be employed, allowing more scans to be completed in a shorter period of time.

Synthetic preparations for specific research compounds are given in the references contained in the results and discussion section that follows. It should be noted that the variable temperature ^{31}P measurements (down to -80 °C) presented some difficulty with sample spinning likely resulting from friction due to adventitious moisture condensation on the cooled spinner. Spinning was stopped for these coldest temperature samples with no obvious detrimental effect on the observed ^{31}P signals whose lines are inherently broader than typical ^{13}C and ^{1}H signals. The preparations of scorpionate chelates used here have been reported previously: KTp* (*16*) and KTp′ (*17*).

Deuterated solvents were obtained from Acros Organics, Sigma-Aldrich, or Cambridge Isotope Laboratories Inc. and used without further purification unless otherwise noted. When protio-solvents were used, the spectrometer was not locked but was gradient shimmed on the solvent protio signal. ^{13}C NMR spectra were referenced to solvent-carbon resonances. ^{11}B NMR spectra were referenced to $[NBu_4]BF_4$ (-2.0 ppm). In some cases internal references were added as sealed capillary tubes (1 M H_3PO_4(aq) for ^{31}P NMR spectra). Cost savings realized in using protio in place of deuterated solvents is also a lesson for students to learn in times of diminishing operating funds and more competitive grant atmospheres. For example, routine NMR spectra recorded in DMSO usually utilized protio DMSO than the much more costly DMSO-d_6, which was reserved for final published spectra.

Heterogeneous samples using cross-linked polystyrene with PEG tethers (Novasyn TG amine resin HL from Nova Biochem) required time to prepare and load. A typical run involved about 100 mg of beads scooped into the mouth of the NMR tube and gentle tapping of the tube on the bench top, similar to filling a melting point capillary. When the beads had slid several centimeters into the tube, NMR solvent was added to help flush the beads to the bottom of the tube. Bead samples suspended in solvent were allowed to stand for about 50 min, allowing time for them to swell and solvate the PEG-tethered functional groups. Solvent densities caused the bead samples to sink (in acetone-d_6, d = 0.87 g/mL) or float (in DMSO-d_6, d = 1.19 g/mL). Excess solvent was drawn out of the NMR tube via a syringe. Ideal samples had floating beads that dispersed down into clear solvent when the tube was spun.

Results and Discussion

Working with unique sample preparation, deutero versus protio solvents, solid samples, and paramagnetic samples all add to the practical experimental experience of undergraduates. Students training to be scientists should never view spectroscopic measurements as a black box; simply make the sample, record the spectrum, and interpret sterilized results. Automated sequences in modern instrument software further exacerbate this. Real samples, the kind one invariably finds in open-ended research problems, are sometimes too dilute because little material exists, less than pure, or exhibit unusual kinetic or thermodynamic behaviors. Practical NMR experience comes from students preparing and running samples themselves, earning them the confidence that NMR is a dynamic tool, complementary to other experimental methods.

Heteronuclei are usually pivotal centers in molecules. Perhaps the best examples of this are phosphorus donor atoms as chelates in transition metal complexes. Because phosphine ligands bond directly to metal centers, ^{31}P NMR spectra of such systems give detailed geometric and sometimes electronic information.

^{31}P NMR Resolution of *cis/trans* Geometric Isomers

An excellent example is the ^{31}P NMR experiment by Queiroz et al. (*7*) used regularly in our advanced inorganic laboratory. It involves the characterization of cis (thermodynamic) and trans (kinetic) isomers of Ru(dppb)(phen)Cl_2 (dppb = 1,4-diphenylphosphinobutane and phen = 1,10-phenanthroline, Figure 1).

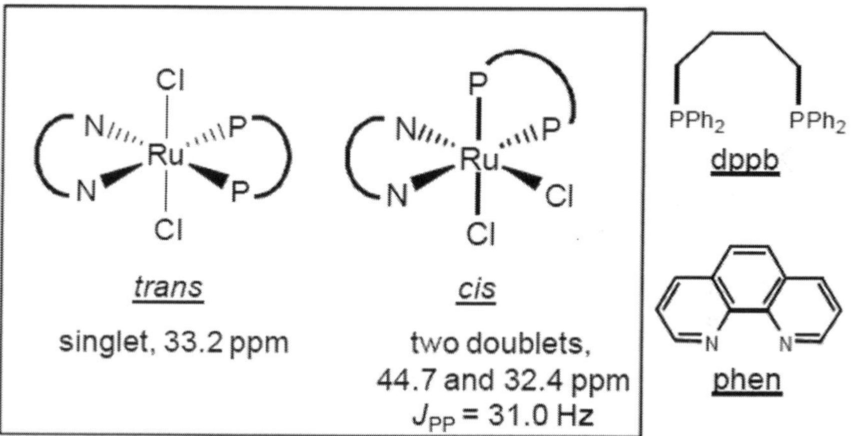

Figure 1. Geometric isomers, cis and trans Ru(phen)(dppb)Cl_2; chelate backbones abbreviated. ^{31}P NMR patterns (in CH_2Cl_2) are indicated.

A first lesson in ^{31}P NMR spectroscopy is provided by the change in chemical shift that occurs with geometrical isomerization. Students note that one of the phosphorus resonances moves upfield of the original trans isomer singlet and one moves downfield. One phosphorus is still trans to N_{phen} but now has its empty *d* orbital rotated 90° relative to the phen pi system. The other phosphorus atom is now trans to a pi-donor chloro ligand. Different electronic phosphorus environments dictate these changes in ^{31}P NMR chemical shifts (*18*).

Students also measure J_{PP} for each distinct phosphorus atom and see that the same value is obtained for each (31 Hz), thus confirming that the now inequivalent phosphorus atoms in the cis isomer are coupled to each other (Figure

1). Observation of strong P-P couplings must mean the two different phosphorus nuclei are bound to the same atom (Ru). This coupling is not observed in the free phosphine chelate, where the atoms are separated by five bonds.

^{31}P NMR of a Fluxional Triphos Chelate

^{31}P NMR measurements of our own phosphine chelate introduced additional experience with variable temperature measurements. Room temperature electronic spectra of [(triphos)NiCysEt]PF_6 confirmed a square planar and correspondingly diamagnetic nickel(II) center in this system (triphos = 1,1,1-tris(diphenylphosphinomethyl)ethane) (*19*). It was puzzling, therefore, that no ^{31}P NMR spectrum was visible at room temperature (see 22 °C spectrum of Figure 2). Literature precedent (*20, 21*) suggested this diamagnetic nickel(II)-phosphine ought to show from one to three ^{31}P resonances. This was a clear case of discordant spectroscopic results from two different methods on a single purified compound. Some reconciliation was needed.

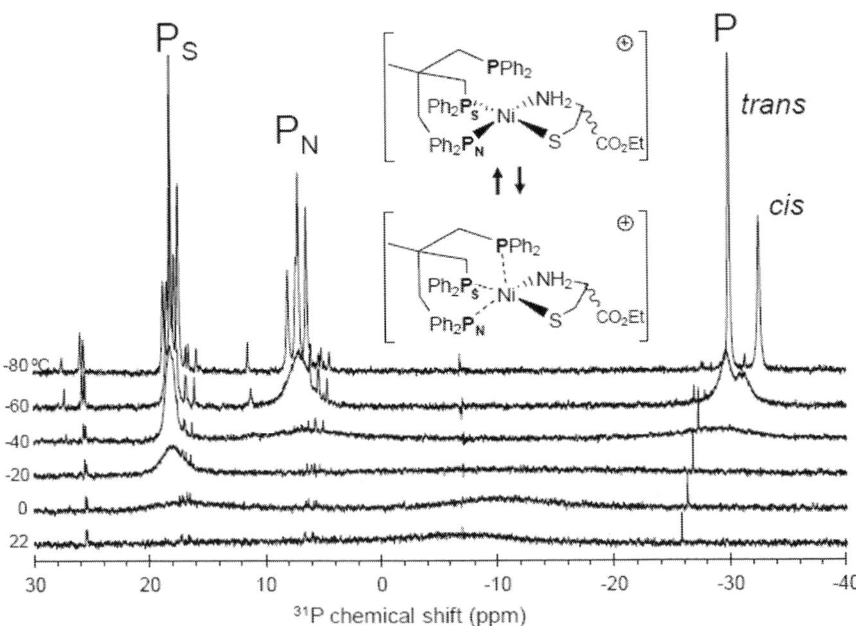

Figure 2. ^{31}P NMR of [(triphos)NiCysEt]PF_6 recorded in acetone-d_6 at increasingly colder temperatures. Evident at -80 °C is the resolution of three distinct resonances assignable to three distinct phosphorus atoms in the (triphos)NiCysEt+ cation. P atom assignments are indicated. Diastereomers result from the use of l-cysteine ethyl ester. Reproduced with permission from reference (19). Copyright 2007 American Chemical Society.

A review of the literature showed that triphos is a characteristically fluxional chelate, especially with d^8 metals prone to square planar coordination spheres. Others have documented this in (triphos)Pt(dtc)$^+$ (dtc = dithiocarbamates) (*22*) and (triphos)$NiCl_2$ (*23*), for which variable temperature ^{31}P NMR spectroscopy confirmed this behavior at low temperature in THF. Not shown is the sharp persistent pentet (integral = 1, δ = -144 ppm, J_{PF} = 708 Hz) due to the free PF_6^- counter ion. These results motivated the same experiment for [(triphos)NiCysEt]PF_6, and these results are summarized in Figure 2; at low temperature (-80 °C) the three expected resonances are clearly resolved with the expected chemical shifts for coordinated and uncoordinated P atoms. Interestingly, the different temperatures at which each resonance was resolved demonstrated the importance of *trans*-P-Ni-S pπ-dπ interactions in these coordination spheres (*19*). This particular result represented a classic "teachable moment" in undergraduate research. What seemed like an intractable problem, irreconcilable spectroscopic measurements on the same compound, resolved itself by use of the chemical literature, that led to a carefully designed NMR measurements, and a self-consistent explanation for a new chemical system.

$^{77}Se/^{31}P$ Hyperfine Splittings

Another heteronucleus, ^{77}Se, that is observable by NMR spectroscopy has proved very useful in identifying specific phosphorus atoms chelating nickel in the complex [dppeNiSeAm]Cl (dppe = 1,2-diphenyl-phosphinoethane; SeAm = 1-amino-ethaneselenate, selenocystamine) (*19*). ^{31}P NMR spectroscopy measurements gave the expected separated doublets at 62.3 and 53.0 ppm (J_{PP} = 41 Hz), one for each phosphorus atom coordinated to nickel. Closer inspection of the spectrum revealed small intensity ^{77}Se satellites associated with each doublet, but each had different values for J_{PSe}. The downfield doublet had satellites with J_{PSe} = 54 Hz vs. J_{PSe} = 40 Hz for the more upfield ^{31}P NMR signal. This work represented a first-ever NMR experience with ^{77}Se for anyone on the project. These results enabled the more downfield ^{31}P, with the larger J_{PSe} value, to be assigned to P-trans-to-Se, owing again to a presumed strong *trans*-P-Ni-E pπ-dπ (E = soft chalcogen donor) effect. The self consistency of this assignment was reinforced by subsequent complexes incorporating the tridentate phosphine, triphos (*vide supra*). The important value of heteroatoms in lending uniquely helpful features to observed spectra is clearly evident in these measurements.

^{11}B NMR Specotrscopy of Scorpionates

Another class of chelates, boron-based scorpionates (*24*, *25*), bear a unique boron heteroatom with NMR chemical shifts and hyperfine splittings very indicative of their chemical environment (Figure 3). ^{11}B NMR spectroscopy is precise and sensitive to the degree of substitution in tetra-substituted boron systems, to the point that this has been employed routinely in undergraduate research applications at our university. This experience motivated the inclusion of experiments from the primary literature in our advanced inorganic laboratory courses dealing with transition-metal catalyzed hydrolysis of amine-borane (*26*), a leading hydrogen-storage material for alternative fuel applications

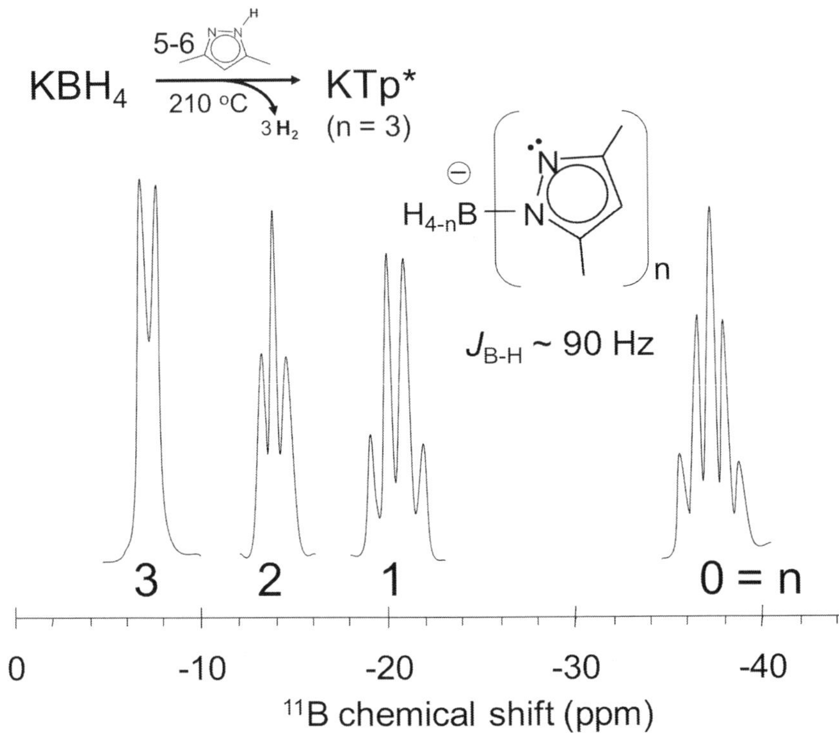

Figure 3. Schematic representations of ^{11}B NMR spectra for BH_4^- ($\delta \sim -37$ ppm) and multiplet patterns when successive hydrides on boron are replaced with pyrazole rings leading to formation of the chelate, Tp-($\delta \sim -7$ ppm).*

Undergraduates in two collaborating research groups in our department have found ^{11}B NMR spectroscopy to be a rapid and certain tool for monitoring the preparation of both well-established and entirely new scorpionates. Synthesis of the new heteroscorpionate KTp′ was monitored and confirmed by this method (*17*). Its preparation involved heating a DMF solution of the potassium salt of the homoscorpionate KTp* in the presence of 1.1 equivalents of benzenetriazole. Over the course of several hours, ^{11}B NMR spectra were recorded for aliquots of the reaction mixture, and a detailed conversion of KTp* (^{11}B $\delta = -7.6$, d $J_{BH} = 90$ Hz) into KTp′ (^{11}B $\delta = -4.6$, d $J_{BH} = 85$ Hz) became apparent. Here, the ability of students to confidently tune the NMR spectrometer probe to ^{11}B, complete the gradient shimming on the protio(CH_3)-DMF signal, and record reliable spectra of a complex mixture was essential. This proven method is being employed in the preparation of a still larger library of heteroscorpionates patterned after KTp′. ^{11}B NMR spectroscopy is the perfect tool for the job, and the students quickly learn the value of this by their own efforts.

Unconventional Hyperfine Splittings in ^{13}C NMR Spectroscopy

Treating hyperfine coupling in ^{13}C NMR spectra is uncommon in undergraduate experiences. While it is the rule to record ^{13}C spectra in *proton*-decoupled mode, ^{13}C NMR spectra are not traditionally recorded in ^{103}Rh ($I = ½$, 100% ab.) or ^{31}P decoupled mode. Similarly, J_{CF} coupling from trifluoromethyl groups, common groups when assessing electronic variability, can add significantly to the complexity of ^{13}C NMR spectra, with J_{CF} ranging from 3 to over 200 Hz for three-bond to one-bond couplings (*27*). Three different systems encountered in undergraduate research projects presented similar examples, and with them the opportunity to practice vigilance for the effect of spin-active nuclei in magnetic resonance measurements.

The first example involved ^{13}C NMR resonances assigned to the ethylene backbones of chelating dppe and cysteinyl thiolate ethyl ester ligands in [dppeNiCysEt]Cl (Figure 4) (*19*).

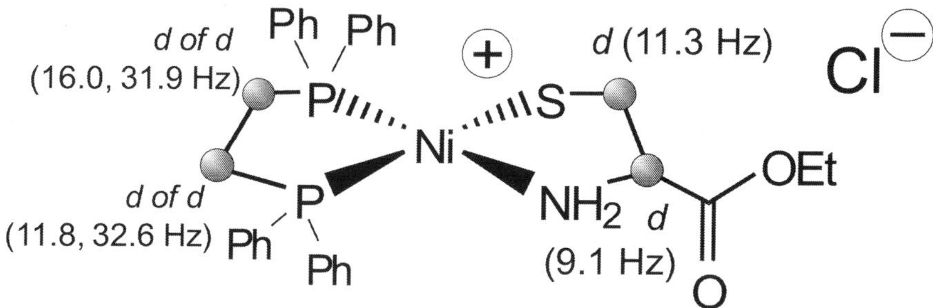

Figure 4. Square planar dppeNiCysEt⁺ cation and J_{CP} splitting patterns observed in proton-decoupled ^{13}C spectra for the $–CH_2CH_2–$ backbone of dppe and for the ethylene backbone of chelating cysteinyl thiolate.

J_{CP} coupling constants ranging from 32 to 11 Hz were observed for the dppe ethylene carbons, reflecting the expected difference between one-bond and two-bond carbon-phosphorus coupling. It should be noted that independent DEPT 135 measurements (not shown) confirmed the $-^{13}CH_2-$ resonances distinct from the α-^{13}CH- resonance of the *l*-cysteine ethyl ester backbone. Surprisingly, we observed substantial $^3J_{CP}$ splittings of 9 to 11 Hz between the Ni-coordinated phosphorus atoms and the cysteinyl ethylene backbone, comparable in magnitude to *two*-bond $^2J_{CP}$ observed for the dppe backbone. This strong coupling was accounted for by long range electronic interaction facilitated by trans-P-Ni-E (E = S or N) dπ-pπ interactions. This hypothesis is supported by the higher value of $^3J_{CP}$ noted for the β-CH_2S carbon atom; sulfur lone pairs should enhance any dπ-pπ interaction compared to the amine group.

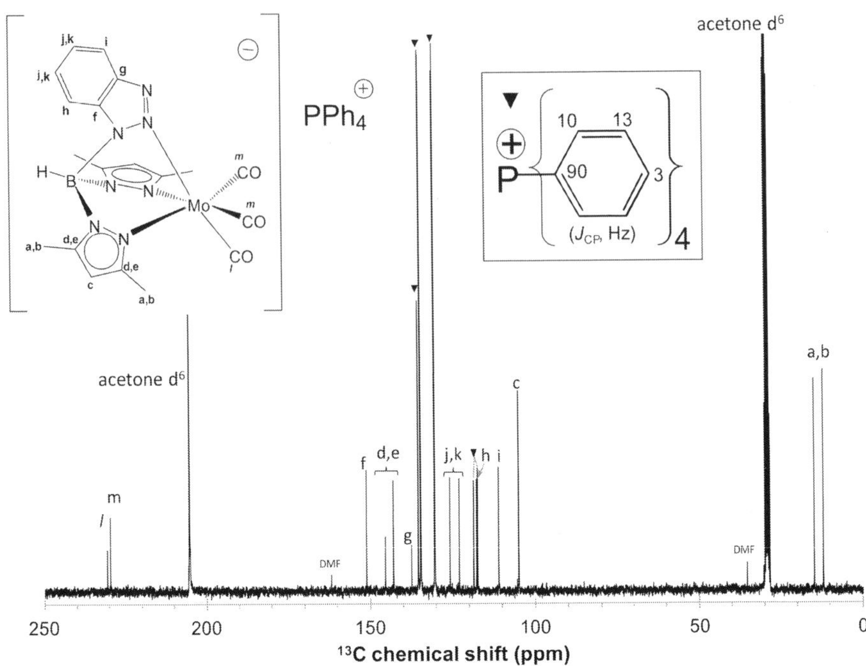

Figure 5. ^{13}C NMR spectrum of [PPh_4][Tp'Mo$(CO)_3$] in acetone-d_6. Assignments are given in the inset figure at left. J_{CP} values for the PPh_4^+ cation are given as an inset figure at right. Black triangles mark these resonances in the spectrum. Reproduced with permission from reference (17). Copyright 2011 American Chemical Society.

Another example of J_{CP} complications in ^{13}C NMR spectra is seen in measurements of the large tetraphenylphosphonium cation. This cation is effective for the precipitation of equally large anions from synthetic reaction mixtures (*28*). The phosphorus heteroatom also offers a site-specific spectroscopic handle; solid state ^{31}P NMR of the closely related triphenylphosphonium cation ($TrPP^+$) experimentally supported theoretical models of halide environments in (TrPP)halide salts (*29*). While phenyl phosphonium cations are effective for these many purposes, it adds complexity to phenyl regions of ^{13}C NMR spectra, especially if other pivotal phenyl carbon groups are also present in the system. Such a case presented itself during the preparation and characterization of [PPh_4][Tp'Mo$(CO)_3$] (Figure 5), where Tp' is a monoanionic heteroscorpionate derivative of Tp* (Figure 3) in which one of the 3,5-dimethylpyrazole (pz*) rings is replaced by a benzotriazole (bzt) ring. There was considerable overlap of pz*, bzt, and PPh_4^+ resonances in the phenyl region of the ^{13}C NMR spectrum of [PPh_4][Tp'Mo$(CO)_3$]. Importantly, each of the ^{13}C NMR resonances of PPh_4^+ could be assigned as neat doublets, albeit with substantially different J_{CP} values (Figure 5), relating to proximity to the phosphorus atom. Other large cations bearing *saturated* groups, like tetraalkylammonium cations, would have yielded a similar salt in this case, but PPh_4^+ introduced desirable selective solubility over NR_4^+ in the target product salt.

A final example of useful hyperfine splittings in $^{13}C\{^1H\}$ NMR spectra is seen in characterization of the new rhodium(I) complex Tp'Rh(cod) (*30*) (Figure 6) in which cod = 1,5-cyclooctadiene. In other settings, a ^{103}Rh-accessible probe (Table 1) would be available, but such is not presently the case at our institution. The alkenyl carbons of cod in this system appear at the expected ^{13}C NMR chemical shift of 81 ppm as a clear doublet with J_{CRh} = 13 Hz. The clarity of this resonance at room temperature supports persistent symmetrical coordination of rhodium(I) by the Tp' chelate, with the benzotriazole group occupying an axial position above a rhodium coordination plane defined by the two pyrazole rings of Tp' and the alkenyl carbon atoms of cod. Variable hapticity of tridentate scorpionates (κ^2 vs κ^3), especially with d^8 second row transition metal ions like rhodium(I), can make solution phase measurements challenging. IR measurements of Tp'Rh(cod) (as KBr pellets; ν(B-H) = 2492 cm^{-1}) support a κ^2 assignment in the solid state. Webster and Hall (*31*) described detailed computational ^{11}B NMR and IR (ν(B-H)) results summarizing an array of systems from the literature in which they were able to definitively correlate ^{11}B chemical shifts and ν(B-H) with κ^n. Others have correlated alkenyl ^{13}C chemical shifts with scorpionate-rhodium hapticity, but the difference is small, only about 3 ppm (*32, 33*).

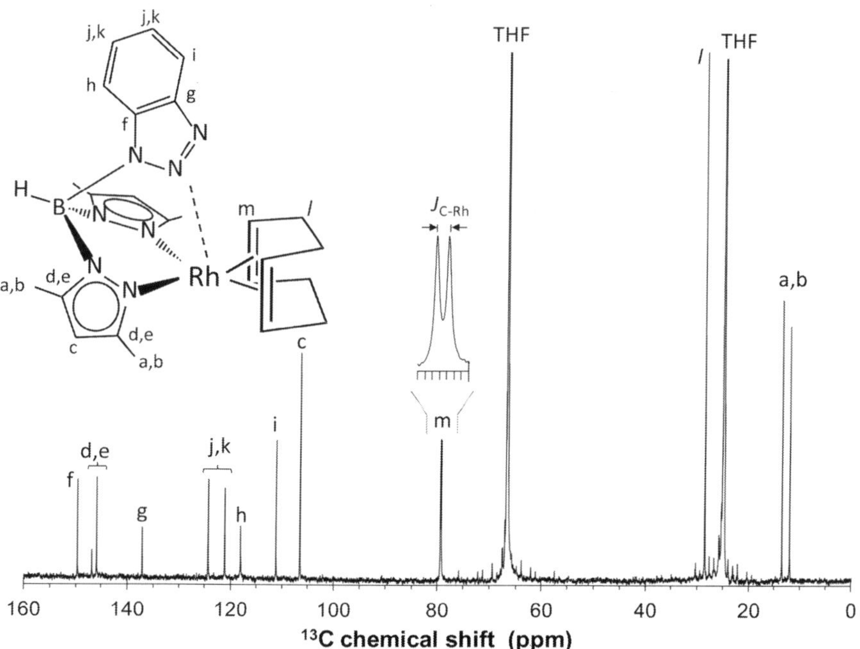

Figure 6. ^{13}C NMR spectrum of Tp'Rh(cod) in protio-THF. Assignments are given in the inset figure at left.

^{27}Al NMR Spectroscopy: A Collaborative Dye Experiment

A final heteroatom attracting attention in our department is ^{27}Al. The aluminum ion is highly abundant in global geology, and its oxides have widespread material and catalytic uses (*34*, *35*). Unfortunately, aluminum possesses limited spectroscopic handles, with ^{27}Al NMR spectroscopy representing one of the few. Aluminum ion (from alum salts) has for centuries been employed as a mordant for fixing colored dyes to fabrics; similar to other Lewis acids like iron(III) and chromium(III) (*36*, *37*). Recently, an organic faculty member in our department has been developing natural organic dye experiments for our organic laboratory in order to replace the current azo-based dye experiment with a greener natural products alternative (*38*). To this end he has enlisted the aid of botonists from our Biology Department to identify good plant species readily available in our campus' extensive Jewel Moore Nature Reserve. Locally, aluminum experiments are also motivated by the role that aluminum ore processing has played historically in the central Arkansas economy. In light of these considerations, solution-phase ^{27}Al NMR experiments were recently incorporated into our advanced inorganic laboratory.

The experiments in this course were designed to introduce students to general chemical shifts of aluminum in solution and to further develop an experiment where equilibrium measurements could differentiate between $[Al^{3+}]_{free}$ and $[Al^{3+}]_{dye}$ in progressive mixtures of dye and $Al^{3+}(aq)$. Several undergraduate ^{27}Al NMR experiments can be found in the chemical literature. One deals with the pH-dependence of $[Al(OH_2)_{6-m}(OH)_m]^{(3-m)+}$ species (*7*), a fundamental property of aqueous Al^{3+} chemistry. Another experiment details the steady progression of ^{27}Al chemical shifts for AlX_4^- tetrahalides as X is systematically varied from fluoride through iodide (*5*). These experiments provide reasonable concentrations to employ and chemical shift ranges to expect. Preliminary ^{27}Al NMR results with dye titrations into samples of $Al^{3+}(aq)$ have yielded several resonances, some sharp and some broad, suggesting possible complications from rapid solution-phase equilibria. While the preliminary results offer a promising "proof-of-concept," ongoing work in future advanced inorganic laboratories will optimize these experiments.

^{13}C NMR Spectroscopy of Resin-Supported Ligands

Heterogeneous solid samples are rarely encountered in undergraduate NMR experiments despite the expanding reach of solid-state NMR measurements (*39*). Short spin-lattice relaxation times (T_1) typically make NMR spectra difficult to record for such samples, and extraordinary instrumentation is required (i.e. MAS probes). Heterogeneous samples, where the spectroscopic handle is removed from the solid lattice by a tether (here PEG was used) allowing it to tumble more freely so that longer relaxation times and meaningful spectra result (*40*, *41*), can circumvent this common limitation.

This type of heterogeneous platform was used in another undergraduate research project to develop a scorpionate ligand anchored to cross-linked polystyrene resin beads via a PEG-amide tether. This heterogeneous ligand was an extension of homogeneous synthesis of the heteroscorpionate, KTp′ (*17*). While the ^{13}C NMR spectrum recorded for this ligand, bead-Tp'K, was straightforward (Figure 7), the effort to prepare the sample and ultimately record the spectrum of this sample was not. This provided an excellent opportunity for undergraduate students to learn to work with these materials with applications ranging from biochemical peptide systems to heterogeneous catalysts.

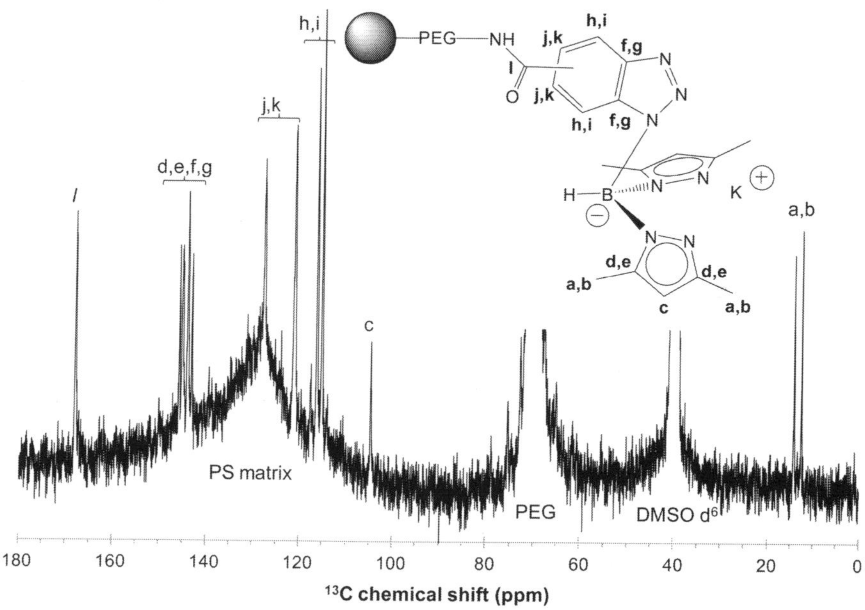

Figure 7. ^{13}C NMR spectrum of KTp′ supported on polystyrene synthesis beads via a PEG tether and through a peptide bond. Reproduced with permission from reference (20). Copyright 2011 American Chemical Society.

Proper solvent choice, careful sample preparation, and long spectra duration were all needed. Lock signals on deuterated solvents were typically more difficult with these resin samples, therefore the less costly solvent $CDCl_3$ was replaced with DMSO-d_6 or acetone-d_6. The presence of six deuterons in these solvents greatly enhanced the solvent lock signal. Generally, DMSO samples gave the best results. In this solvent the beads did float, but spinning caused them to disperse down the length of the tube's spinning vortex. Best sample preparation results were obtained by first packing the dry beads into the NMR tube followed by adding solvent and allowing time for the beads to swell. Typically, solvent was removed from the sample via a syringe, drawing from the clear liquid below the floating beads. Clear spectra resulted from overnight scans (Figure 7) in which one can differentiate sharper resonances assigned to the PEG-tethered ligand versus the very broad resonances of the polystyrene matrix and the PEG $-OCH_2-$ tether. Success with these measurements has motivated still more ambitious NMR measurements on cheaper but more highly cross-linked polystyrene resin beads.

Routine Paramagnetic ^{11}B NMR Spectroscopy

Measurements for paramagnetic samples are now more routinely recorded and in many cases resonances can be assigned and integrated (*42*). The boron heteroatom in paramagnetic Tp*NiX (X = Cl, Br, I, NO_3, BH_4, or a second Tp^R) provided undergraduate researchers the opportunity to develop a unique ^{11}B NMR chemical shift range that was very sensitive to both the identity of X and the metal coordination geometry, Figure 8.

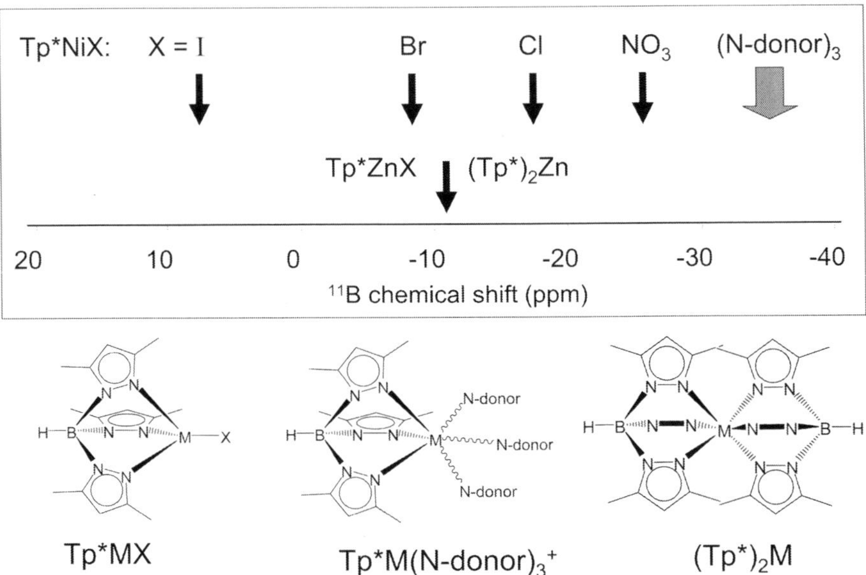

Figure 8. ^{11}B NMR chemical shifts for the B atom of Tp in a variety of nickel(II) and zinc(II) scorpionates.*

The influence of the nickel(II) paramagnet (*S*=1) on the Tp* boron atom is clearly evident. Typical through-space nickel-boron distances in Tp*NiX are ~3 Å (X = Cl, Br, BH_4) (*43*, *44*). Chemical shifts for four-coordinate Tp*NiX geometries varied widely with X. Five-coordinate Tp*NiX geometries (or fluxional variations) gave chemical shifts near -25 ppm, and six-coordinate octahedral Tp*NiX cases gave chemical shifts near -36 ppm over a wide range of N-donors (including a second Tp*, 3 $NCCH_3$, or 3 NH_3). In contrast, the boron chemical shift in the closed-shell zinc(II) cases, Tp*ZnX, is insensitive to X, such that the boron chemical shifts of Tp*ZnX (X= Cl and I) and $Zn(Tp^*)_2$ only differed by 0.5 ppm.

The broad utility of scorpionates in paramagnetic transition metal complexes has motivated the development of meaningful ^{11}B chemical shift libraries of these complexes. The nickel(II) examples in Figure 8 demonstrate the usefulness of this method for rapidly identifying metal-scorpionate coordination geometries in reaction mixtures. Others have reported similar trends in paramagnetic cobalt(II) examples (*45*). The combination of such trends and the general excellent sensitivity of ^{11}B NMR specotrscopy has made this tool a quick and routine confirmatory measurement for existing and newly prepared metal scorpionates in our work. For example, this trend was useful in identifying *in situ* formation of Tp'$NiNO_3$ (^{11}B $\delta \sim 25$ ppm), a product of this new heteroscorpionate but one that proved difficult to isolate from reaction mixtures (*17*).

A final example of the use of NMR in metal scorpionates is seen in comparison spectra recorded for Tp*$NiBH_4$ and Tp*$ZnBH_4$ (Figure 9) (*3*). Measurements were first recorded for Tp*$NiBH_4$ using a conventional (*diamagnetic*) chemical shift window of 50 to -100 ppm. This traditional window yielded only one broad resonance near -26 ppm; no discernible resonance assignable to the nickel-coordinated BH_4 could be identified. In contrast, the zinc(II) case yielded two resonances (1:1 integral), a doublet for the boron of Tp* (-10.7 ppm) and second pentet (-50.2 ppm) for the boron of BH_4 coordinated to zinc. These results were initially and reluctantly interpreted to mean that the strong paramagnetism of the high spin nickel(II) center precluded observation of a signal for the BH_4 due to effective spin-spin relaxation.

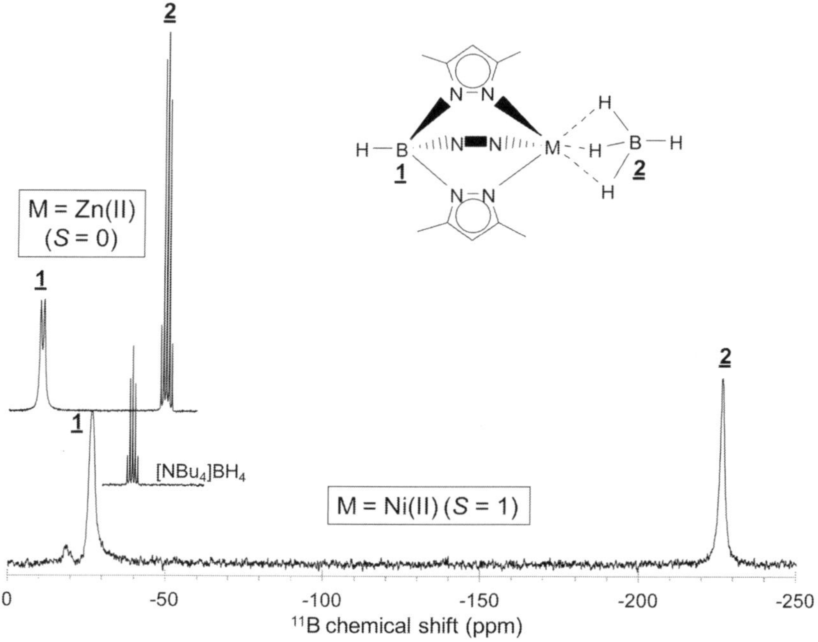

Figure 9. ^{11}B NMR spectrum of TpMBH_4 in CD_2Cl_2. Distinct resonances for each boron atom in the molecules are evident. In Tp*$ZnBH_4$: J_{BH} (Tp*) = 110 Hz and J_{BH} (BH_4) = 83 Hz. The spectrum of "free" BH_4^- (from the salt [NBu_4]BH_4) is given as an inset for comparison.*

A second experiment recorded the Tp*$NiBH_4$ spectrum but with a window twice as wide as typically employed for ^{11}B samples. This measurement revealed the signal eventually assigned to nickel-coordinated BH_4^- (near -230 ppm).

Importantly, these measurements confirm the borohydride ion to be closely (covalently) associated with the zinc(II) and nickel(II) centers in Tp*MBH_4 as evidenced by the considerable shift for the coordinated BH_4-ion versus a "free" BH_4-ion seen in [NBu_4]BH_4 (Figure 9). Independent experiments involving BD_4/BH_4 exchange confirmed that these signals indeed resulted from metal-coordinated versus free unassociated BH_4^- in solution (*3*). An important lesson from this work is the need to employ unconventional wide shift windows to observe all resonances from a single molecule and not to presume such signals cannot be observed since the spin-spin effects from paramagnetic systems on NMR signals can sometimes be extreme and unpredictable.

Conclusions

FT-NMR measurements on a variety of heteronuclear, heterogeneous, and paramagnetic systems have all served to provide undergraduates with a rich experience that adds to traditional ^1H and ^{13}C NMR spectroscopy applications in organic offerings. Students trained in these methods earn valuable hands-on experience with modern instrumentation and become versatile chemical problem solvers. While this is a contribution to a series devoted to NMR measurements, NMR remains *complementary* to and not exclusive of other characterization methods whenever possible. Multiple points of coincidence, including solid state and solution phase measurements (e.g.,. IR and UV-vis spectroscopies and chromatography) remain the best approach to thoroughly characterize a chemical system. Perhaps this is the most valuable lesson we can teach our undergraduate students from these experiences.

Acknowledgments

Dr. Chip Detmer and Dr. Ashok Krishnaswami provided valuable technical assistance and travel support was provided from Robert DiPasquale (all from JEOL). The NMR spectrometer purchased for this work was funded by a grant from the National Science Foundation (Grant CCLI 0125711 to J.M.M. and R.M.T.). This material is based upon work supported by the National Science Foundation under CHE-0717213 (P.J.D. and R.M.T.) and by the donors of the American Chemical Society Petroleum Research Fund (39644-B3 to P.J.D.). I am also grateful to Profs. Richard M. Tarkka and Jerald M. Manion for their expertise and many helpful and informative NMR discussions. The very capable work of the many undergraduates responsible for some or all of the spectra recorded is gratefully acknowledged including Brian Besel, Jared Evanov, Kristin Thorvilson, Adam Corken, Ariel Marshall, Chris Sutton, Davis Duong, Bonnie Hong, and Stacey LeLievre.

References

1. *ACS Guidelines and Evaluation Procedures for Bachelor's Degree Programs*; American Chemical Society: Washington, DC, 2008; p 6.
2. Quote attributed to M. J. Molina, 1995 Nobel Laureate, in Van Houten, J. *J. Chem. Educ.* **2002**, *79*, 1182–1188.
3. Desrochers, P. J.; Sutton, C. A.; Abrams, M. L.; Ye, S; Neese, F.; Telser, J.; Ozarowski, A.; Krzystek, J. *Inorg. Chem.* **2012**, *51*, 2793–2805.
4. Mebi, C. A.; Trujillo, J. J.; Rosenthal, B. L.; Bowman, R. B.; Noll, B. C.; Desrochers, P. J. *Transition Met. Chem.* **2012**, *37*, 645–650.
5. Davis, C. M.; Dixon, B. M. *J. Chem. Educ.* **2011**, *88*, 309–310.
6. Curtin, M. A.; Ingalls, L. R.; Campbell, A.; James-Pederson, M. *J. Chem.Educ.* **2008**, *85*, 291–293.
7. Queiroz, S. L.; de Araujo, M. P.; Batista, A. A.; MacFarlane, K. S.; James, B. R. *J. Chem. Educ.* **2001**, *78*, 87–88.
8. Basuli, F.; Bailey, B. C.; Brown, D.; Tomaszewski, J.; Huffman, J. C.; Baik, M.-H.; Mindiola, D. J. *J. Am. Chem. Soc.* **2004**, *126*, 10506–10507.
9. Chatterjee, P. B.; Crans, D. C. *Inorg. Chem.* **2012**, *51*, 9144–9146.
10. Borer, L. L.; Russell, J. G.; Settlage, R. E.; Bryant, R. G. *J. Chem. Educ.* **2002**, *79*, 494–497.
11. Boivin, S.; Outurquin, F.; Paulmier, C. *Tetrahedron Lett.* **2000**, *41*, 663–666.
12. Silbestri, G. F.; Flores, J. C.; de Jesús, E. *Organometallics* **2012**, *31*, 3355–3360.
13. Maslow, A. H. *The Psychology of Science, A Reconnaissance*; Henry Regnery Co.: Chicago, IL, 1966.
14. Uzun, S. S.; Sen, S. *J. Phys. Chem. B* **2007**, *111*, 9758–9761.
15. Du, L. –S.; Stebbins, J. F. *J. Phys. Chem. B* **2003**, *107*, 10063–10076.
16. Trofimenko, S. *J. Am. Chem. Soc.* **1967**, *89*, 3170.
17. Desrochers, P. J.; Besel, B.; Corken, A. L.; Evanov, J. R.; Hamilton, A. L.; Nutt, D.; Tarkka, R. M. *Inorg. Chem.* **2011**, *50*, 1931–1941.
18. Garrou, P. E. *Chem. Rev.* **1981**, *81*, 229–266.
19. Desrochers, P. J.; Duong, D.; Marshall, A. S.; Lelievre, S. A.; Hong, B.; Brown, J. R.; Tarkka, R. M.; Manion, J. M.; Holman, G.; Merkert, J. W.; Vicic, D. A. *Inorg. Chem.* **2007**, *46*, 9221–9233.
20. Busby, R.; Hursthouse, M. B.; Jarrett, P. S.; Lehmann, C. W.; Malik, K. M. A.; Phillips, C. *J. Chem. Soc., Dalton Trans.* **1993**, 3767–3770.
21. Autissier, V.; Clegg, W.; Harrington, R. W.; Henderson, R. A. *Inorg. Chem.* **2004**, *43*, 3098–3105.
22. Colton, R.; Tedesco, V. *Inorg. Chim. Acta* **1992**, *202*, 95–100.
23. Kandiah, M.; McGrady, G. S.; Decken, A.; Sirsch, P. *Inorg. Chem.* **2005**, *44*, 8650–8652.
24. Pettinari, C. *Scorpionates II: Chelating Borate Ligands*; Imperial College Press: London, U.K., 2008.
25. Trofimenko, S. *Scorpionates: The Coordination Chemistry of Polypyrazolborate Ligands*; Imperial College Press: London, U.K., 1999.
26. Kalidindi, S. B.; Indirani, M.; Jagirdar, B. J. *Inorg. Chem.* **2008**, *47*, 7424–7429.

27. Tidwell, C. P.; Bharara, P.; Rudeseal, G.; Rudeseal, T.; Rudeseal, F. H., Jr.; Simmer, C. A.; McMillan, D.; Lanier, K.; Fondren, L. D.; Folmar, L. L.; Belmore, K. *Molecules* **2007**, *12*, 1389–1398.
28. Rojas, R.; Valderrama, M.; Wu, G. *Inorg. Chem. Comm.* **2004**, *7*, 1295–1297.
29. Burgess, K. M. N.; Korobkov, I.; Bryce, D. L. *Chem. Eur. J.* **2012**, *18*, 5748–5758.
30. Tp'Rh(cod) from J. R. Evanov and P. J. Desrochers, unpublished results. Compounds of the class $Tp^{R}Rh$(diene) are very active catalysts for the polymerization of phenylacetylene. See for example: Katayama, H.; Yamamura, K.; Miyaki, Y.; Ozawa, F. *Organometallics* **1997**, *16*, 4497–4500.
31. Webster, C. E.; Hall, M. B. *Inorg. Chim. Acta* **2002**, *330*, 268–282.
32. Adams, C. J.; Anderson, K. M.; Charmant, J. P. H.; Connelly, N. G.; Field, B. A.; Hallett, A. J.; Horne, M. *Dalton Trans.* **2008**, 2680–2692.
33. Bucher, U. E.; Currao, A.; Nesper, R.; Rüegger, H.; Venanzi, L. M.; Younger, E. *Inorg. Chem.* **1995**, *34*, 66–74.
34. Cai, W.; Yu, J.; Anand, C.; Vinu, A.; Jaroniec, M. *Chem. Mater.* **2011**, *23*, 1147–1157.
35. Suchanek, W. J.; Garcés, J. M.; Fulvio, P. F.; Jaroniec, M. *Chem. Mater.* **2010**, *22*, 6564–6574.
36. Sun, Y.; Men, Q. *Adv. Mater. Res.* **2012**, *441*, 92–95.
37. Pozzi, F.; Lombardi, J. R.; Bruni, S.; Leona, M. *Anal. Chem.* **2012**, *84*, 3751–3757.
38. R. M. Tarkka, private communication.
39. Dybowski, C.; Bai, S. *Anal. Chem.* **2008**, *80*, 4295–4300.
40. Cardona, C. M.; Jannach, S. H.; Huang, H.; Itojima, Y.; Leblanc, R. M.; Baker, G. A.; Brauns, E. B.; Gawley, R. E. *Helv. Chim. Acta* **2002**, *85*, 3532–3558.
41. *High-Throughput Synthesis: Principles and Practice*; Sucholeiki, I., Ed.; Marcel Dekker Inc.: New York, 2001; p 24.
42. Reinaud, O. M.; Rheingold, A. L.; Theopold, K. H. *Inorg. Chem.* **1994**, *33*, 2306–2308.
43. Tp*NiCl and $Tp^{*}NiBH_4$: Desrochers, P. J.; LeLievre, S. A.; Johnson, R. J.; Lamb, B. T.; Phelps, A. L.; Cordes, A. W.; Gu, W.; Cramer, S. P. *Inorg. Chem.* **2003**, *42*, 7945–7950.
44. Tp*NiBr: Desrochers, P. J.; Telser, J.; Zvyagin, S. A.; Ozarowski, A.; Krzystek, J.; Vicic, D. A. *Inorg. Chem.* **2006**, *45*, 8930–8941.
45. Myers, W. K.; Duesler, E. N.; Tierney, D. L. *Inorg. Chem.* **2008**, *47*, 6701–6710.

Physical Chemistry and Biochemistry

Chapter 12

Substituent Interactions in Aromatic Rings: Student Exercises Using FT-NMR and Electronic Structure Calculations

James B. Foresman*,[1] and Donald D. Clarke[2]

[1]Department of Physical Sciences, York College of Pennsylvania, York, Pennsylvania 17405
[2]Department of Chemistry, Fordham University, Bronx, New York 10458
***E-mail: JForesma@ycp.edu**

The examination of chemical shifts allows students the opportunity to measure both the dramatic and subtle electronic effects that occur when one or more substituents are added to a benzene ring. Some of these effects are not obvious and cannot be predicted by simple empirical additivity models. The goal of the exercises presented in this chapter is to demonstrate the importance of these effects and how to correctly assign ^{13}C NMR spectra based on two-dimensional NMR techniques and quantum mechanical calculations.

Introduction

Proton spectra for typical organic molecules can usually be assigned by deductions based on the splitting of the signal due to nearby hydrogen atoms and the magnitude of the shift. By comparison, proton decoupled ^{13}C NMR spectra are much less obvious. For aromatic carbons, additivity parameters (*1–3*) derived from the spectra of mono-substituted benzenes are traditionally used, but these can lead to incorrect assignments when signals are close to one another and when there is interaction between substituents. By using other NMR experiments, it is often possible to unambiguously assign the ^{13}C NMR spectrum of a sample by correlating it with its proton spectrum. For example, the two-dimensional NMR experiment CH-COSY (*4*) shows cross peaks that correlate ^{1}H nuclei that are directly bonded to ^{13}C nuclei. Students can compare assignments made using ^{1}H, ^{13}C, and CH-COSY to those predicted by additivity parameters to see if there are electronic interactions among the substituents which may not be simply an additive effect.

It is also constructive to introduce students to more sophisticated models for predicting ^{13}C NMR spectra. For example, electronic structure programs such as Gaussian (*5*) have the ability to simulate NMR spectra from first principles of quantum mechanics. This modeling can be done in either gas-phase or solution. Calculated quantities can include all of the subtle electronic interactions occurring among aromatic substituents, and may offer explanations for cases in which the order of peaks are incorrectly determined from additivity parameters. An examination of the output from a program such as Gaussian also gives students a better understanding of the physical origin of the property of chemical shift. Orbital pictures can help students visualize how electrons sometimes act in opposition to an applied magnetic field, while at other times their motion can add to the field through paramagnetic interactions.

The electronic effects exposed through these student exercises are not just important in assigning spectra, but in many cases also govern chemical reactivity. This chapter will focus on the practical issues associated with calculating ^{13}C chemical shifts and comparing them to experimental spectra. Case studies are presented that would be useful to instructors who would like to incorporate two-dimensional NMR and electronic structure calculations into the undergraduate curriculum.

Background

Nuclei have an intrinsic spin angular momentum (**I**) just like electrons. The quantum mechanical eigenvalue equations associated with this observable can be written:

$$\hat{\mathbf{I}}^2\phi = I(I+1)\hbar^2\phi \qquad (1)$$

$$\hat{\mathbf{I}}_z\phi = m_I\hbar\phi \qquad (2)$$

The nuclear spin quantum number, I, can take on both integer and fractional values depending on the identity of the nucleus. For a ^{13}C nucleus, I has a value of ½. The values possible for m_I are restricted to:

$$m_I = -I, (-I+1), (-I+2), \ldots, +I \qquad (3)$$

The spin eigenfunctions ($\varphi = \alpha,\beta,\gamma$) can be viewed as mathematical functions describing the various spin states and do not need to be specified in order to interpret an NMR spectrum. In the absence of a magnetic field, these would all be degenerate in energy. The total spin angular momentum squared is given by equation 1 and the z-component is given by equation 2. These two operators are said to commute, therefore both values can be known simultaneously. The other components of the spin angular momentum (x and y) are not known with any amount of precision, and any attempts to measure one component will cause the other two to become less defined, a consequence of quantum mechanics. As a result, the system is typically described as a vector precessing about the z-axis. The vector's length (the total spin angular momentum) and its projection onto one axis are known (z-component), but since it is spinning around the z-axis the x and y components are uncertain during the course of a given experiment.

In a magnetic field, the nuclear spin states no longer possess the same energy. There are different energy levels possible (two for nuclei that have spin ½), and transitions among them can be detected just like in other forms of spectroscopy. The transition energy (**Δε**) is proportional to the strength of the magnetic field (**B_o**), the magnetogyric ratio of the nucleus (expressed in terms of a "**g** factor"), and the electronic environment of the nucleus (expressed in terms **σ**, the absolute shielding). Using these definitions and the values of the nuclear magneton (**μ_N**) and Planck's constant (***h***), the transition energy can be derived in frequency units (**B_0** in Tesla):

$$\Delta\varepsilon = h\nu = g_I \times \mu_N \times B_O \times (1-\sigma) \qquad (4)$$

$$\nu[MHz] = 7.6226 \times g_I \times B_O \times (1-\sigma) \qquad (5)$$

Spectrometers are named according to the frequency associated with a bare (**σ**=0) proton resonance, so using a value of 5.5854 for the **g** factor of a ^{1}H nucleus, we see that a 300 MHz instrument would be associated with a magnet of strength 7.05 Tesla. A ^{13}C nucleus has a **g** factor of 1.4042, so its bare frequency in that same spectrometer would occur in the vicinity of 75 MHz. More importantly, the ^{13}C nuclei of benzene and those of tetramethylsilane would be found at only slightly lower frequencies (4.5 and 14.2 kHz below the bare nucleus, respectively). All of the important electronic information that distinguishes typical carbon atoms, one from another, is contained within this narrow 15 kHz wide range on a 300 MHz spectrometer.

To discuss results from different instruments in a consistent fashion, it is useful to divide by the frequency of the bare nucleus (**ν_{spect}**) and define a unitless quantity called the chemical shift, δ, measured relative to a standard such as tetramethylsilane (TMS):

$$\delta[ppm] = \frac{(\nu - \nu_{TMS})}{\nu_{spect}} \times 10^6 \qquad (6)$$

Figure 1 shows chemical shift values in ppm for several carbon nuclei relative to TMS. These are gas-phase estimates based on Gaussian 09 (*5*) calculations using a standard model (*6*) that will be discussed later. In this case the bare ^{13}C nucleus is shown as a C^{6+} cation. The absolute shielding for C^{6+} is zero, so the value found here is simply the absolute shielding, **σ**, calculated for TMS. The terms downfield and upfield are artifacts from the days when the spectrum was obtained by varying the field strength, B_o, and holding the receiver frequency constant. For example, the magnetic field would have to be increased to observe those nuclei that were more shielded. All modern spectrometers now vary the receiver frequency and use a fixed high-field value for **Bo**. In what follows, we will attempt to rationalize what we see in Figure 1. This explanation has been presented previously (*7*) and is useful to include as a pre-lab discussion.

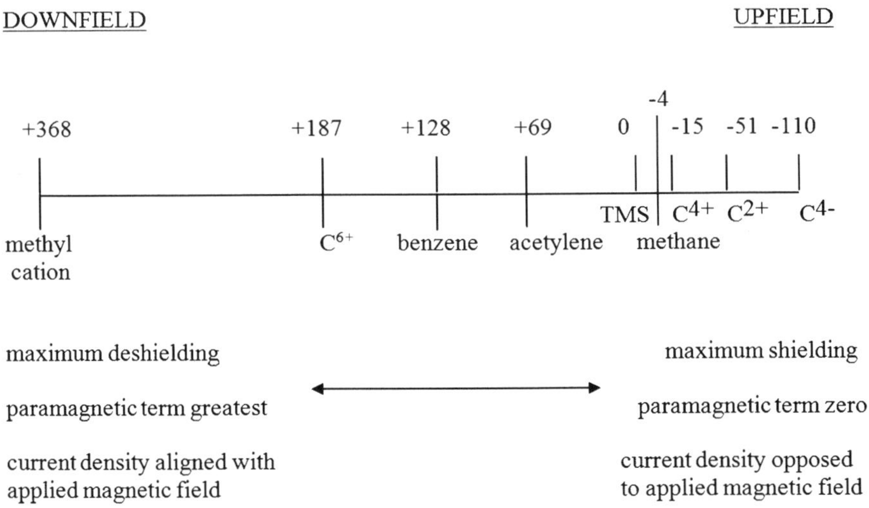

Figure 1. Conventions, terminology, and values for ^{13}C chemical shifts.

The 10 electrons in C^{4-} would represent two completely filled orbital shells around the atom, thus placing it at the extreme right in Figure 1 (*8*). Electrons in filled shells or filled subshells act to both stabilize the atom and shield the nucleus. The magnetic field generated by these electrons would act in opposition to the applied magnetic field in the spectrometer. The component of the chemical shift due to electrons behaving in this way is referred to as a "diamagnetic contribution," and any atom that has at least a filled 1s shell of electrons will exhibit this.

Electrons that occupy incompletely filled orbital shells (valence electrons responsible for bonding) can also create magnetic fields that shield the nucleus (or that might align with the applied magnetic field), increasing the field at the nucleus. Consider another extreme example, the methyl cation. An estimate of the absolute shielding, **σ**, for the ^{13}C atom in this planar molecule is –181 ppm. Therefore, it is found that many units to the left of C^{6+} and is significantly *deshielded* compared to TMS. This is explained by considering the orbital hybridization. Since the carbon in methyl cation is sp^2 hybridized this means that there is an empty p_z orbital perpendicular to the molecular plane. A magnetic field applied in one of the in-plane directions will cause a circulation of electrons, moving them from the sp^2 orbitals into the unoccupied p_z orbital, adding to the applied magnetic field instead of opposing it. This is the paramagnetic contribution to the chemical shift. Since the magnetic field of the instrument is applied in only one direction and the orientation of the molecules in the sample is random, only a rotational average of all directions is observed. Therefore in this case, the paramagnetic term is the greatest contributing factor towards the chemical shift value for the methyl cation.

In general, most carbon atoms will fall between these two extreme cases. For example, a carbon found in a benzene ring would be affected by roughly half as much paramagnetism as methyl cation since its p*z* orbital is half occupied. Substituents present on the ring would have an effect on the occupation of that orbital and its energy relative to the in-plane orbitals. Therefore, chemical shifts will be directly controlled by the identity and placement of substituents on the ring. These effects are the subject of the laboratory investigation our students are asked to complete.

Electronic Structure Calculations of Chemical Shift Values

A simulation of NMR spectra from first principles (i.e., starting without empirical parameters) begins with the calculation of the absolute nuclear magnetic shielding at each nucleus. This is the second derivative of the energy, once with respect to the external magnetic field, B_o, and once with respect to the magnetic moment of that nucleus (*9*). Since both the external field and the moment are expressed with x, y, and z components, this is a 3 x 3 tensor matrix. The absolute shielding, σ, at each atom is the average of the diagonal (xx, yy, zz) components of this tensor. This is referred to as the isotropic shielding. There are several variations of this numerical procedure which reflect different solutions to the gauge problem, with the results dependent upon where you choose the origin for the atomic coordinates. The most popular of these is the Gauge Including Atomic Orbital (GIAO) method initially proposed by Ditchfield (*10*). This procedure involves the use of atomic basis functions that each includes the field as part of their definition. We exclusively use the GIAO method in this chapter when referring to NMR shielding calculations.

Students will need some advice as to which theoretical method to use in computing the NMR spectrum. First, the geometry of the molecule should be optimized at a reliable level of theory. Second, the NMR shielding calculation should be done with a fairly large basis set. It is also important to use a method that accounts somewhat for electron correlation (such as MP2 or Density Functional Theory). It is known that Hartree-Fock calculations typically do not reproduce experimental spectra. In what follows, we have consistently used the 6-311+G(2d,p) basis set as a reasonable compromise between accuracy and the amount of time available to students to complete the exercise.

In searching for a method to suggest, we first looked for models that reproduced the value of 60.6 ppm for the gas-phase absolute shielding of ^{13}C in benzene. This is the 300 K experimental value (*11*) corrected to 0 K by removing the thermal and vibrational contributions (*12*). Table 1 lists the results obtained from Gaussian 09 (*5*) using several popular density functionals.

Table 1. Gas-phase isotropic shielding of a ^{13}C nucleus in benzene[a]

Functional	*σ [ppm]*
B3LYP	50.2
B3PW91	53.8
CAM-B3LYP	50.3
M06	42.7
O3LYP	58.2
ω-B97X	56.7
TPSSh	59.1
Expt	60.6

[a] Theoretical values are calculated using the 6-311+G(2d,p) basis set and the B3LYP 6-311+G(2d,p) optimized geometry for benzene.

Based on the success of the TPSSh functional, we did a further calibration study comparing measured chemical shifts, δ, of aromatic carbons relative to TMS. Figure 2 compares the computed chemical shifts of 125 aromatic carbons.

The molecules in the test set included mono-, di-, and tri-substituted benzene compounds. To calculate the chemical shift for a particular carbon, a calculation on TMS is performed first and then the shift is computed as the difference:

$$\delta_C = \sigma_{TMS} - \sigma_C \tag{7}$$

Geometries were optimized using the B3LYP density functional and included the effect of the solvent ($CDCl_3$ or DMSO-d_6) by using a polarizable self-consistent reaction field continuum model (the default solvent procedure in Gaussian 09). The NMR calculations used the same solvent model. Based on the systematic improvement seen in the TPSSh results, our students use the B3LYP solvated geometries and the TPSSh NMR calculations done in the presence of the solvent. This was also the model used to generate the values in Figure 1, only in that case the solvent was not included. We are not claiming to have found the ultimate theoretical method for NMR calculations, just one that works for the present class of compounds. In fact, it would be instructive for students to experiment with this choice and discuss their own findings in groups.

The Curious Case of 2-Nitroaniline

Samples of 2-nitroaniline [CAS 88-74-4], available from Aldrich, can be used without purification. Students should wear gloves at all times when handling the amine since the orange solid is corrosive and should be handled with care. Spills should be cleaned up immediately. To prepare an NMR sample students should add 300 milligrams of the solid into an NMR tube and mix it with 1.0 mL of $CDCl_3$ which contains 0.03% TMS. The tube can be slightly heated by placing it in a beaker of hot tap water in order to fully dissolve the solid.

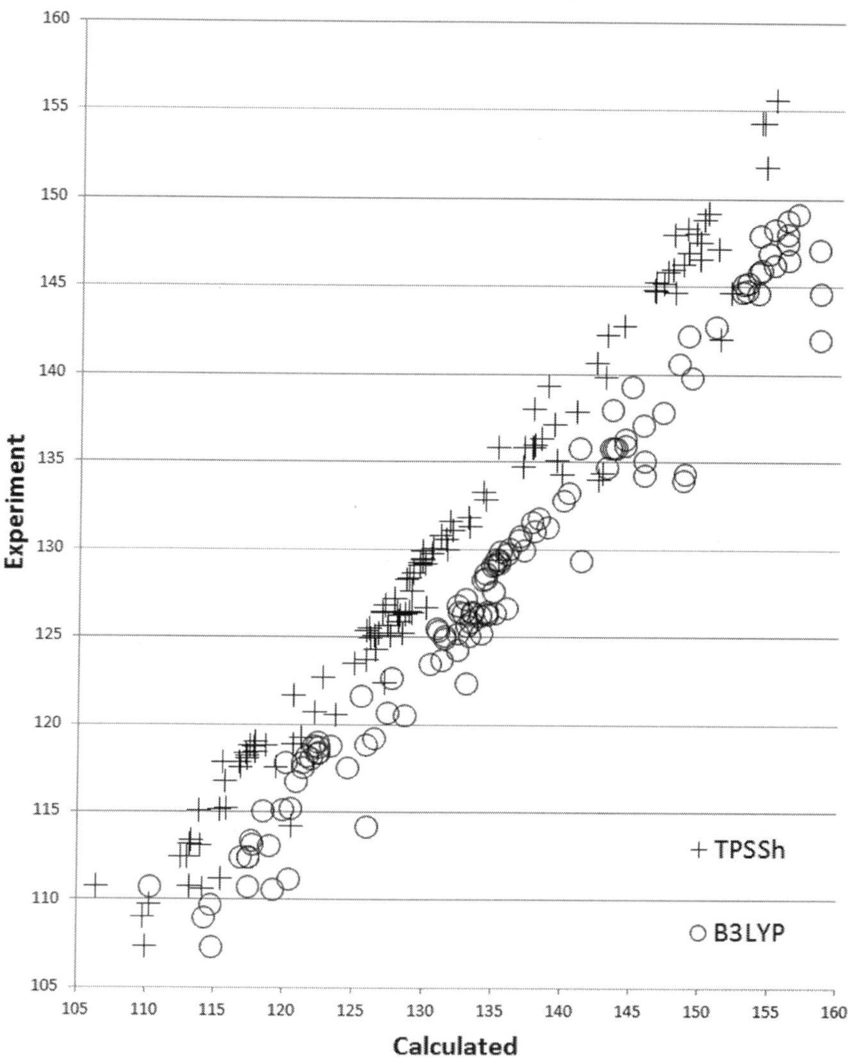

Figure 2. ^{13}C shifts relative to TMS for 125 aromatic carbon atoms measured and computed in $CDCl_3$ or DMSO-d_6 using (+) TPSSh and (o) B3LYP functionals.

A 300 MHz FT-NMR spectrometer is used to obtain all the spectra needed for the analysis. First students obtain the ^{13}C NMR spectrum by using the standard proton-decoupled experiment with 1024 scans. At this point it is instructive for the students to make reasonable guesses as to which signal corresponds to which carbon. A calculation based on additivity may assist in this. Figure 3 shows the ^{13}C NMR chemical shifts of aniline, nitrobenzene, and 2-nitroaniline relative to benzene.

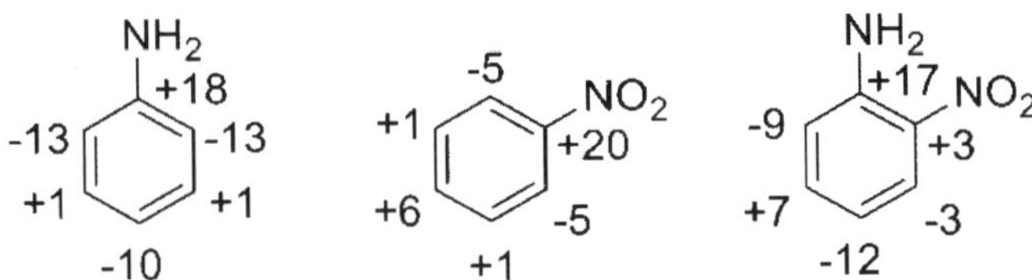

Figure 3. Experimental ^{13}C chemical shifts [ppm] of aniline, nitrobenzene, and 2-nitroaniline, relative to benzene (measured in $CDCl_3$).

In what follows, we will refer to particular carbon atoms by their position relative to these figures, with C_1 being at the top and continuing clockwise around the ring. At first glance, it seems that effects from the two substituents simply add to produce the net result in 2-nitroaniline, however C_4 and C_6 have anomalous values and students would have these two carbon atoms switched based on additivity values alone.

Table 2. Experimental and calculated ^{13}C chemical shifts relative to TMS

Carbon Atom	*TPSSh*[a]	*Expt*[b]
1	146.99	145.16
2	133.01	131.79
3	128.05	125.90
4	115.69	116.76
5	136.96	135.82
6	117.86	119.00

[a] basis set is 6-311+G(2d,p). [b] in $CDCl_3$.

To further support their assignments, the class is asked to compute them using the standard model described in the previous section. So that they do not need to repeat the calculation of TMS, the value of 187.40 ppm is provided. Results are shown in Table 2.

It is at this point that they realize the reversal in C_4 and C_6. Could this be a failure of the electronic structure calculation? The only way to know for sure is to obtain the assignments by using two-dimensional NMR experiments. Figure 4 shows the output of the CH-COSY experiment (*4*).

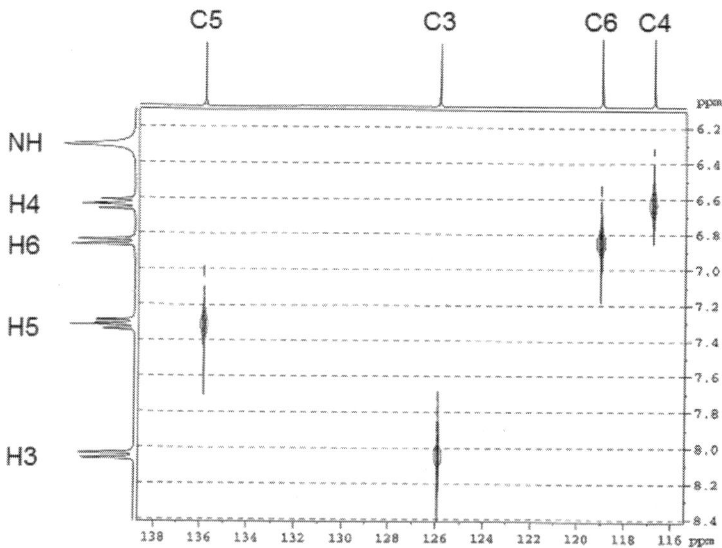

Figure 4. 2-D CH-COSY spectrum of 2-nitroaniline in $CDCl_3$.

The ^{13}C NMR spectrum is at the top and shows only the signals corresponding to carbons that have directly attached hydrogen atoms. Comparison with the ^{13}C proton decoupled spectrum verifies that the C_1 and C_2 carbons were correctly assigned as carbons bearing substituents. To assign the other four peaks, students make correlations with the proton spectrum shown on the left. Reading the ^{13}C peaks from left to right, the signal at 136 ppm correlates with a triplet in the proton spectrum, meaning that it corresponds to C_4 or C_5. Since it is the less shielded of these two, it must correlate to C_5. The next ^{13}C peak at 126 ppm correlates with a doublet, meaning that it corresponds to C_3 or C_6. Again, since it is less shielded, it must be C_3. Finally we come to the remaining two carbon signals whose assignments are more contentious. The ^{13}C peak at 119 ppm corresponds to C_6 since it matches with a doublet in the proton spectrum. Likewise, the final ^{13}C peak at 116 ppm is assigned to C_4 since it correlates with a triplet in the proton spectrum. This data is quite surprising given the fact that the ordering of C_4 and C_6 is reversed from that in the spectrum of aniline. While the predictions based on additivity parameters are very close in value to the experimental chemical shifts, they lead to incorrect assignments for C_4 and C_6.

There is nothing magical about our standard model in terms of making the qualitative prediction about which carbon is most upfield in this example. For example, other density functionals and basis sets have been tested, and though the chemical shift values may deviate more from experiment, the order of C_4 and C_6 is maintained.

Students can explore several aspects of the calculation in their discussion of the exercise. First, there is the question of molecular geometry. Gas-phase calculations predict that this molecule is non-planar (the amine nitrogen is slightly pyramidal). In solution, the calculation predicts a planar structure. There is a slight increase in chemical shift (decrease in shielding) due to this structural change. In contrast, aniline is predicted to be a non-planar structure in both gas-phase and solution. As a result the C-NH_2 bond length decreases in going from aniline to 2-nitroaniline. With a greater orbital interaction possible between the p_z orbitals of both C_6 and the amine nitrogen, it is reasonable that the paramagnetic term at C_6 would be greater in 2-nitroaniline and therefore C_6 is moved downfield.

Second, one can look at the shielding tensors as reported in the calculation and compare them to aniline. The z-axis is perpendicular to the benzene ring for both molecules. The x and y tensor components are not comparable since the molecular axis will be rotated, but we can combine these into an in-plane component for comparison. Table 3 summarizes the results from our standard model. Notice how the diamagnetic contribution is large, and is coming from the out-of-plane diagonal tensor, σ_{zz}. Students can interpret this as resulting from electrons that are in sp orbitals resisting that direction of the magnetic field which would circulate electrons in the plane of the molecule. Contrasting with this is the sum of the components coming from in-plane directions of the magnetic field. These resist the field much less and, in some cases, add to the field (negative values). This component is increased at C_4 and decreased at C_6 in going from aniline to 2-nitroaniline, suggesting that the origin of the reversal in peak order is due to a change in paramagnetism at these two carbons.

Table 3. Calculated components of the shielding tensor [ppm]

	Aniline			*2-nitroaniline*		
	$\sigma_{xx}+\sigma_{yy}$	σ_{zz}	σ_{iso}	$\sigma_{xx}+\sigma_{yy}$	σ_{zz}	σ_{iso}
C_1	-36.5	149.7	37.7	-28.5	149.4	40.3
C_2	55.3	165.3	73.5	45.9	117.0	54.3
C_3	-1.9	175.3	57.8	2.2	175.6	59.3
C_4	32.4	177.9	70.1	40.2	174.6	71.6
C_5	-1.9	175.3	57.8	-22.4	173.4	50.3
C_6	55.3	165.3	73.5	50.0	158.3	69.4

Looking at the isotropic values of the shielding tensor, we see that the net effect of these contributions is that C_4 is moved 1.5 ppm upfield while C_6 is moved 4.1 ppm downfield, resulting in a reversal of order.

We have extended this exercise to include the toluidines (placing methyl substituents in various locations on the ring) and replacing NH_2 and NO_2 with other electron donating and electron withdrawing groups on the ring so that students working in a given semester will have a wide variety of aromatic ^{13}C assignments to make. In comparing their studies, they will discover some situations where the substituent effects are more additive than others. A representative sample of these is given in Table 4.

Table 4. ^{13}C chemical shift differences for the C_4 and C_6 nuclei based on additivity and actual values

C_1	C_2	C_5	*Additivity*	*Actual*
H	H	H	0.0	0.0
NH_2	NO_2	CH_3	3.3	0.3
NH_2	NO_2	Cl	3.3	-0.3
NH_2	CH_3CO	H	3.3	-1.6
NH_2	NO_2	H	3.3	-2.2
$NH(COCH_3)$	NO_2	H	3.8	1.1
$N(CH_3)_2$	NO_2	H	4.0	-0.2
$NH(CH_3)$	NO_2	H	4.7	1.8
OCH_3	OH	H	6.7	10.6
OCH_3	CN	H	6.7	9.5
OCH_3	NO_2	H	6.7	6.6
OCH_3	NH_2	H	6.7	3.5

Keeping the substituent the same at C_1 and varying the ones at C_2 and C_5 will result in no change in the C_4 and C_6 shift difference according to simple additivity, since these positions are equivalent relative to C_2 and C_5. In reality, there are measurable differences, as shown here, including cases of reversal in order. These are confirmed using 2-D NMR and electronic structure calculations.

^1H-Coupled ^{13}C NMR Spectra

Another extension of the project described above is to have students interpret a completely ^1H-coupled ^{13}C NMR spectrum obtained with a modified pulse program which shows splitting of signals due to ^1H nuclei (*13*). For aromatic carbons, a proton that is attached will couple most strongly, however protons that are three bonds away will also produce measureable splitting (more so than the protons that are just two bonds away). This general guideline is usually enough to provide students with the means to interpret the spectra, however an electronic structure calculation can also be used to predict these coupling constants and guide the students in their assignments.

One example where this kind of *J*-resolved spectrum is helpful in assigning the correct ^{13}C signals is 5-methyl-2-nitroaniline.

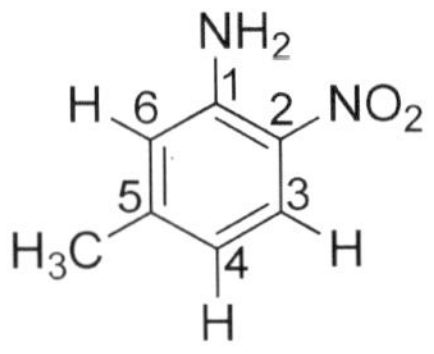

NH2, H, 6, 1, 2, NO2, 5, 3, H3C, 4, H, H

Figure 5. 5-methyl-2-nitroaniline.

Table 5. Nuclear spin J_{CH} couplings calculated for 5-methyl-2-nitroanilinea

Atom pair	*Coupling [Hz]*
$^{2}J(C_1\text{-NH})$	4.1
$^{2}J(C_1\text{-}H_6)$	1.5
$^{3}J(C_1\text{-}H_3)$	6.8
$^{2}J(C_5\text{-}H_4)$	2.7
$^{2}J(C_5\text{-}H_6)$	0.2
$^{3}J(C_5\text{-}H_3)$	10.3
$^{2}J(C_5\text{-}CH_3)$	6.0

[a] using the TPSSh model

Here, the two signals that are too similar to distinguish based on additivity alone are the ring carbons with the amine and methyl substituents (C_1 and C_5). In the ^{1}H-coupled ^{13}C NMR spectrum these signals would have very different splitting patterns. Table 5 lists the two and three bond coupling constants computed for these carbon atoms using an efficient procedure (*14*) available in Gaussian 09 (*5*).

From this information, and ignoring splitting values of less than 4 Hz, it would be predicted that C_5 will appear as a doublet of quartets, whereas C_1 will appear as a doublet of doublets. Figure 6 shows the ^{1}H-coupled ^{13}C spectrum in the region where these two carbon atoms are found.

Even though the predicted coupling information is not entirely observable on a 300 MHz NMR spectrometer, it can be confirmed that it is C_1 that is the upfield signal.

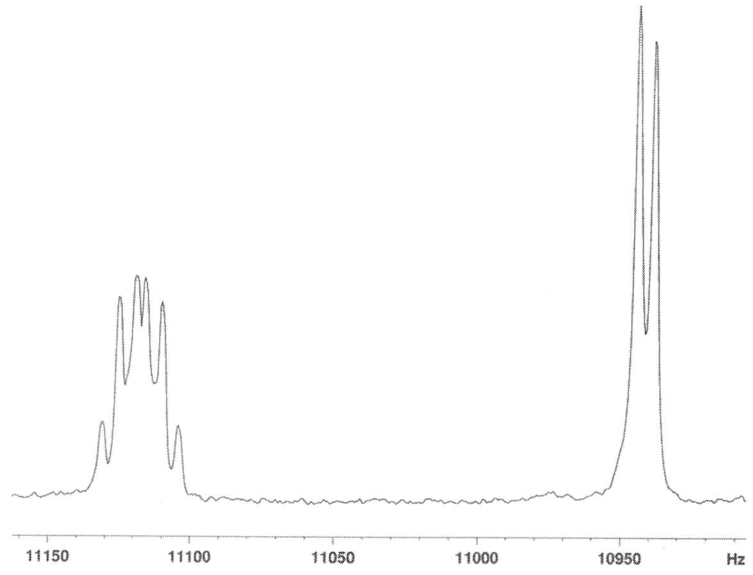

Figure 6. A selected region of the ^{1}H-coupled ^{13}C spectrum of 5-methyl-2-nitroaniline (taken using a 300 MHz NMR spectrometer).

Indanes

Another class of compounds whose ^{13}C NMR assignments are non-intuitive are 5-substituted indanes, Figure 6.

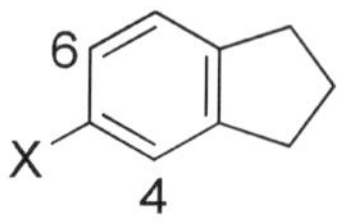

Figure 7. A 5-substituted indane.

We have considered cases where X is equal to H, NO_2, OH, OCH_3, NH_2, and found that in all cases the relative ordering of C_4 and C_6 in the ^{13}C NMR spectrum changes from what you would predict from that of *o*-xylene, where the alicyclic ring is replaced by methyl groups at the two positions. That is to say C_6 should be found upfield from C_4, but instead the reverse is true. This can be confirmed by both two-dimensional NMR experiments and electronic structure calculations. Examination of the latter again points to a difference in the paramagnetic term for these two carbons. This has been explained in the literature as rehybridization induced by the strain created by the five-membered ring (*15*).

Hydrocarbon Systems

2,2,4-trimethyl-1,3-pentanediol is another case that we sometimes ask students to investigate, Figure 7. This compound is mentioned in Silverstein and Webster (*1*) as having a coincidental spectral equivalency. The claim is that the two carbons labeled here as (c) and (d) are not chemically equivalent, but in certain instances are accidentally chemically equivalent, though this is both field strength and solvent dependent.

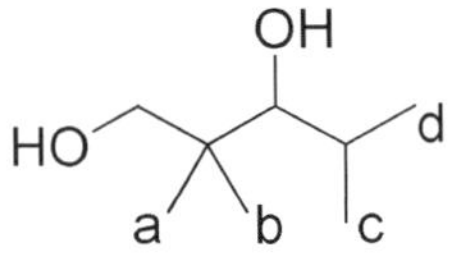

Figure 8. 2,2,4-trimethyl-1,3-pentanediol.

However, on a 300 MHz NMR instrument all eight signals can be fully resolved, though the carbon atoms at positions (b) and (d) are nearly equivalent. This can be shown using the two-dimensional CH correlation NMR experiment since the carbon at position (b) will be linked to a proton that is a singlet (the isobutyl group has no hydrogen), while the one at (d) is linked to a doublet (the isopropyl group has one hydrogen). The diastereotopicity of the methyl carbon atoms is due to their proximity to the stereocenter and steric effects (*16*). Electronic structure calculations can investigate this effect, however these calculations are much more complicated since the computed spectra for several rotational isomers of the molecule must be considered and these must be Boltzmann averaged in order to simulate the experimental spectrum.

Practical Considerations

For all the compounds mentioned in this chapter, we recommend to students that they prepare samples for study by dissolving 300 mg in 1.0 mL of $CDCl_3$. If solubility is a problem, we encourage students to first slightly warm the tube by placing it in a beaker of hot water. If that is not successful, they can try dividing the concentration in half. Only as a last resort do we suggest using DMSO-d_6 as the solvent, and then only under close supervision of the instructor. All NMR vials are prepared in the hood with students wearing gloves.

Conclusion

These experiments are used in a two-week laboratory experiment associated with the Physical Chemistry course taken by undergraduates (third and fourth year chemistry majors are enrolled in the class). Students sign up in pairs for times outside of the normal lab meeting time to use the NMR. They already have experience with the NMR instrument from Organic Chemistry, but two-dimensional NMR experiments are not covered in that earlier course. Students will also know how to use Gaussian 09 and have heard lectures on density functional theory as well as traditional quantum chemistry topics. Students have access to Gaussian 09 through a web-based interface known as WebMO (*17*) which includes an easy to use molecular editor for creating input and visualization tools to view the output. The WebMO interface allows students to submit jobs to a server and not have to wait for a computation to finish on a lab computer. The sign convention for chemical shifts should be discussed with the students at the start of the lab, since this is often a point of confusion. The spectra that emerge from the instrument will show positive chemical shifts with respect to TMS. The more positive the shift, the less shielded the nucleus is compared to the corresponding nucleus in TMS. In a Gaussian calculation, the isotropic shielding (**σ**) is reported as an absolute quantity; the larger the value, the greater the shielding. Therefore, to convert the Gaussian data it must be *subtracted from* the TMS reference value. Rather than have the students compute the value for TMS, that value is simply provided in the lab handout. For a shielding calculation done using the TPSSh functional with the large 6-311+G(2d,p) basis set and the B3LYP 6-311+G(2d,p) geometries, the TMS value is 186.95 (gas-phase), 187.40 ($CDCl_3$), and 187.56 (DMSO-d_6). Here we turn the solvent on in the calculation by using the default reaction field procedure (keyword SCRF) in Gaussian 09.

The objective of this laboratory experiment is to have students take a known compound and assign its ^{13}C NMR spectrum. They use experimental techniques, rules of additivity for aromatic carbons, predictions from Gaussian calculations, and intuition from their prior experience to achieve this. In writing their laboratory report they must reflect on all of this information, compare and contrast assignments, and write a conclusion which explains any anomalies. These experiments provide a good opportunity to practice their technical writing skills in that they must present evidence, interpret it, and reach a conclusion that is precise since the assignments are absolute. By working together in pairs and comparing their progress with other students who have similar compounds, they also learn how collaboration is important in solving chemical problems.

References

1. Silverstein, R. M.; Webster, F. X. *Spectrometric Identification of Organic Compounds*, 6th ed.; Wiley: New York, 1998; Chapter 5.
2. Cooper, J. W. *Spectroscopic Techniques for Organic Chemists*; Wiley: New York, 1980; p 181.
3. Wehrli, F. W.; Wirthlin, T *Interpretation of Carbon-13 NMR Spectra*; Heyden and Son: New York, 1976; p 47ff.
4. Also referred to as HETCOR. Our Bruker instrument uses the name HCCOSW for the composite procedure. For a description, see: Braun, S.; Kalinowski, H.-O.; Berger, S. *150 and More Basic NMR Experiments*; Wiley-VCH: New York, 1998; p 375.
5. M. J. Frisch; G. W. Trucks; H. B. Schlegel; G. E. Scuseria; M. A. Robb; J. R. Cheeseman; G. Scalmani; V. Barone; B. Mennucci; G. A. Petersson; H. Nakatsuji; M. Caricato; X. Li; H. P. Hratchian; A. F. Izmaylov; J. Bloino; G. Zheng; J. L. Sonnenberg; M. Hada, M. Ehara; K. Toyota; R.

Fukuda; J. Hasegawa; M. Ishida; T. Nakajima; Y. Honda; O. Kitao; H. Nakai; T. Vreven; J. A. Montgomery, Jr.; J. E. Peralta; F. Ogliaro; M. Bearpark; J. J. Heyd; E. Brothers; K. N. Kudin; V. N. Staroverov; T. Keith; R. Kobayashi; J. Normand; K. Raghavachari; A. Rendell; J. C. Burant; S. S. Iyengar; J. Tomasi; M. Cossi; N. Rega; J. M. Millam; M. Klene; J. E. Knox; J. B. Cross; V. Bakken; C. Adamo; J. Jaramillo; R. Gomperts; R. E. Stratmann; O. Yev; A. J. Austin; R. Cammi; C. Pomelli; J. W. Ochterski; R. L. Martin; K. Morokuma; V. G. Zakrzewski; G. A. Voth; P. Salvador; J. J. Dannenberg; S. Dapprich; A. D. Daniels; O. Farkas; J. B. Foresman; J. V. Ortiz; J. Cioslowski; Fox D. J. *Gaussian 09*, Revision C.01; Gaussian, Inc.: Wallingford, CT, 2010.

6. Foresman, J. B. Unpublished results. Geometries were optimized using the B3LYP 6-311+G(2d,p) theoretical model. NMR shielding calculations were then performed on these using the same basis set and the TPSSh density functional.
7. Wiberg, K. B.; Hammer, J. D.; Keith, T. A.; Zilm, K. *J. Phys. Chem. A* **1999**, *103*, 21–27.
8. This is by no means the most extreme case for nuclear shielding. Carbon tetraiodide, CI_4, is reported to have a ^{13}C signal that appears at −290.2 ppm with respect to TMS. This particular value would be difficult to calculate since relativistic effects need to be considered. See Fukawa, S.; Hada, M.; Fukuda, R.; Tanaka, S.; Nakatsuji, H. *J. Comput. Chem.* **2001**, *22*, 528–536.
9. Cheeseman, J. R.; Trucks, G. W.; Keith, T. A.; Frisch, M. J. *J. Chem. Phys.* **1996**, *104*, 5497–5509.
10. Ditchfield, R. *Mol. Phys.* **1974**, *27*, 789.
11. Jameson, A. K.; Jameson, C. K. *Chem. Phys. Lett.* **1987**, *134*, 461.
12. Harding, M.; Metzroth, T.; Gauss, J. *J. Chem. Theory Comput.* **2008**, *4*, 64.
13. For example, see: Braun, S.; Kalinowski, H.-O.; Berger, S. *150 and More Basic NMR Experiments*; Wiley-VCH: New York, 1998; p 120. For our Bruker instrument, students simply load the default ^{13}C proton decoupled experiment and change the pulse program to zggd30. This will allow NOE enhancement by having the decoupler turned on for the relaxation delay, but it is off for the entire acquisition period.
14. Gaussian keyword NMR=Mixed, see: Deng, W.; Cheeseman, J. R.; Frisch, M. J. *J. Chem. Theory Comput.* **2006**, *2*, 1028–37.
15. Stanger, A.; Tkachenko, E. *J. Comput. Chem.* **2001**, *22*, 1377–1386.
16. Kroschwitz, J. I.; Winokur, M.; Reich, H. J.; Roberts, J. D. *J. Am. Chem. Soc.* **1969**, *91*, 5927–5928.
17. Schmidt, J. R.; Polik, W. F. *WebMO: Web-Based Computational Chemistry*; http://www.webmo.net, Hope College: Holland, MI, 2000 (accessed December 2012).

Chapter 13

Using NMR Spectroscopy To Elucidate the Effect of Substituents on Keto-Enol Equilibria

Anderson L. Marsh*

Department of Chemistry, Lebanon Valley College, 101 N. College Avenue, Annville, Pennsylvania 17003
***E-mail: marsh@lvc.edu**

In the physical chemistry laboratory, students are generally tasked with using quantitative results to explain trends in chemical reactivity. In one such experiment at Lebanon Valley College, students use proton nuclear magnetic resonance (^1H NMR) spectroscopy to understand the effects of substituents on keto-enol tautomerization equilibria in a series of substituted β-diketones. The results illustrate to the students that additional effects beyond those that involve electron donation or electron withdrawal, such as resonance effects, need to be considered. Furthermore, the experiment allows students to use NMR spectroscopy as a quantitative tool rather than simply as a qualitative aid in structure characterization.

Introduction

In the laboratory curriculum at Lebanon Valley College, undergraduate chemistry and biochemistry majors are introduced to nuclear magnetic resonance (NMR) spectroscopy in the spring semester of their first year, where it is used to characterize synthesized aspirin. Laboratories during the second year focus on structural characterization of more complex organic and inorganic compounds, which is especially useful for those students interested in working on a synthetic research project. It is not until the analytical and physical chemistry laboratory sequences that students are introduced to the quantitative aspects of NMR spectroscopy.

A widely used experiment in the physical chemistry laboratory involves the determination of keto-enol equilibrium constants for various diketones (*1–4*). Students use ^{1}H NMR spectroscopy to find the amount of each tautomer in a solution, since separate resonance signals are observed for protons in the keto and the enol species. Among the variables examined as affecting the equilibrium distribution are concentration (*2, 3*), solvent (*2–4*), and temperature (*2, 3, 5*). Kinetic variations have been described (*5, 6*), as well as chemical structure effects with a series of β-diketones and β-ketoesters (*1, 2, 4*). It is also noted that the substituents on the α-carbon of 2,4-pentanediones may affect the keto-enol equilibria (*1*). A reaction equilibrium involving a 3-substituted-2,4-pentanedione is shown in Scheme 1. In the initial version of this laboratory experiment developed at Lebanon Valley, students performed this experiment using the methyl and chloro substituted forms in addition to 2,4-pentanedione (*7*). The experiment has recently been updated to include the addition of several other 3-substituted-2,4-pentanediones, along with another quantitative aspect that involves substituent constants.

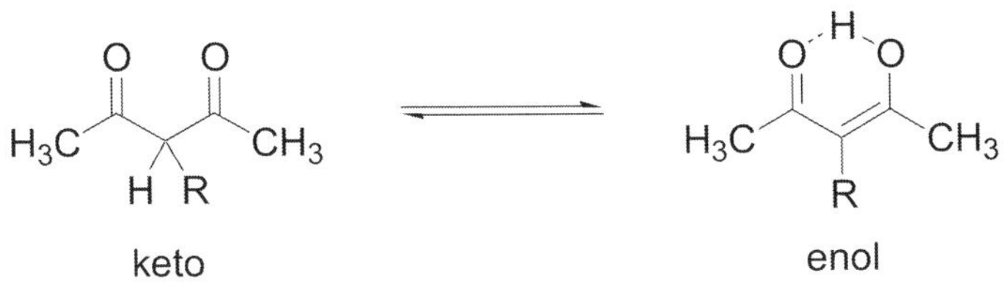

keto ⇌ enol

Scheme 1. The tautomeric equilibrium of a 2,4-pentanedione derivative with a substituent on the α-carbon

Experimental Methods

Deuterated chloroform containing tetramethylsilane (TMS), 2,4-pentanedione, 3-chloro-2,4-pentanedione, 3-methyl-2,4-pentanedione, 3-ethyl-2,4-pentanedione, and 3-acetyl-2,4-pentanedione, also known as triacetylmethane, were purchased from Sigma-Aldrich and used without further purification. The diketones 3-butyl-2,4-pentanedione and 3-phenyl-2,4-pentanedione were obtained from TCI America and used without further purification. The diketone 3-phenylazo-2,4-pentanedione was obtained from Alfa Aesar and used without further purification.

Students were given a handout referencing background information and published methods necessary to plan their own procedures to carry out the experiment. Prior to the laboratory period students were expected to locate and read the references. At the beginning of the experiment, students calculated volumes or masses needed to prepare solutions at a mole fraction of 0.001 of each 3-substituted-2,4-pentanedione in deuterated chloroform in standard NMR tubes and wrote out a detailed procedure in their laboratory notebooks. Students then worked in groups of two to four individuals to prepare their solutions. After these were prepared, student groups acquired standard ^{1}H NMR spectra using a Bruker

Avance 300 MHz FT-NMR spectrometer. The program uses a 30° excitation pulse, a 4.4 s acquisition time, and a 2 s relaxation delay over 16 scans (*6*). The temperature of the probe is maintained at 297 K.

Because the cost of several of the diketones is prohibitive to using amounts that yield a 0.2 mole fraction in solution, a mole fraction of 0.001 was used. A prior report demonstrated that 2,4-pentanedione solutions of this mole fraction yielded similar results for equilibrium constants found when using solutions of 0.2 mole fraction (*4*). For the hydrogen, methyl, and chloro substituents, no significant variation from our previously published results was observed (*6*).

Results and Discussion

Results from student NMR data are summarized in Table 1. Chemical shifts are reported relative to tetramethylsilane (TMS).

Table 1. NMR shift assignments, peak integrations, and experimentally determined values of the keto-enol equilibrium constant K_c from student data

	keto CH_3		*enol CH_3*		
compound	*shift (ppm)*	*int*	*shift (ppm)*	*int*	K_c
2,4-pentanedione	2.2	1.39	2.1	6.71	4.83
3-methyl-2,4-pentanedione	2.2	8.93	2.1	6.70	0.750
3-ethyl-2,4-pentanedione	2.2	15.83	2.1	7.36	0.465
3-butyl-2,4-pentanedione	2.3	11.36	2.2	6.21	0.550
3-chloro-2,4-pentanedione	2.4	0.48	2.3	6.07	13
3-acetyl-2,4-pentanedione	2.3	1.12	2.3	6.12	5.46
3-phenyl-2,4-pentanedione	--	--	1.8	5.87	--

Student groups assigned the chemical shifts in their spectra using integrated peak intensities, as well as results found in the literature (*8–12*). The spectra for the pentanediones with the hydrogen, methyl, ethyl, butyl, chloro substituents on the α-carbon presented no problems for the student groups to analyze. Because the enol species is the sole tautomer existing in solution for the pentanedione with the phenyl substituent, the spectrum was easier for the student groups to assign than if a mixture of tautomers existed. The spectrum for the pentanedione with the acetyl substituent was the most challenging for the student groups to evaluate, particularly since the peak for the protons on the three methyl groups in the keto form overlaps with the peak for two of the methyl groups on the enol form (*11*).

Equilibrium constants were calculated by the student groups from peak integrations for the protons in the keto and enol methyl groups bonded to the carbonyls using

$$K_c = \frac{[\text{enol}]}{[\text{keto}]} \quad (1)$$

where [enol] and [keto] represent the corresponding peak integrations. As seen from the data in the table, the equilibria for the pentanediones with the alkyl substituents (methyl, ethyl, and butyl) favor the keto tautomer, whereas the equilibria for the pentanediones with the hydrogen, chloro, and acetyl substituents favor the enol tautomer. For the pentanedione with the phenyl substituent, the equilibrium lies completely with the enol tautomer. Once again, these observations made by the student groups are supported by previous literature results (*8–11*).

In explaining the findings, the student groups initially argued that electron donating or electron withdrawing effects were responsible. For example, one would expect the alkyl substituents to be electron donating, which should lead to stabilization of the keto form (*13*). As mentioned above, this result is in fact what the student groups observed. On the other hand, electron withdrawing substituents, such as chloro or acetyl, should result in the enol being favored, as observed by the student groups. In order to quantify this inductive argument, the student groups compared equilibrium constants to Hammett substituent constants, the results of which are summarized in Table 2.

Table 2. Comparison of keto-enol equilibrium constants K_c with Hammett substituent constants

compound	K_c	σ_{para}
2,4-pentanedione	4.83	0.00
3-methyl-2,4-pentanedione	0.750	–0.14
3-ethyl-2,4-pentanedione	0.465	–0.15
3-butyl-2,4-pentanedione	0.550	–0.16
3-chloro-2,4-pentanedione	13	0.47
3-acetyl-2,4-pentanedione	5.46	0.24
3-phenyl-2,4-pentanedione	--	–0.01

The student groups researched Hammett constants in the literature and decided that the para substituent constants would be the most suitable for comparison (*14*). As observed with the data in the table, there is some correlation between the magnitude of the para substituent constant and the magnitude of the keto-enol equilibrium constant. Only in the case of the phenyl substituent is there no agreement. Still, Hammett plots of log K_c versus σ_{para} constructed by the student groups showed less than ideal correlation coefficients from the linear regression analysis. On the other hand, resonance forms could yield a greater stabilization of the enol tautomer. This resonance induced stabilization would explain the predominance of the enol form for the phenyl substituted pentanedione, as well as the chloro substituted pentanedione, especially if one considers contributions from the *d* orbitals on the chlorine atom (*10*).

Conclusions

Using NMR spectroscopy as a quantitative tool has allowed students in the Physical Chemistry laboratory course at Lebanon Valley College to examine structure-reactivity relationships for equilibria involving keto and enol tautomers in solution. Each student group designs and plans experimental procedures, collects data, and analyzes their findings. The experiment allows them to think about qualitative concepts learned during organic chemistry in a more quantitative manner that is traditional to physical chemistry. While interpretation of the NMR spectra may be challenging to some of the student groups, the solutions may be prepared and the spectra may be collected easily during a normal four hour laboratory period.

The laboratory experiment has recently been revised even further to a guided inquiry version where students are required to predict a trend in the equilibrium constants prior to performing the experiment in the laboratory. Interestingly, the student groups this time again invoked an electron donating/withdrawing argument in developing a hypothesis. Further updates may include having the students use the Taft substituent constants, which account for both polar and steric effects, as a means of quantifying their observations with regards to the equilibrium constants (*15*). In addition, student groups could perform the experiments using solvents of different polarities and pool their data before deriving conclusions. A computational component could also be introduced in order to more fully understand the effect of resonance structures. Finally, independent projects performed by students in the course could involve the synthesis of a substituted pentanedione for which the equilibrium constant has not been determined.

References

1. Garland, C. W.; Nibler, J. W.; Shoemaker, D. P. Experiment 42: NMR Determination of Keto-Enol Equilibrium Constants. In *Experiments in Physical Chemistry*, 8th ed.; McGraw-Hill: New York, 2009; pp 466−474.
2. Drexler, E. J.; Field, K. W. *J. Chem. Educ.* **1976**, *53*, 392–393.
3. Grushow, A.; Zielinski, T. J. *J. Chem. Educ.* **2002**, *79*, 707–714.
4. (a) Cook, A. G.; Feltman, P. M.; *J. Chem. Educ.* **2007**, 84, 1827−1829. (b) Cook, A. G.; Feltman, P. M.; *J. Chem. Educ.* **2010**, 87, 678−679.
5. Koudriavtsev, A. B.; Linert, W. *J. Chem. Educ.* **2009**, *86*, 1234–1237.
6. Nichols, M. A.; Waner, M. J. *J. Chem. Educ.* **2010**, *87*, 952–955.
7. (a) Manbeck, K. A.; Boaz, N. C.; Bair, N. C.; Sanders, A. M. S.; Marsh, A. L. *J. Chem. Educ.* **2011**, 88, 1444−1445. (b) Manbeck, K. A.; Boaz, N. C.; Bair, N. C.; Sanders, A. M. S.; Marsh, A. L. *J. Chem. Educ.* **2012**, 89, 421.

8. Rogers, M. T.; Burdett, J. L. *J. Am. Chem. Soc.* **1964**, *86*, 2105–2109.
9. Tanaka, M.; Shono, T.; Shinra, K. *Bull. Chem. Soc. Jpn.* **1969**, *42*, 3190–3194.
10. Yoshida, Z.; Ogoshi, H.; Tokumitsu, T. *Tetrahedron* **1970**, *26*, 5691–5697.
11. Dreyfus, M.; Garnier, F. *Tetrahedron* **1974**, *30*, 133–140.
12. Spectral Database for Organic Compounds home page; http:// http://riodb01.ibase.aist.go.jp/sdbs/cgi-bin/direct_frame_top.cgi (accessed September 2012).
13. Zuffanti, S. *J. Chem. Educ.* **1945**, *22*, 230–234.
14. Perrin, D. D.; Dempsey, B. Substituent Constants for the Hammett and Taft Equations. In *pK_a Prediction for Organic Acids and Bases*; Chapman and Hall: New York, 1981; pp 109–126.
15. Emsley, J.; Lewina, Y. Y. M.; Bates, P. A.; Motevalli, M.; Hursthouse, M. B. *J. Chem. Soc., Perkin Trans. 2* **1989**, 527–533.

Chapter 14

NMR-Based Kinetic Experiments for Undergraduate Chemistry Laboratories

Eric J. Kantorowski,* Bijan D. Ghaffari, Allee Macrorie, Kellan N. Candee, Jennifer M. Petraitis, Melanie M. Miller, Gayle Warneke, Michelle Takacs, Vanessa Hancock, and Zoe A. Lusth

Department of Chemistry and Biochemistry, California Polytechnic State University, San Luis Obispo, California 93407
***E-mail: ekantoro@calpoly.edu**

The dehydration reaction of 1,1-diphenylpropan-1-ol and a series of 1,1-diphenylethan-1-ol derivatives can be followed by ^1H NMR spectroscopy. The reaction conveniently occurs at room temperature in $CDCl_3$ using trichloroacetic acid or trifluoroacetic acid. The kinetic experiments are suitable for a variety of courses and center on the use of NMR spectroscopy to follow reaction progress, establish reaction order with respect to each component, examine temperature dependence, and explore structure-reactivity relationships. Many of the kinetic studies described can be investigated using a 60 MHz FT-NMR although higher-field instruments are necessary for accurate initial rate studies for establishing the order of the reaction.

Introduction

While kinetic experiments generally appear at each level of undergraduate chemistry curricula, they are more of a rarity in the organic chemistry teaching laboratory. This is due in part to the initial emphasis placed on techniques (e.g., extraction, TLC, distillation), hands-on exposure to instrumentation (e.g., GC, IR, NMR), and the gradual shifting of focus towards reactions which integrate the previous skills. Consequently, in the setting of the organic teaching lab, time is not generally made for kinetics. A recent survey indicated that kinetics earned a lower ranking compared with some of the aforementioned topics covered in the organic teaching lab (*1*).

Despite its expense, NMR spectroscopy is regarded as one of the most important hands-on techniques for students to learn (*1*). However, its primary use remains rooted in structure elucidation of unknowns and/or for confirmation and characterization of synthesis products. The substantial investment in such instrumentation makes it desirable to exploit its capabilities as well as broaden its use across multiple levels of the curriculum. Several reports have described the use of NMR spectroscopy as a tool for investigating kinetics (*2–6*) in the teaching lab.

At our institution, students enrolled in organic chemistry have access to a permanent-magnet FT-NMR (60 MHz, Anasazi upgrade) and a medium-field FT-NMR (300 MHz). The experiments described herein are adaptable to make use of both of these instruments. We have successfully introduced the first of the four experiments into our sophomore-level organic laboratory course. The remaining, more complex kinetic experiments, have only recently been fully elucidated and are now poised for formal introduction into our curriculum, particularly in our advanced organic and physical chemistry laboratories. Nevertheless, the data and results contained in this chapter were produced exclusively by undergraduate students (from non-majors enrolled in the organic laboratory through to majors involved in research projects). In this regard, multiple laboratory sections have contributed critical pieces to the collective data set and student researchers have refined much of the work demonstrating the viability of these experiments in the hands of undergraduates.

Monitoring Dehydration of 1,1-Diphenylpropan-1-ol by ^1H NMR Spectroscopy

Acid-catalyzed dehydration of an alcohol is a standard experiment found in many organic chemistry lab courses. A common approach involves heating an alcohol (e.g., cyclohexanol) in the presence of an inorganic acid and collecting the lower-boiling alkene by distillation as it is formed. The process can also be applied to higher-molecular weight compounds and the dehydration of 1,1-diphenylpropan-1-ol (**1**) to 1,1-diphenylprop-1-ene (**2**) works well in this respect (Figure 1). We have found that this reaction provides a remarkable opportunity for students to follow the dehydration process by ^1H NMR spectroscopy.

The dehydration is initiated by reversible protonation of the alcohol providing oxonium ion **3**, and loss of water leads to doubly-benzylic tertiary carbocation **4**. It is reasonable to consider that relief of backstrain also promotes this process. Deprotonation of **4** by either the conjugate base of HA or water provides alkene **2**. Students will recognize that this has all of the hallmarks of an E1 reaction. A discussion of both carbocation stability and the thermodynamic drive to a conjugated product fittingly attend this reaction.

We have observed that many organic acids are capable of initiating the dehydration of **1** at room temperature in a variety of available NMR solvents. Trichloroacetic acid (TCA), trifluoroacetic acid (TFA), *p*-toluenesulfonic acid

(p-TsOH), acetic acid, and sulfuric acid were investigated. TCA and TFA provided the most convenient reaction rates. Heating was required for acetic acid and p-TsOH, and each contributed undesirable signals in the proton spectrum.

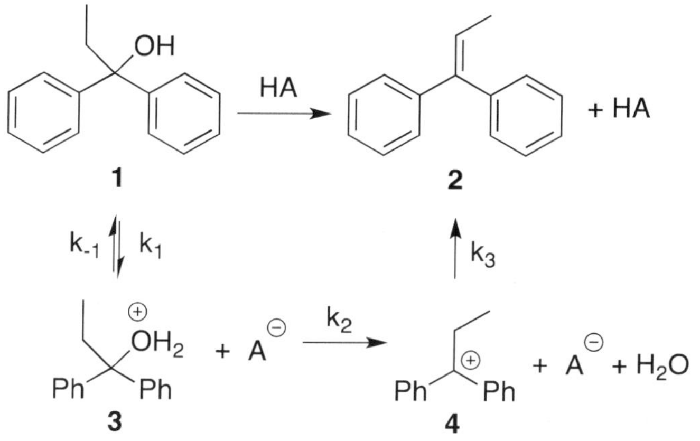

Figure 1. Postulated mechanism of the dehydration of 1,1-diphenylpropan-1-ol.

$CDCl_3$ was selected for the NMR solvent in these studies due in part to its prevalence in undergraduate labs as well as its low cost. Other solvents either obscured critical ^{1}H NMR signals associated with the starting alcohol (acetone-d_6, DMSO-d_6) or required heating (CD_3OD). Ultimately, the combination of TCA and $CDCl_3$ was found to be the most experimentally convenient.

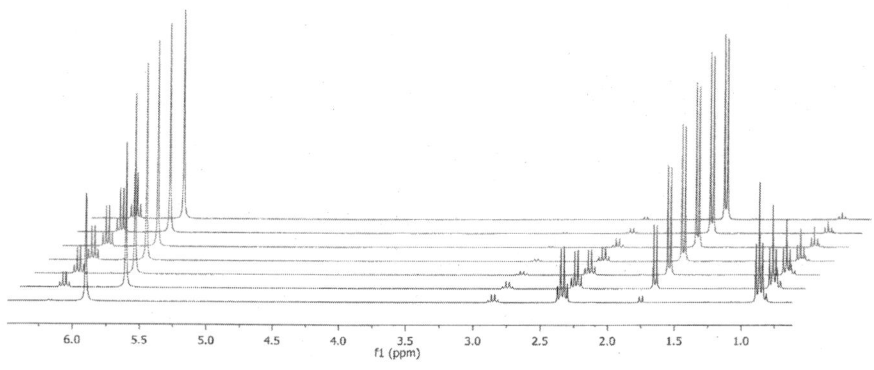

*Figure 2. Dehydration of **1** → **2** in $CDCl_3$ with TCA (300 MHz). The front slice is at 1 min and each subsequent slice is at 12 min intervals. The aromatic region has been omitted.*

A representative stacked plot of the dehydration is shown in Figure 2. Each signal is resolved, with the signals associated with the ethyl group of alcohol **1** appearing near 0.9 and 2.3 ppm and those of alkene **2** at 1.7 and 6.2 ppm. The methylene group of the intermediate can be seen as a quartet at 2.8 ppm. Spectroscopic evidence points to oxonium **3** as the most likely intermediate (*7, 8*). The singlet appearing near 5.8 ppm is due to the collective exchangeable protons present during the course of the reaction (e.g., alcohol **1**, TCA, and water). This signal experiences a slight upfield shift in the early stages of the reaction and then steadies its location; it reliably avoids overlap with the vinyl signal of the alkene at 300 MHz but is occasionally an issue at 60 MHz.

Observing the reaction with a 60 MHz FT-NMR spectrometer also provides a satisfactory stacked plot (Figure 3). Fortuitously, each of the signals remains resolved. Although less apparent compared with the 300 MHz spectra, the quartet of the intermediate is still observable. The aromatic region has also been included and noticeably displays the conversion from **1** to **2** as evidenced by changeover to the upfield signals of the more highly conjugated aromatic system.

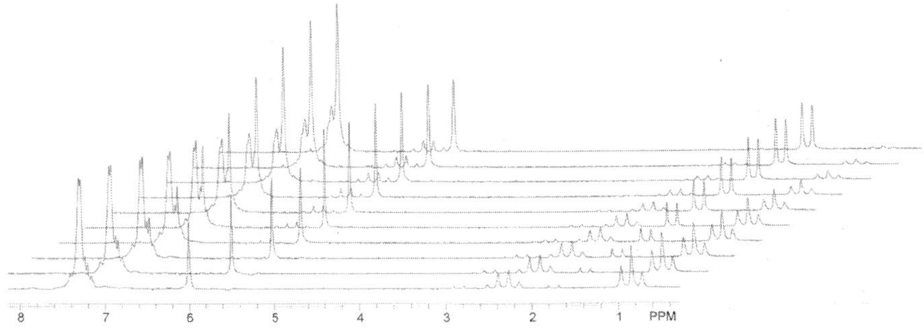

*Figure 3. Dehydration of **1** → **2** in $CDCl_3$ with TCA (60 MHz) at 1.0, 1.5, 2.5, 4.0, 6.0, 8.5, 11.5, 15.0, 19.0, 23.5 min (front to back).*

In calculating the relative concentrations of each component from the collected spectra it is recommended that the methylene quartet of the alcohol (2.3 ppm), the methyl doublet of the alkene (1.7 ppm), and the methylene quartet of the intermediate (2.8 ppm) be used. Although it is possible to use the vinyl quartet (6.2 ppm) to quantify alkene **2**, the methyl doublet offers a stronger signal (3H vs. 1H). The proximity of the semi-mobile –OH cluster can also interfere with the vinyl signal at 60 MHz.

Caution must also be exercised if the triplet at 0.87 ppm is utilized. At 300 MHz it is immediately clear that it is actually two triplets that overlap, that of the starting alcohol and the intermediate (Figure 4). The relative concentration of the alcohol or the intermediate can be found using this signal provided that 1.5x the value of the corresponding methylene integral value of the undesired signal is subtracted. The intermediate does not appear when TFA is used as the acid or when TCA is used with acetone-d_6.

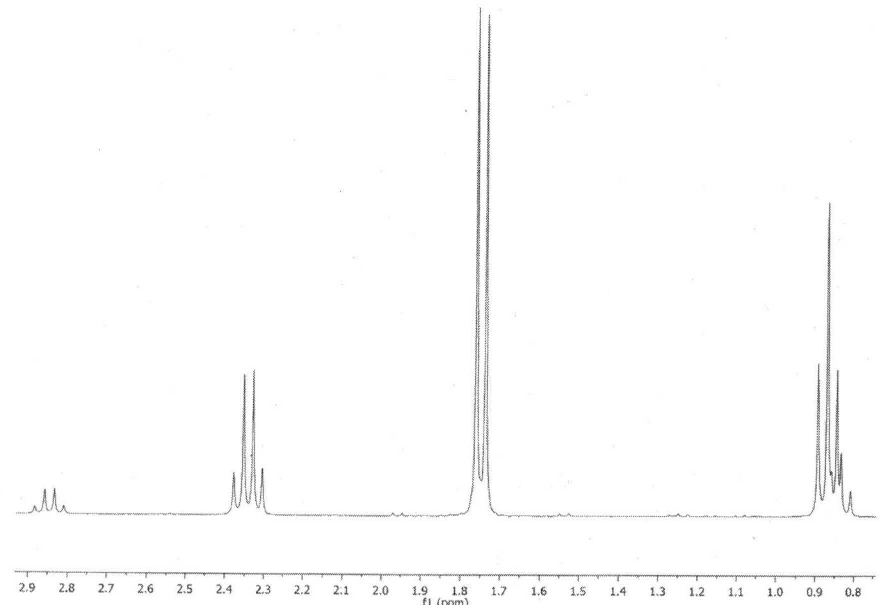

*Figure 4. One slice of the ¹H NMR (300 MHz) spectral data set from Figure 2. The quartet of the intermediate is present at 2.85 ppm and its associated methyl triplet is buried within the triplet of alcohol **1** at 0.85 ppm.*

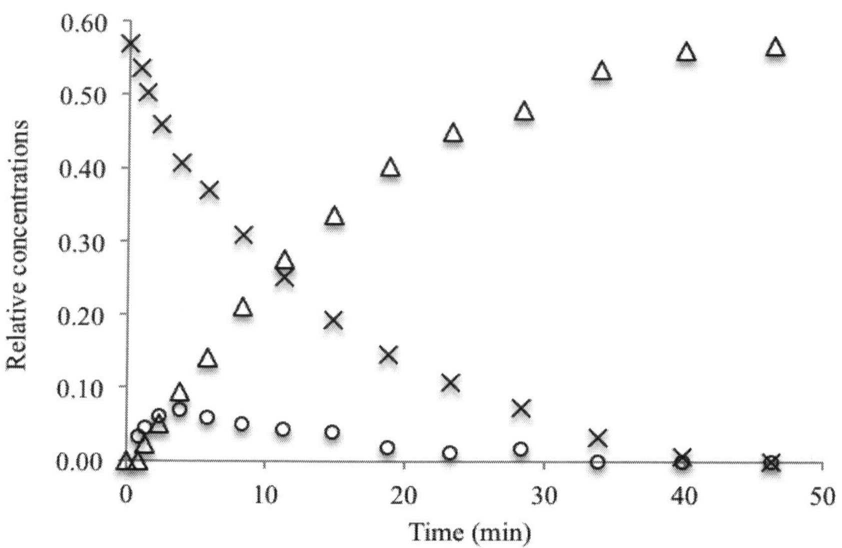

*Figure 5. Relative concentrations of alcohol **1** (X), alkene **2** (Δ), and intermediate **3** (O). Each data point is from a single acquisition obtained on a 60 MHz FT-NMR.*

Figure 5 shows the relative concentration changes over time for a mixture of **1** (0.57 M) and TCA (0.37 M) in 1.0 mL of $CDCl_3$ (containing 1% v/v TMS) at 22 °C. Relative concentrations were determined by either using TMS as an internal standard or taking the aromatic region to be constant. The latter method is viable as the only signals evident are those of **1**, **2**, and **3**. Despite recording this reaction at 60 MHz with only a single acquisition for each slice, the plot clearly shows a smooth trend for each component including the prompt appearance and subsequent steady decay behavior of the intermediate.

Experimental Details

1,1-Diphenylpropan-1-ol (**1**) is conveniently prepared by the reaction between phenyl Grignard and propiophenone. It may be recrystallized from 95% ethanol or by trituration using available hydrocarbon solvents (e.g., 60-90 ligroin, hexanes, etc.). This latter method conveniently removes the slight excess of ketone employed in the reaction as well as any unreacted bromobenzene. It also permits accelerated drying compared to the 95% ethanol and exposes students to liquid-solid extraction. This reaction reliably provides yields of 70-85% across a range of reaction scales (0.5-50 mmol).

The O<u>H</u> signal of **1** consistently appears near 2.1 ppm which, on a 60 MHz spectrometer, causes overlap with the methylene quarter centered at 2.3 ppm. A D_2O shake is instructive for students as it cleans up the distorted quartet and draws attention to the change in the integration value. Additionally, the sterically impacted alcohol appears as a *sharp* stretch in the FTIR spectrum (3540 cm^{-1}) characteristic of compounds prohibited from intermolecular H-bonding. 1,1-Diphenylpropan-1-ol (**1**): ^{1}H NMR δ 7.41-7.19 (10H, m), 2.30 (2H, q, J = 7.1 Hz), 2.05 (1H, s), 0.86 (3H, t, J = 7.1 Hz) ppm; ^{13}C NMR δ 146.9, 128.1, 126.7, 126.1, 78.4, 34.4 ppm.

The dehydration of the alcohol to **2** can be performed on a preparatory scale by mixing it with 6 M aq. HCl and heating it over a steam bath for approximately 1 h. Alkene formation is evidenced by formation of an oily upper layer. Upon cooling, the alkene solidifies and can be isolated by vacuum filtration. Alternatively, extraction with dichloromethane can be employed. Prop-1-ene-1,1-diyldibenzene (**2**): ^{1}H NMR δ 7.28-6.13 (10H, m), 6.16 (1H, q, J = 7.0 Hz), 1.75 (3H, d, J = 7.0 Hz) ppm; ^{13}C NMR δ 142.9, 142.4, 140.0, 130.0, 128.1, 128.0, 127.2, 126.8, 126.7, 124.1, 15.67.

For intervallic monitoring of the dehydration reaction by NMR spectroscopy it is most convenient for students to work in pairs. The alcohol and TCA are individually dissolved in $CDCl_3$ (99.8 atom % D, containing 1% v/v TMS) (0.5 mL each). The kinetics program package on the spectrometer is readied, the two solutions are mixed thoroughly (t = 0), transferred to an NMR tube and placed in the spectrometer for the appropriate duration. As expected, the retrieved sample displays a slightly hazy appearance due to formation of water in $CDCl_3$. Each time slice is then processed and analyzed. Table 1 provides some approximate times for the dehydration reaction to go to completion. These times were estimated from the concentration versus time curves (e.g., Figure 5).

Table 1. Approximate Completion Times for Various Amounts of 1[a]

Alcohol ***1***		
mass (g)	*mmol*	*t (min)*
0.10	0.47	6.5
0.12	0.57	10.0
0.14	0.66	15.0
0.16	0.75	23.0

[a] Each reaction is performed with 0.10 g (0.61 mmol) of TCA and 1.0 mL of $CDCl_3$.

Dehydration Kinetics and Establishing Reaction Order

Inspection of Table 1 reveals an unexpected oddity associated with this dehydration. The reaction rate is *inversely* proportional to the concentration of alcohol. First-order rate plots (ln [**1**] vs. t) consistently provided a slight curvature (downward) further refuting the simple E1 process shown in Figure 1. The analysis can be deceptive in that the least-squares curve fit is excellent through three and sometimes four half-lives (with $r^2 > 0.990$ in many instances). That this reaction obligingly goes to completion with no indications of decomposition allows this feature to be exposed. The possibility of a second-order process is easily dismissed based on poor fit in the second-order rate plot ($[\mathbf{1}]^{-1}$ vs. t). Inspection of other NMR solvents and other acids gave rise to similar complexities in that first- and second-order rate plots also deviated from linearity.

The inability to achieve a linear fit does not detract from the earlier described use of NMR spectroscopy as a tool to monitor the progress of a reaction. It also serves as an important reminder that despite any preconceived ideas we might have about a given mechanistic pathway, the kinetics can only be established through empirical investigation. It now becomes clear that the original reaction scheme presented in Figure 1 is incorrect or incomplete. Unable to derive the reaction order from the integration method, it became necessary to investigate initial rates (*9*). Such a study would be a good fit for an undergraduate physical chemistry laboratory experiment.

For the initial rate study, a stock solution of TCA was prepared and the alcohol concentration was varied across several kinetic runs. After mixing, an acquisition was taken immediately and concentrations were determined. The concentration of TCA and **1** were carefully selected to ensure that the reaction would not progress beyond 5%. One set of initial rate data is provided in Table 2. A plot of ln ($\Delta 1/\Delta t$) against $\ln[\mathbf{1}]_0$ results in a line providing the order (gradient) and ln k (which can be extracted from the y-intercept) for the reaction (Figure 6).

Table 2. Initial Rate Data for Determining Alcohol Order[a]

$[\mathbf{1}]_0$ (M)	$[\mathbf{1}]_t$ (M)	$\ln [\mathbf{1}]_0$	$\Delta[\mathbf{1}]$ (M)	$\ln (\Delta[\mathbf{1}]/\Delta t)$
0.3786	0.3614	-0.9714	0.0172	-8.1066
0.4724	0.4575	-0.7500	0.0149	-8.2493
0.5681	0.5548	-0.5655	0.0133	-8.3633
0.6605	0.6485	-0.4148	0.0119	-8.4715

[a] Performed with [TCA] = 0.1839 M; Δt = 57 s.

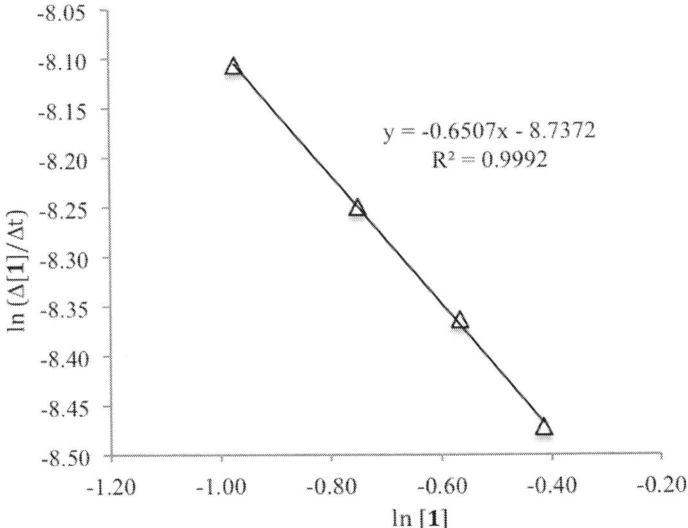

*Figure 6. Initial rate plot to determine the order of **1**, and Table 2. (from data in Table 2)*

Three repetitions provided an average order of -0.67 with respect to the alcohol. This negative fractional order demands a revision of the proposed mechanism. It is proposed that the initially formed oxonium ion **3** experiences a bifurcated pathway (Figure 7). One path leads to the expected carbocation **4** which is subsequently deprotonated, irreversibly producing the alkene. In competition with this is the possibility that **3** elects to aggregate with the starting alcohol, presumably a more favorable arrangement with respect to solvation despite the considerable steric congestion incurred. It is certainly conceivable to invoke an expanded network involving the acid, its conjugate base, and, as the dehydration progresses, water as well.

For the dehydration reaction to commence requires that this complex dissociate, giving rise to a negative order (*10*). It holds that the addition of more alcohol would then increase the proportion of this aggregate that is diverted into the cul-de-sac which thereby retards the reaction. The fractional aspect of the substrate order likely represents a composite of the involvement of intermediate **3** and any aggregates it may form with **1** (i.e., dimers, higher oligomers). Wherein the rate-determining step for this elimination still depends on the formation of carbocation **4** (k_2), the rate expression has become more complex now that the available amount of intermediate **3** is adjusted by k_4/k_{-4} in the revised reaction profile.

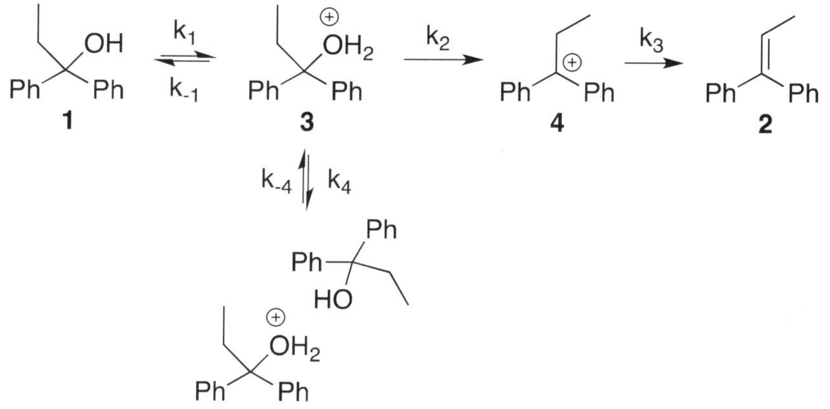

Figure 7. Expanded reaction manifold for dehydration of alcohol ***1****.*

Table 3. Initial Rate Data for Determining TCA Order[a]

$[1]_0$ (M)	$[1]_t$ (M)	$\ln [TCA]_0$	$\Delta[1]$ (M)	$\ln (\Delta[1]/\Delta t)$
0.5662	0.5586	-0.9714	0.0076	-8.9281
0.5662	0.5504	-0.7500	0.0158	-8.1897
0.5662	0.5370	-0.4148	0.0292	-7.5780

[a] $\Delta t = 57$ s.

The initial rates method was also employed to determine the order with respect to TCA. A stock solution of **1** was prepared and the TCA concentration was varied across several kinetic runs. As before, these kinetic runs were carefully designed to observe the reaction at less than 5% progress. One set of initial rate data is provided in Table 3 and graphically displayed in Figure 8. The average of three experiments provided an order of 1.92. An explanation for this order is not entirely clear, but may be intimately connected with the aforementioned aggregation event for the alcohol.

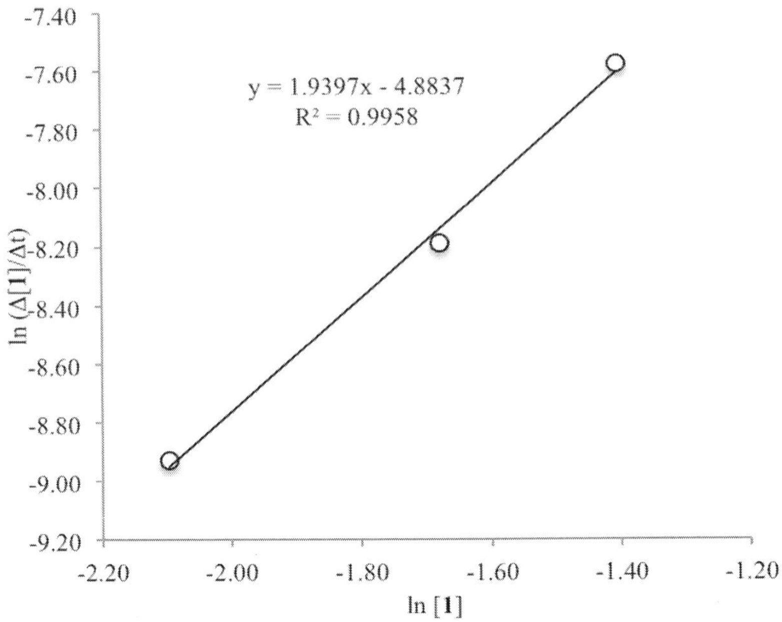

Figure 8. Initial rate plot to determine the order of TCA, and Table 3. (from data in Table 3)

Table 4. Summary of Initial Rate Studies

	*order of **1***		*order of TCA*	
Set	*slope*	*y-int*	*slope*	*y-int*
1	-0.73	-8.82	1.94	-4.88
2	-0.62	-8.82	1.83	-5.01
3	-0.65	-8.73	1.98	-4.76
average	-0.67	-8.79	1.92	-4.88

A summary of the three initial rate studies for alcohol **1** and TCA is provided in Table 4. All of the initial rate studies described were performed on a 300 MHz NMR spectrometer and a single-scan experiment was used for each run. At 60 MHz, the signal to noise ratio using only a single scan resulted in the introduction of significant error.

Establishing the reaction order with respect to the alcohol and TCA allows the rate constant to be calculated equation (1). Information from the initial rate data used for determining the order with respect to **1** produces $k = 4.50 \times 10^{-3}$ M s^{-1}; from the TCA analysis $k = 5.20 \times 10^{-3}$ M s^{-1}. These two values are in good agreement considering that obtaining k from an intercept inherently introduces high error.

$$\ln \text{rate} = \ln k - 0.67 \ln [\mathbf{1}] + 2.00 \ln [\text{TCA}] \qquad (1)$$

Experimental Details

The initial rate experiments were performed using volumetric glassware. For determination of the order with respect to the alcohol, a stock solution of TCA in $CDCl_3$ (99.8 atom % D, containing 1% v/v TMS) was prepared (0.1839 M). The chosen amount of alcohol was measured in a vial and dissolved in 0.50 mL $CDCl_3$. A 0.50 mL aliquot of the stock TCA solution was added to the alcohol ($t = 0$), mixed well, and added to an NMR tube. The sample was placed in the spectrometer and a single scan was taken. The probe temperature was maintained at 23.0 ± 0.1 °C for all runs. In order to bypass the relatively lengthy (~5 min) spectrometer preparations (e.g., shimming, etc.) for each kinetic run, a dummy sample containing similar concentrations of **1** and TCA was prepared and used to calibrate the instrument moments ahead of each set.

For determination of the order with respect to TCA, a stock solution of alcohol **1** in $CDCl_3$ was prepared (0.5662 M), and experiments were performed in a similar fashion using varying amounts of TCA.

Arrhenius Analysis Using VT-NMR

The dehydration of 1,1-diphenylpropan-1-ol (**1**) is also well suited for a variable temperature (VT) study. We examined the time required for the reaction to go to completion across a temperature range of 26.0 – 50.0 °C (Table 5). Completion times were estimated from each of the concentration versus time plots. The "rule of thumb" often told to students is that for each 10 °C increase in temperature the rate of the reaction approximately doubles is nicely demonstrated by this study. The Arrhenius plot of the data from Table 5 shows excellent correlation (Figure 9). From the analysis, the energy of activation is found to be 59.8 kJ/mol (14.3 kcal/mol).

Table 5. Temperature Dependence for Dehydration of 1

T (K)	*1000/T*	$t_{100\%}$ *(min)*	$1/t_{100\%}$	$ln\ (1/t_{100\%})$
299.0	3.34	90.0	1.11E-02	-4.50
303.0	3.30	65.0	1.54E-02	-4.17
308.0	3.25	45.0	2.22E-02	-3.81
313.0	3.19	30.0	3.33E-02	-3.40
318.0	3.14	19.5	5.13E-02	-2.97
323.0	3.10	16.0	6.25E-02	-2.77

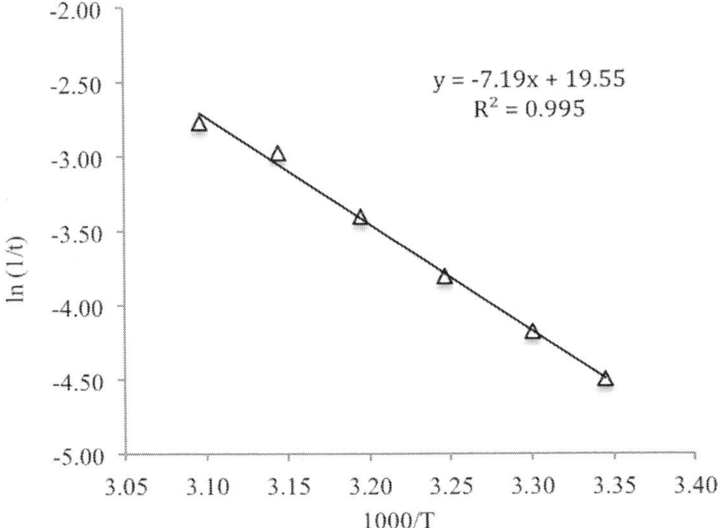

Figure 9. Arrhenius plot of the data in Table 5.

Experimental Details

Stock solutions of alcohol **1** and TCA in $CDCl_3$ (99.8 atom % D, no TMS) were prepared (0.5657 M and 0.3672 M, respectively). The spectrometer probe was equilibrated to the desired temperature, the reactants (0.50 mL each) were mixed, transferred to an NMR tube, and the sample was placed in the spectrometer. Scans were taken at appropriately spaced intervals in order to track the reactions to completion.

While the sample temperature for each kinetic run was maintained below the boiling point of $CDCl_3$ (b.p. = 61 °C), it was well above that for TMS (b.p. = 26 °C). As a precaution, $CDCl_3$ free of TMS was used for these studies. This required use of the aromatic region to serve as an internal standard.

Hammett Study

This dehydration reaction is also an excellent candidate for a structure-reactivity relationship investigation in an advanced organic chemistry lab. Preparation of substituted derivatives sets the stage for a Hammett study. In this respect we chose to prepare a series of 1,1-diarylethan-1-ol derivatives (**5**, Figure 10). The availability of a wide selection of substituted acetophenones and aryl bromides makes alcohol **5** an appealing substrate. The opportunity to provide students with ownership over a specific target and then to collaboratively pool the resulting kinetic data adds great value to this experiment.

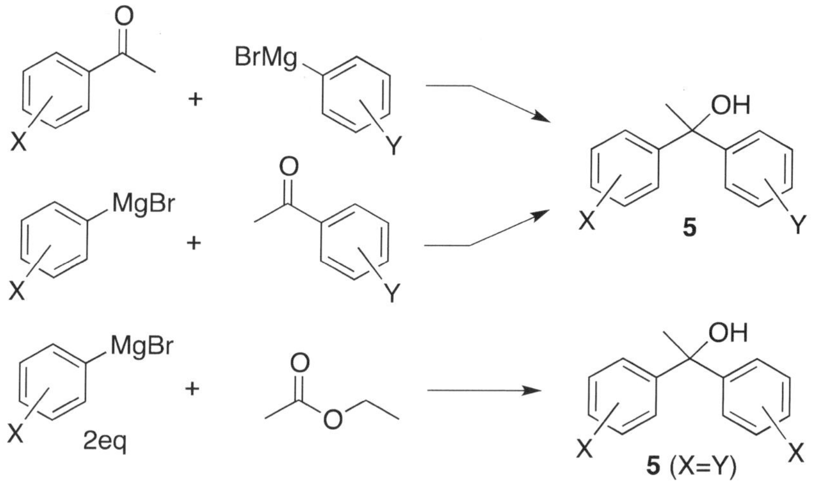

Figure 10. Synthetic routes to 1,1-diarylethan-1-ol derivatives.

We limited our study to para-substituted derivatives including *p*-OMe, *p*-Me, H, *p*-Ph, and *p*-Cl as possibilities for X and Y (Table 6). All of the resulting alcohols were purified by chromatography and characterized by ^1H and ^{13}C NMR, FTIR, and GC-MS. Of the nine compounds used for the study, only three were crystalline with the others being viscous oils.

Several noteworthy discussion points emerge from the syntheses. That *p*-chlorophenyl magnesium bromide can be made from 1-bromo-4-chlorobenzene is an excellent example of chemoselectively. 1,1-Bis(4-chlorophenyl)ethanol can be prepared either by treating *p*-chloroacetophenone with *p*-ClC_6H_4MgBr or by using two equivalents of the same Grignard with ethyl acetate. This dichloro alcohol product also challenges students to think about how the isotopic distribution affects the appearance of the molecular ion peak in mass spectrum. Lastly, the opportunity to have different students prepare the same compound by alternate routes creates valuable discussion regarding efficiency in terms of cost and yield.

The dehydration of **5** to **6** (Figure 11) was carried out using TFA in $CDCl_3$, but TCA can also be used if preferred. With TFA, the intermediate **3** is not observed; preliminary studies suggest that the reaction is third order with respect to TFA. The progress of the dehydration is readily followed by decay of the methyl singlet from alcohol **5** and growth of the vinyl protons in alkene **6**. In compound **6**, when $X \neq Y$, the vinyl protons are nonequivalent and appear as two singlets which students frequently mistake for a doublet. The observed chemical shifts are shown in Table 6.

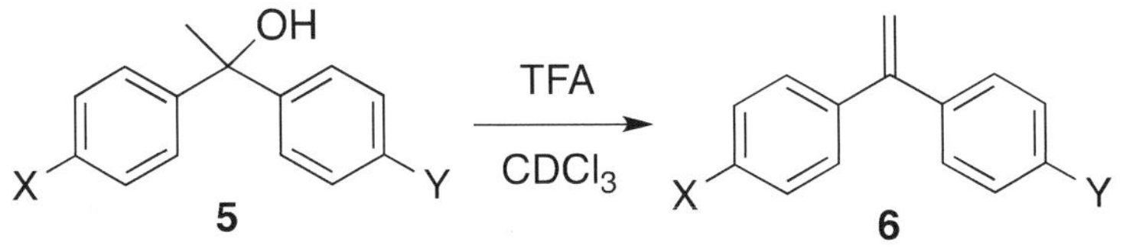

Figure 11. Dehydration of 1,1-diarylethan-1-ol.

Table 6. Chemical Shifts of Key Signals for 5 → 6.

Alcohol	*X*	*Y*	*Alcohol* δ *(methyl)*[a,b]	*Alkene* δ *(vinyl)*[a]	
5a	*p*-MeO	H	1.92	5.37	5.33
5b	*p*-Me	H	1.93	5.42	5.40
5c	*p*-MeO	*p*-Cl	1.88	5.39	5.33
5d	*p*-Ph	H	1.99	5.51	5.46
5e	H	H	1.94	5.45	
5f	*p*-Me	*p*-Cl	1.87	5.46	5.42
5g	*p*-Cl	H	1.92	5.49	5.47
5h	*p*-Cl	*p*-Cl	1.91	5.44	

[a] In ppm relative to TMS; [b] methyl refers to the CH_3 group attached at C1.

For this experiment it proved challenging to find the time when the reaction was 100% complete. A considerable amount of subjectivity contributes to the error when attempting to extrapolate from the concentration versus time curve. For simplicity, the rate of the reaction was taken to be the time required for the reaction to reach ~98% completion.

Table 7 lists the individual σ values for each substituent, the composite value (Σσ) (*11*, *12*), and the time required for ~98% completion. It is clear that the dehydration rate is accelerated by electron-donating groups (**5a**, **5b**) compared against the unsubstituted derivative (**5e**), and slowed by electron-withdrawing groups (**5g**, **5h**). As part of this study, we also prepared 1,1-bis(4-methoxyphenyl)ethan-1-ol (**5**, X = Y = OMe), but found its reaction time essentially instantaneous under these conditions. As expected, difficulties were encountered during its preparation and purification due to its pronounced acid sensitivity.

In the cases of **5c** and **5f**, each equipped with one electron-donating and one electron-withdrawing group, the composite σ value (Σσ) holds up well against the observed dehydration rate. Similarly, the additive effect of two chlorines properly places **5h** as the slowest of the series. The exceedingly slow dehydration of **5h** (~28 h) also allowed us to observe that the dehydration shows no evidence of decomposition in the spectrum even after more than one day.

Table 7. Sigma Values and Reaction Times for Alcohols 5a-5h with TFA

Alcohol[a]	*X*	σ_X	*Y*	σ_Y	$\Sigma\sigma$	*t (min)*[b]	*log (1/t)*
5a	*p*-MeO	-0.27	H	0.00	-0.27	3	-0.48
5b	*p*-Me	-0.17	H	0.00	-0.17	22	-1.34
5c	*p*-MeO	-0.27	*p*-Cl	0.24	-0.06	16	-1.20
5d	*p*-Ph	-0.01	H	0.00	-0.01	35	-1.54
5e	H	0.00	H	0.00	0.00	57	-1.76
5f	*p*-Me	-0.17	*p*-Cl	0.24	0.04	200	-2.30
5g	*p*-Cl	0.24	H	0.00	0.21	220	-2.34
5h	*p*-Cl	0.24	*p*-Cl	0.24	0.42	1700	-3.23

[a] The alcohol concentration was 0.25 M for each kinetic run; [b] The time indicated is at ~98% completion.

Since the substituents are capable of stabilizing the intermediate carbocation directly through resonance, it was necessary to also inspect the σ^+ values (σ^+ for *p*-OMe = -0.78, *p*-Me = -0.31, *p*-Ph = -0.18, *p*-Cl = +0.11). The resulting Hammett plots (Figure 12) yield ρ values of -3.76 (ρ) and -2.11 (ρ^+). The former value shows good correlation ($R^2 = 0.914$) and is consistent with a benzylic carbocation intermediate involved in the rate-determining step. The ρ^+ value, however, shows greater scatter with a magnitude atypical of such an intermediate.

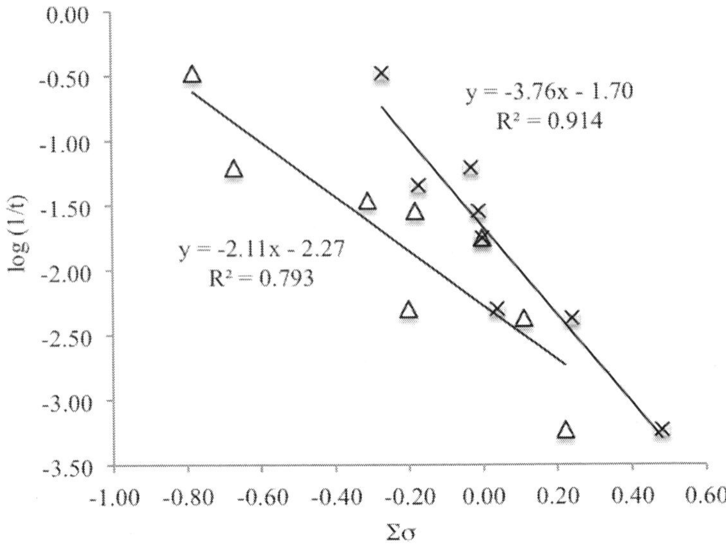

Figure 12. Hammett plot for dehydration of ***5a*** *–* ***5h*** *using σ (X) and σ⁺(Δ).*

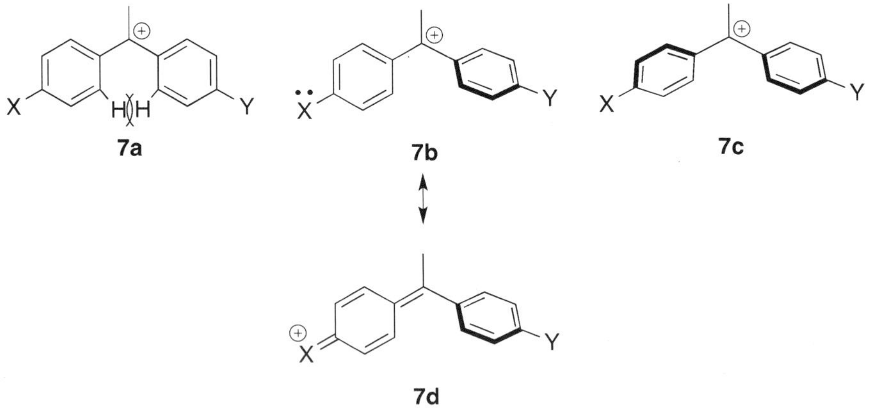

X H H Y
7a

X Y
7b

X Y
7c

X Y
7d

Figure 13. Some conformational options for intermediate 7.

An explanation for this deviation comes from examination of possible conformations of intermediate carbocation **7** (Figure 13). Conformation **7a**, with both aromatic rings coplanar, experiences repulsion between the ortho hydrogen atoms. Twisting one ring (e.g., **7b**) alleviates this interaction, but sacrifices (partially or entirely) resonance interaction with the carbocation by that aryl group. Twisting both rings results in a propeller-like conformation (e.g., **7c**), but at the expense of severely reduced or complete loss of resonance stabilization by both aryl groups. Whether **7b** or **7c** dominates is unclear at present, but the better fit obtained by using σ values implies that **7c** may have a greater contribution. It is also a consideration that one of these conformations is promoted in importance depending on the nature of X and Y. If **7b** is the preferred conformer, then it might be expected that a better correlation would be realized using σ^+ for X and σ for Y. A similar situation has been reported in a related system (*13*). We are currently investigating the interplay of these factors.

From the available Hammett data (either ρ or ρ^+), students have the opportunity to contribute to the study by selecting unexplored combinations. This confers greater ownership of the project, making the student an invested researcher.

Experimental Details

Alcohols **5a** – **5h** are prepared by standard Grignard reactions. Typical yields range from 55 – 90%. The alcohol products are purified by column chromatography with 5 – 25% ethyl acetate in heptanes as an eluent depending on the specific substrate. 1-(4-Methoxyphenyl)-1-phenylethanol (**5a**): ^{1}H NMR δ 7.41-7.22 (7H, m), 6.83 (2H, d, *J* = 8.8 Hz), 3.78 (3H, s), 2.17 (1H, br s), 1.92 (3H, s) ppm; ^{13}C NMR δ 158.1, 148.4, 140.4, 127.8, 127.1, 126.5, 125.7, 113.2, 75.6, 54.8, 30.7 ppm. 1-Phenyl-1-(p-tolyl)ethanol (**5b**): ^{1}H NMR δ 7.60-7.10 (9H, m), 2.32 (3H, s), 2.15 (1H, s), 1.93 (3H, s) ppm; ^{13}C NMR δ 148.0, 145.0, 135.9, 128.5, 127.7, 126.4, 125.7, 125.6, 75.7, 30.4, 20.7 ppm. 1-(4-Chlorophenyl)-1-(4-methoxyphenyl)ethanol (**5c**): ^{1}H NMR δ 7.33-7.22 (6H, m), 6.82 (2H, d, *J* = 8.9 Hz), 3.77 (3H, s), 2.23 (1H, s), 1.88 (3H, s) ppm; ^{13}C NMR δ 158.6, 146.9, 139.7, 132.5, 128.1, 127.2, 127.1, 113.5, 75.5, 55.2, 30.9 ppm. 1-([1,1′-Biphenyl]-4-yl)-1-phenylethanol (**5d**): ^{1}H NMR δ 7.58-7.23 (14H, m), 2.12 (1H, s), 1.99 (3H, s) ppm; ^{13}C NMR δ: 147.9, 147.0, 140.6, 139.6, 128.7, 128.1, 127.2, 127.0, 126.9, 126.7, 126.3, 125.8, 76.0, 30.7 ppm; m.p. = 71-73 °C. 1,1-Diphenylethanol (**5e**): ^{1}H NMR δ 7.42-7.20 (10H, m), 2.14 (1H, s), 1.94 (3H, s) ppm; ^{13}C NMR δ 148.0, 128.1, 126.9, 125.8, 76.2, 30.8 ppm; m.p. = 78-79 °C. 1-(4-Chlorophenyl)-1-(p-tolyl)ethanol (**5f**): ^{1}H NMR δ: 7.32-7.09 (8H, m), 2.31 (3H, s), 2.18 (1H, s), 1.87 (3H, s) ppm; ^{13}C NMR δ 146.7, 144.5, 136.7, 132.4, 128.8, 128.0, 126.2, 125.7, 75.6, 30.6, 20.9 ppm. 1-(4-Chlorophenyl)-1-phenylethanol (**5g**): ^{1}H NMR δ 7.38-7.25 (9H, m), 2.10 (1H, s), 1.92 (3H, s) ppm; ^{13}C NMR δ 147.3, 146.5, 132.5, 128.1, 128.0, 127.2, 127.0, 125.7, 77.7, 30.5 ppm. 1,1-Bis(4-chlorophenyl)ethanol (**5h**): ^{1}H NMR δ 7.33 (4H, d, *J* = 9.1 Hz), 7.28 (4H, d, *J* = 9.1 Hz), 2.12 (1H, s), 1.91 (3H, s) ppm; ^{13}C NMR δ 146.0, 133.0, 128.3, 127.2, 75.5, 30.7 ppm; m.p. = 64-65 °C.

For intervallic monitoring of the dehydration reaction the alcohol is dissolved in 1.0 mL $CDCl_3$ (99.8 atom % D, containing 1% v/v TMS) and transferred to an NMR tube. TFA (10 μL) is added to the NMR tube using a micropipette (t = 0) in a fume hood, mixed thoroughly by several inversions of the capped NMR tube, and placed in the spectrometer for the appropriate duration.

Conclusion

The dehydration reactions of 1,1-diphenylpropan-1-ol and 1,1-diarylethan-1-ol derivatives provide experiments that broadly apply at various levels in the chemistry curriculum. The ability to directly monitor these reactions impresses upon students that NMR spectroscopy is more than a tool for structural elucidation. The opportunity to spectroscopically observe an intermediate during the course of an organic reaction is an added advantage attending this set of experiments. Finally, this reaction plainly demonstrates the pitfalls of cavalierly categorizing reactions as straightforward processes (e.g., E1, E2, etc.). Despite the availability of numerous examples to influence our predictions regarding the kinetics, empirical evidence remains the ultimate arbiter.

Acknowledgments

We would like to thank Drs. John Hagen, John Marlier, and Derek Gragson for helpful discussions.

References

1. Martin, C. B.; Schmidt, M.; Soniat, M. *J. Chem. Educ.* **2011**, *88*, 1630–1638.
2. Gallaher, T. N.; Gaul, D. A.; Schreiner, S. *J. Chem. Educ.* **1996**, *73*, 465–467.
3. Turner, J. *J. Chem. Educ.* **1992**, *69*, 242–244.
4. Socrates, G. *J. Chem. Educ.* **1967**, *44*, 575–576.
5. Ruston, G. T.; Burns, W. G.; Lavin, J. M.; Chong, Y. S.; Pellechia, P.; Shimizu, K. D. *J. Chem. Educ.* **2007**, *84*, 1499–1501.
6. Koudriavtsev, A. B.; Linert, W. *J. Chem. Educ.* **2009**, *86*, 1234–1237.
7. Olah, G. A.; White, A. M.; O'Brien, D. H. *Chem. Rev.* **1970**, *70*, 561–591.
8. Olah, G. A.; Prakash, G. K. S.; Liang, G.; Westerman, P. W.; Kunde, K.; Chandrasekhar, J.; Schleyer, P. v. R. *J. Am. Chem. Soc.* **1980**, *102*, 4485–4492.
9. Canle, M; Maskill, H.; Santaballa, J. A. In *The Investigation of Organic Reactions and their Mechanisms*; Maskill, H., Ed.; Blackwell Publishing: Ames, IA, 2006; pp 50−58.
10. Deprez, N. R.; Sanford, M. S. *J. Am. Chem. Soc.* **2009**, *89*, 11234–11241.
11. Hart, H.; Sedor, E. A. *J. Am. Chem. Soc.* **1967**, *89*, 2342–2347.
12. Hansch, C.; Leo, A.; Taft, R. W. *Chem. Rev.* **1991**, *91*, 165–195.
13. Hegarty, A. F.; Lomas, J. S.; Wright, W. V.; Bergmann, E. D.; Dubois, J. E. *J. Org. Chem.* **1972**, *37*, 2222–2228.

Chapter 15

Physical Chemistry Laboratory Projects Using NMR and DFT-B3LYP Calculations

A. C. Bagley, C. C. White, M. D. Mihay, and T. C. DeVore*

Department of Chemistry and Biochemistry, James Madison University, Harrisonburg, Virginia 22807
***E-mail: devoretc@jmu.edu**

Five laboratory exercises that use the NMR spectrum of methanol and DFT-B3LYP-GIAO calculations are presented. DFT calculations explain the different chemical shifts observed for the OH proton in the liquid, the vapor, and varying concentrations of methanol in solution. The equilibrium constant for the methanol dimer is determined from the change in the OH chemical shift observed in the vapor phase NMR spectrum as the temperature is changed. The hydrogen bond energy between methanol and common NMR solvents and the activity coefficient at infinite dilution are determined using the OH chemical shift observed for infinitely dilute solutions. A procedure for measuring the mass susceptibility and magnetic moment of O_2 is also presented.

Introduction

Since NMR spectroscopy is a versatile technique that can be applied across the undergraduate curriculum, the number of exercises designed to use NMR spectroscopy is rapidly increasing (*1*). Modern high field NMR spectrometers can detect small amounts of material, making it relatively easy to obtain NMR spectra of very dilute solutions and even vapor molecules. This means that classic physical chemistry NMR experiments, such as measuring the keto-enol equilibrium constant for 2,4-pentanedione (*2–4*) or measuring reaction dynamics (*4, 5*), and traditional non-spectroscopy experiments, such as determining the enthalpy of vaporization, can be done using NMR spectroscopy (*2*). Five laboratory projects are presented that use NMR spectroscopy of vapor phase methanol and/or dilute methanol solutions are presented here, of which four also use DFT-B3LYP-GIAO calculations.

Two of these exercises are extensions of the NMR experiment to measure the enthalpy of vaporization presented previously (*2*). These exercises were developed to answer questions students had about the NMR spectrum of methanol that showed the OH chemical shift moving from up-field of the CH_3 proton chemical shift in the vapor to down-field of the CH_3 chemical shift in the liquid. In response, an exercise that uses the DFT-B3LYP-GIAO method to calculate the NMR spectrum of methanol molecules and clusters was developed. The results of these calculations confirm that the formation of methanol clusters produces the chemical shifts observed. It was also very apparent to the students that the chemical shifts measured for the methanol vapor differ from those calculated by ~2 ppm, thus leading to a second laboratory exercise. This exercise uses the difference between the measured and calculated chemical shift of the methanol vapor molecule to determine the mass susceptibility and magnetic moment of O_2 using a procedure slightly modified from the NMR method for measuring magnetic susceptibility developed by Evans (*6–9*). The small changes in the vapor phase OH chemical shift as the temperature or the partial pressure of the methanol is changed can be used to measure the enthalpy of dimerization for methanol. The procedure for doing this is presented as the third laboratory exercise. The fourth and fifth experiments presented were developed from undergraduate research projects that used NMR spectroscopy to investigate hydrogen bonding. The fourth experiment uses the measured chemical shift of the methanol OH proton in infinitely dilute solutions to estimate the methanol-solvent hydrogen bond energies. The fifth experiment uses the changes in the OH chemical shifts as a function of concentration to determine the activity coefficients for the methanol-acetone system. The two main advantages of using methanol for these exercises are (i) methanol only has two different types of protons giving a simple NMR spectrum and (ii) since it only contains six atoms with a total of eighteen electrons, it is possible to do high level DFT calculations with the modest hardware/software typically available for use in the teaching laboratories.

Experimental

NMR spectra of neat liquids and vapor molecules are obtained using capped 5 mm OD sample tubes fitted with a stem coaxial insert (Wilmad LabGlass). The insert contains D_2O which serves as the lock solvent. Following the suggestion of Gottleib et al. (*10*), the HDO that is always present in the D_2O is used as a secondary standard to determine the chemical shifts relative to TMS. The outer tube is filled with degassed neat liquid to measure the liquid spectrum. Simply pouring the liquid out of the outer tube usually leaves enough liquid in the tube to produce the vapor spectra. The liquid can also be added to a clean dry tube with a pipette or syringe if more control of the sample size is desired. Measuring spectra at several temperatures clearly establishes the gas phase bands since the vapor intensity increases while the liquid intensity from condensation on the walls decreases relative to the intensity of the HDO peak in the inner tube as the temperature is increased. Since all gas phase spectra were collected in air, the chemical shifts of the gas phase molecules are shifted downfield by ~2 ppm by the paramagnetic oxygen present in the NMR tube. Although the liquid droplets are not degassed, the NMR spectrum of them is not changed significantly from the NMR spectrum determined for the degassed neat liquid.

Solutions were prepared using a microliter syringe to add the solute (usually in 1 μL increments) to 1 mL of the deuterated solvent placed in a standard 5 mm OD NMR tube (Norell). Since the sample signals have roughly the same intensity as the impurity peaks for the most dilute samples, spectra of the neat solvent were also taken to establish the location of "impurity peaks" (usually caused by incomplete deuteration of the solvent) (*10*). The solvent doubled as the lock solvent and in many cases contained TMS which was used to calibrate the spectrum. While it is relatively easy to degas these samples to remove traces of oxygen, we generally do not have students do this as part of the laboratory exercises.

Although most of the spectra were obtained using a Bruker Spectrospin 400 NMR (Avance DRX-400) equipped with a variable temperature 5 mm broadband auto tune probe with a Z gradient, a 300 MHz Bruker DPX equipped with a 5 mm variable temperature probe and a Bruker Ultrashield 600 Plus (Avance 600) equipped with multi-nuclear variable temperature probes have also been successfully used. Generally, 8-64 scans, with a receiver gain of 1300 and a spectral width set between 15 ppm - 30 ppm is adequate to obtain the spectra. A vaporization temperature between 295 K and 330 K as measured by the variable temperature probe is usually sufficient to identify the vapor phase molecules. NMR spectra of dilute solutions are usually obtained with the temperature set at 300 K.

Theoretical calculations were done using the DFT-B3LYP-GIAO method available in the Gaussian 03 PC software package. Density functional theory was chosen since it requires less computer time and less sophisticated computer systems to achieve high level results than most of the other high accuracy methods (*11–15*). The DFT-B3LYP method, which uses the Becke exchange functional and the LYP correlation functional to parameterize the calculation, is among the most commonly used DFT methods (*11–15*). The gauge including atomic orbital method (GIAO) uses basis functions that have explicit field dependence to determine the NMR spectrum (*14, 15*). Since Cheeseman et al. have reported that the DFT-B3LYP-GIAO method provides a reasonable trade-off between cost and accuracy, and there are over one hundred examples using this method in the recent chemical literature, we used it for these experiments (*14*). Calculations done using the 6-31G^{++} basis set, usually considered to be the minimum basis set needed for reliable calculations, can easily be done in a 3 hour class period for methanol starting with an arbitrary structure and for methanol clusters or methanol-solvent interactions starting with previously minimized structures. Minimization of floppy structures like the methanol dimer can take considerably longer than 3 hours starting from an arbitrary structure even for this level of theory. Better calculations using the 6-311G^{++} (3df, 3pd) basis set have also been done, but these generally require more than three hours using our PC based system. These calculations are started and allowed to process on a dedicated computer until they finished. The calculations are generally done in three steps. First, the energy is minimized by changing the molecular geometry to determine a stable structure for the molecule. Next, vibrational frequencies are calculated to establish that this structure has no imaginary frequencies indicating that it is at least a local minimum on the potential curve. Finally, the NMR spectra are calculated using the GIAO method and the chemical shifts are determined relative to the values calculated for TMS using the same level of theory. While it is possible to calculate the chemical shifts relative to chemical shifts stored in the software, this changes the values by as much as 1 ppm. Better agreement with experiment is obtained using the same level of theory. While many other basis sets and other approaches such as Hartree Foch, MP-2 or other DFT methods are available in the Gaussian 03 package, we have not used them. Additional information about these approaches for calculating an NMR spectrum can be found in Cheeseman et al. (*14*), Krykunov et al. (*16*), or Loibl et al. (*17*).

The Effect of Hydrogen Bonding on the OH Chemical Shift

The NMR spectra observed for an equilibrium mixture of neat and vapor phase CH_3OH are presented in Figure 1. One of the more striking features is that the OH proton is downfield relative to the CH_3 protons in the liquid while it is upfield from the CH_3 peak in the vapor.

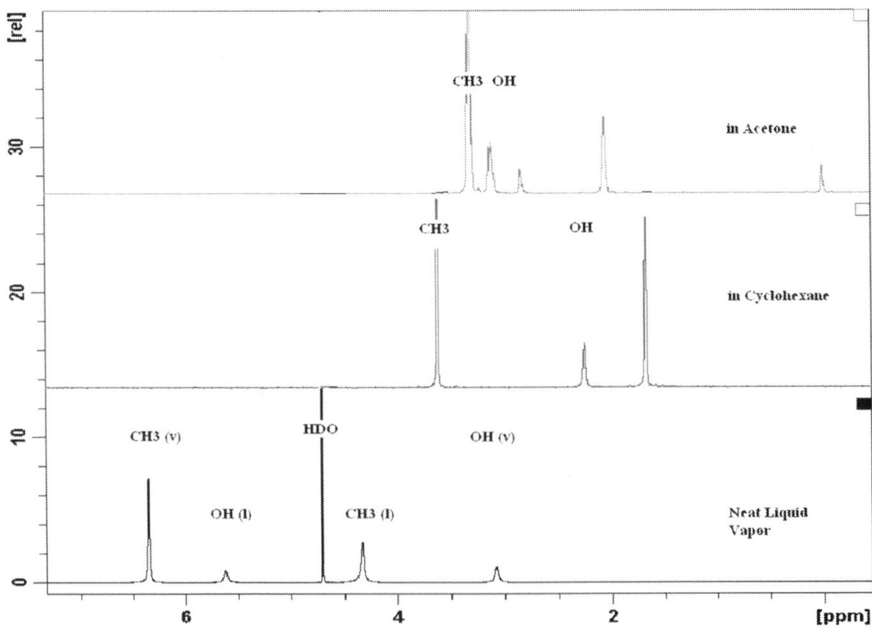

Figure 1. NMR spectra of methanol neat liquid and vapor in air, dissolved in cyclohexane (χ = 0.01), and dissolved in acetone (χ = 0.01).

Students can use DFT calculations to confirm that the different OH chemical shifts result from hydrogen bonded methanol clusters present in the liquid. A comparison of the chemical shifts calculated for methanol molecules, the linear dimer, and ring trimer (see Figure 2) using the 6-31G^{++} basis set and the 6- 311G^{++} (3df,3pd) basis set are compared to the measured experimental chemical shifts in Table 1.

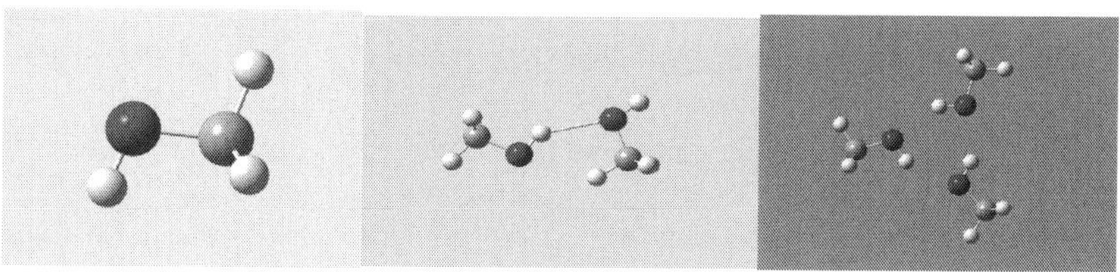

Figure 2. The minimized structures determined for methanol molecules, dimers, and trimers.

Table 1. Comparison of measured and calculated chemical shifts for methanol

Method	*δ CH_3*	*δ OH*	*Δδ (CH_3-OH)*
Neat Liquid	3.5	5.0	-1.5
Vapor	5.6	2.3	3.4
Monomer[a]	3.63	-0.14	3.77
Monomer[b]	3.60	-0.23	3.83
Dimer[a]			
H- bonded	3.59	3.35	0.24
Free OH	3.63	0.28	3.35
Dimer[b]			
H- bonded	3.59	3.35	0.24
Free OH	3.59	0.32	3.27
Trimer[a]	3.53	5.11	-1.58
Trimer[b]	3.52	4.22	-0.70

[a] Calculated using DFT-B3LYP- GIAO with a 6-31G++ basis set. [b] Calculated using DFT-B3LYP- GIAO with a 6-311G++(3DF,3PD) basis set.

Both sets of calculations show that the OH chemical shift moves from approximately 0 ppm for the methanol molecule to a value greater that 4 ppm for the ringed trimer. The chemical shifts in the linear dimer show that the hydrogen bonded OH are downfield from the OH that is not bonded, confirming that the hydrogen bonding is producing the change.

It probably should be pointed out to the students that these are not the only clusters that are present in the neat liquid methanol, but this set provides enough information to confirm that the chemical shift for the OH peak moves downfield as the amount of hydrogen bonding increases.

The Mass Susceptibility and Magnetic Moment of O_2

As shown in Table 1, the chemical shifts for the CH_3 and the OH protons in the gas phase spectrum collected in air are shifted downfield by ~ 2 ppm relative to the chemical shifts calculated for methanol molecules in vacuum. This is from deshielding by the oxygen molecules in the air in the sample tube. The difference between the observed and calculated chemical shifts can be used to measure the mass susceptibility and the magnetic moment for oxygen using the method developed by Evans (*6–9*). This can be done by modifying the physical chemistry laboratory exercise for the determination of paramagnetic susceptibility using NMR spectroscopy exercise developed by Garland et al (*9*). In this exercise the mass susceptibility χ_g is given by:

$$\chi_g = \frac{3}{2\pi m}\frac{\Delta f}{f} + \chi_o + \frac{\chi_o(d_o - d_s)}{m} \qquad (1)$$

where m is the concentration of the paramagnetic substance expressed in grams per milliliter; Δf, the change in chemical shift, is the difference between f_{obs}, the value for the chemical shift in air and f, the value for the chemical shift in vacuum ($\Delta f = f_{obs} - f$); and χ_o is the mass susceptibility of the pure solvent. The final term in equation (1) accounts for differences in the density of the solutions. Since this difference is usually small, this term is usually set equal to zero for laboratory exercises (*9*). Assuming the solvent is nitrogen, χ_o is -5.4×10^{-9}. The value for m is determined by multiplying the density by the mass fraction of oxygen and the molar mass of oxygen. For dry air it is ~0.0105 g/ ml at STP. Determining m is one of the major sources of error in this experiment since the density of air depends on the air pressure and the relative humidity. Since the chemical shift observed for the OH proton is a function of the temperature and the methanol concentration, the chemical shift of the CH_3 protons is used for this experiment. There are three possible methods for determining the unperturbed chemical shift for the CH_3 protons, f. The most convenient is to assume that the chemical shifts observed for the CH_3 protons are independent of the species present and use the value observed for the neat liquid as f. Since the DFT calculations indicate that the chemical shift for the CH_3 protons in the methanol monomer differ from those in the ringed trimer by less that 0.1 ppm, this is a reasonable approximation and is accurate enough for this laboratory exercise. This assumption also has the advantage that the change in the chemical shift can be measured directly since, as shown in Figure 1, the liquid and the vapor spectrum can be observed simultaneously. A second approach is to use the calculated chemical shift for the monomer as f. The third method is to obtain f from the NMR spectrum of methanol in a dilute solution of a non-polar, non-bonding solvent such as cyclohexane. Extrapolation to zero concentration produces the chemical shift of "isolated molecules" surrounded only by solvent molecules. If research quality results are wanted, these samples can be degassed to remove traces of dissolved oxygen and corrected for the diamagnetic chemical shift produced by the solvent. Establishing the value of *f* is the second source of error in this experiment.

Once the change in the chemical shift (Δf) and *f* are established, the mass susceptibility (χ_g) can be calculated from equation (1). Multiplying χ_g by the molar mass gives the molar susceptibility χ_M and the magnetic moment can then be calculated from the molar susceptibility using equation 2 (*6–9*).

$$\mu = 2.84\,(\chi_M\,T)^{1/2} \qquad (2)$$

Typical results obtained for methanol at 300 K for each approach, which were then used to determine *f*, are presented in Table 2. The results obtained are generally within 5% of the literature value when methanol is used as the probe (*7*). We have also tried acetone, chloroform, and cyclohexane as the probe. While each gives measurements of the same order of magnitude as the literature value, they generally have larger errors than found when using methanol as the probe.

Table 2. The values determined for the molar susceptibility and the magnetic moment for oxygen based on the relative chemical shift for methanol

Quantity	*Observed*	*Liquid*	*Solution*	*Calculated*
δ CH_3	5.80	3.68[a]	3.62	3.64
$\Delta f/f$		2.12[b]	2.18	2.16
$\chi_g(*10^6)$	106.3[c]	105	109	108
$\chi_M(*10^3)$	3.40[c]	3.38	3.50	3.47
μ	2.86[c]	2.85	2.90	2.89
H- bonded	3.59	3.35	0.24	

[a] For methanol droplets on the walls of the NMR tube during the vapor experiment.
[b] Measured directly from the NMR spectrum. [c] From Reference (*7*) at 293 K.

Determination of the Enthalpy of Dimerization for Methanol Vapor

In contrast to the CH_3 chemical shifts, the corrected chemical shift for the OH proton does not agree well with the calculated value for the methanol molecule. While some of this difference undoubtedly results from error in the calculations, some of it results from the presence of dimers and higher clusters in the vapor. It has long been known that methanol vapor in equilibrium with the liquid contains small amounts of dimer and larger species (most often assigned as the tetramer) (*18–23*). Since the exchange rate is fast on the NMR time scale, the NMR spectra of the individual species are not resolved and the observed chemical shift is a weighted average of the chemical shifts for all of the molecules present (*23–26*). This average chemical shift can still be used to determine the equilibrium constant for the vapor phase dimerization reaction.

If the partial pressure of the larger clusters is assumed to be negligible, methanol vapor would contain a large partial pressure of molecules (P_M) in rapid equilibrium with a small partial pressure of dimers (P_D).

$$D \Leftrightarrow 2\,M \qquad (3)$$

If the pressures are in bars, the equilibrium expression for reaction (3) is:

$$K_P = P_M^2 / P_D \qquad (4)$$

According to Dalton's Law of partial pressures, the pressure of the system (P) is the sum of the partial pressures of the monomer (P_M) and dimer (P_D) and the

$$P = P_M + P_D \qquad (5)$$

observed chemical shift is the weighted average of the chemical shifts for the monomer (δ_M) and the dimer (δ_D) (*26*).

$$\delta = f_M\,\delta_M + f_D\,\delta_D \qquad (6)$$

where: $f_M = P_M / P$ and $f_D = P_D / P$

Combining equation (5) with equation (6) and rearranging gives

$$\delta = \delta_D + (\delta_M - \delta_D)\,P_M / P \qquad (7)$$

If the chemical shifts for the monomer and dimer are known, P_M can be established from the measured chemical shift and the vapor pressure of methanol. The vapor pressure can be determined from the temperature by using vapor pressure tables (*27*) if the liquid and vapor are in equilibrium or by totally vaporizing a small measured amount of the liquid and using the ideal gas law to calculate the pressure. The latter method is recommended for this exercise since using a fixed amount of methanol produces larger changes in the OH chemical shift as the temperature is changed. Using the calculated chemical shifts given in Table 1 as the values for δ_M and δ_D, P_M can be determined from the pressure of methanol and the measured chemical shift using equation (7). Once P_M is determined, it can be used to calculate P_D using equation (5). From these, the equilibrium constant can be determined using equation (4) and the Gibbs Energy calculated using

$$\Delta G = -RT \ln K_p \qquad (8)$$

By measuring ΔG at several temperatures, ΔH and ΔS can also be obtained

$$\Delta G = \Delta H - T\,\Delta S \qquad (9)$$

Typical results using a fixed mass of methanol and the chemical shifts from the calculations using the 6-31G++ basis set are presented in Figure 3.

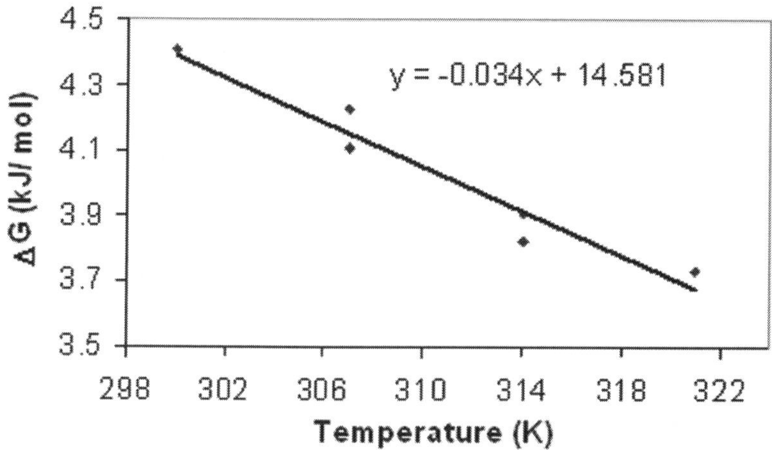

Figure 3. The determination of ΔH = 14.6 kJ mol $^{-1}$ and ΔS =34 J K^{-1} mol^{-1} for the dimer-monomer equilibrium of methanol determined using the chemical shifts calculated using the DFT-B3LYP-GIAO method with the 6-31G++ basis set for a constant amount of methanol vapor in the tube.

The duplicate points were obtained by systematically increasing the temperature of the sample and then systematically decreasing the temperature to check for systematic temperature measurement errors. The temperatures in Figure 3 were measured by the variable temperature NMR probe, though the probe can be calibrated using the methanol thermometer if more accurate temperature measurements are desired (*28*). The values determined are 14.58 kJ mol^{-1} for ΔH and 34 J K^{-1} mol^{-1} for ΔS. These ΔH values are consistent with the previously reported values of ΔH of 13.4 and 16.8 kJ mol^{-1} from vapor pressure measurements (*18–20*), 13.5 kJ mol^{-1} from thermal conductivity measurements (*21*), 19.2 kJ mol^{-1} from photo-ionization measurements (*27*). A value of 17.6 kJ mol^{-1} can be estimated by assuming that the vaporization of ethanol vapor requires breaking approximately 2 hydrogen bonds/molecule vaporized (*29*), though high-level theoretical calculations predict a slightly higher value (~ 25 kJ mol^{-1}) (*30*). The entropy change determined is approximately ½ the value of ~ 69 J K^{-1} mol^{-1} reported previously, indicating that there is an error in the measurements using this method (*18–21*). The students should also recognize that their calculations indicate there is considerably more dimer present in the equilibrium mixture than reported from the vapor pressure measurements (*18–20*). While some of this error results from neglecting the larger clusters in the model used to make the calculation, the largest error arises from the uncertainty in the values used for the chemical shifts of the component species. Better results can be obtained by combining the measured chemical shift with partial pressure data from the literature (*18–20*) to determine better values for δ_M and δ_D using Equation 7. The values determined were $\delta_D = 2.4$ ppm and $\delta_M = 0.22$ ppm. These chemical shifts and the data presented in Figure 3 give $\Delta H = 15.9$ kJ mol^{-1} and $\Delta S = 65$ J K^{-1} mol^{-1} with the equilibrium vapor determined to be ~97 % monomer at 300 K.

Methanol in Dilute Solutions

The NMR spectra observed for a dilute solution of methanol dissolved in acetone and in cyclohexane are presented in Figure 1. The chemical shifts observed for the OH proton as a function of the mole fraction of methanol are presented in Figure 4.

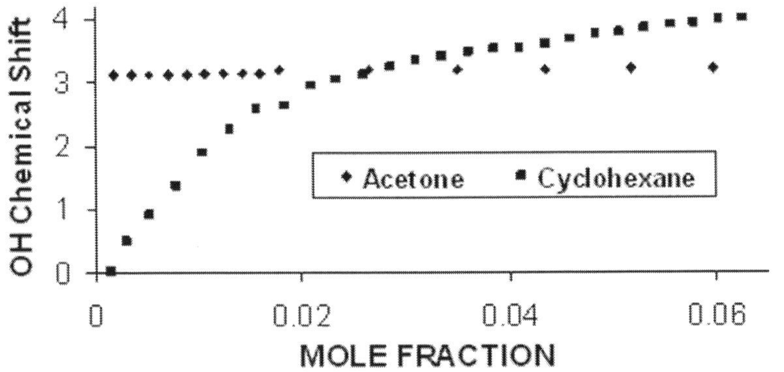

Figure 4. The observed OH chemical shift as a function of the mole fraction methanol in deuterated acetone and deuterated cyclohexane.

In cyclohexane, the OH proton peak rapidly moves downfield non-linearly as the mole fraction of methanol increases. Only a small and approximately linear downfield shift is observed in acetone. The chemical shifts of the infinitely dilute solutions determined by extrapolating the observed chemical shift to zero mole fraction are also dramatically different. The intercept in cyclohexane (-0.055 ppm) is similar to the value calculated for isolated vapor phase molecules suggesting only a weak interaction with the solvent while the intercept observed for acetone (3.12 ppm) indicates a stronger interaction with the solvent. Both observations can be explained by hydrogen bonding. Since cyclohexane does not interact strongly with methanol, the methanol molecules are only effectively isolated in very dilute solutions. As the concentration increases, dimers and larger clusters form readily, producing the rapid downfield chemical shift as the concentration increases. Since methanol can hydrogen bond to acetone, the OH chemical shift is significantly shifted downfield relative to the gas phase value even in infinitely dilute solutions. The solvent interaction inhibits the formation of clusters and the chemical shift changes more slowly. In principle, the OH chemical shift can be used to determine the equilibrium constants for the monomer-dimer equilibrium in solution and these types of investigations have been reported previously (*26*). However, the OH stretching region in the IR spectrum of methanol shows the presence of multiple species even in dilute solutions (*31*). The multiple equilibria complicate the analysis enough that the simple model used above does not work well. The complete analysis requires sophisticated fitting procedures generally not available in undergraduate laboratories. As a result, we have not had students attempt to measure equilibrium constants for these solutions. However, there are two relatively simple exercises using dilute solutions that can be done by typical physical chemistry laboratory students and these are given below.

Methanol–Solvent Interaction Energies

The chemical shift at infinite dilution could be used to estimate the hydrogen bond energy between the methanol and the solvent if a relationship between the chemical shift and the bond energy can be established. For this exercise, a linear relationship was assumed and determined by setting the hydrogen bond energy of the isolated molecules equal to zero and the hydrogen bond energy for the molecules in the neat liquid equal to the enthalpy of vaporization. From this relationship, and using the chemical shifts calculated for the molecules using the 6-311G^{++}(3df,3pd) basis set and for the neat liquid, -0.2 and 5.15 ppm respectively, the relationship between the hydrogen bond energy and the chemical shift was determined to be

$$E_{HBE} = 6.6 \times \delta_{OH} + 1.3 \quad (10)$$

The hydrogen bond energies determined using this equation for several common solvent systems are given in Table 3.

Wendt et al. found that the calculated binding energy between methanol and one solvent molecule provides a reasonable estimate of the H bond energy for methanol in solution (*25*). This model has been used as part of this exercise by having the students calculate the OH chemical shifts and the binding energy for the hydrogen bonded interaction between a methanol and a solvent molecule. The structures are built with the OH proton in methanol hydrogen bonding to the O, N, or Cl in the solvent molecule and the energy is minimized. The hydrogen bond energy is determined by subtracting the energy of the methanol-solvent pair from the energy calculated for methanol and the solvent molecule. This produces a positive number that represents the hydrogen bond energy. To simplify the calculations, ethane is used in place of cyclohexane and dimethyl ether is used in place of THF. The $CHCl_3$-methanol complex is more stable when the H on $CHCl_3$ bonded to the oxygen on the methanol than when the H on methanol is bonded to the chlorine on $CHCl_3$. An average of the two structures is used since it gives better approximation to the interactions expected to occur in the solution. The NMR spectrum is then calculated and compared to the observed spectrum to test the model. The results obtained for the 6-311G^{++} (3df.3pd) calculations are compared to the experimental values determined using Equation 10 in Table 3. The agreement is probably too good considering the simplicity of the model used to determine the experimental values and that zero point and other similar corrections normally used in the best calculations were not done. There are also undoubtedly other interactions occurring in the solution that are being ignored in the calculations. A possible example of these other interactions would be the ring currents in toluene. This is one possible reason why the calculated and measured values do not agree as well for this system since the results for the hydrogen bond energies determined using the 6-31G^{++} basis set are ~ 50% larger than the values calculated using the 6-311G^{++} (3df.3pd) basis set.

Table 3. Comparison of the experimentally determined (obs) OH Chemical Shifts (in ppm) and hydrogen bond energy (in kJ mol^{-1}) for methanol in solution and those calculated using DFT-B3LYP with the 6-311G^{++}(3df, 3pd) basis set

Solvent	*δ OH*		*H Bond Energy*	
	Obs	Calc	Obs	Calc
Vapor	(-0.2)[a]	-0.23	0	0
Cyclohexane	-0.055	-0.15	0.96	0.97
Toluene	-0.16	-0.30	0.26	8.45
Chloroform	0.92	0.38	7.39	9.67
CH_3CN	2.17	2.17	15.7	18.9
THF	3.02	3.04	21.2	20.4
Acetone	3.12	4.18	21.9	23.8
DMSO	4.09	3.47	28.3	29.1
Liquid	5.15[a]	----	35.3[a]	----

[a] assumed.

Methanol–Acetone Activity Coefficients at Infinite Dilution

Xu et al. (*32*) coupled the observed OH chemical shift ($\delta_{i,obs}$) with the local composition model presented by Deng et al. (*33*) to measure activity coefficients at infinite dilution. The solution is assumed to only contain two species: the molecule of interest i surrounded by solvent molecules j and molecule of interest i only surrounded by other i molecules. The observed chemical shift is given by the weighted average of the chemical shifts for these species

$$\delta_{i,obs} = \delta^{P}_{i}\ \Phi_{ii} + \delta^{S}_{i}\ \Phi_{ji} \tag{11}$$

where δ^{P}_{i} and δ^{S}_{i} are the chemical shifts for the pure substance i and for i in an infinitely dilute solution respectively. Φ_{ii} and Φ_{ji} are the local volume fractions of i surrounded by i molecules and of i surrounded by the solvent j. The value for δ^{S}_{i} can be determined by extrapolating $\delta_{i,obs}$ to zero concentration and δ^{P}_{i} is taken as the chemical shift observed for the neat liquid. The probability of finding a molecule j around a central molecule i can be defined in terms of the bulk mole fractions and the interaction energies ε_{ji} and ε_{ii} for i-j and i-i interactions respectively. Defining Λ_{ji} in terms of the interaction energies and molar volumes (v_i) gives

$$\Lambda_{ji} = v_j^L / v_i^L \exp [- (\varepsilon_{ji} - \varepsilon_{ii}) / RT] \quad (12)$$

The local volume fractions can be expressed in terms of the bulk mole fractions and Λ_{ji}

$$\Phi_{ii} = x_i / (x_i + \Lambda_{ji} x_j) \text{ and } \Phi_{ji} = \Lambda_{ji} x_j / (x_i + \Lambda_{ji} x_j) \quad (13)$$

Substitution of equation (13) into equation (11) gives an equation that can be used to determine Λ_{ji}.

$$\delta_{i,obs} = \delta^P_i x_i / (x_i + \Lambda_{ji} x_j) + \delta^S_i \Lambda_{ji} x_j / (x_i + \Lambda_{ji} x_j) \quad (14)$$

The activity coefficient at infinite dilution (γ_i) is then given by

$$\ln (\gamma_i) = - \ln (\Lambda_{ji}) + 1 - \Lambda_{ij} \quad (15)$$

and the difference in interaction energies can be calculated using equation (12).

The methanol –acetone solution is a good solution for measuring the activity coefficient using this method in the physical chemistry lab. As shown in Figure 4, the chemical shift changes in a nearly linear fashion with concentration. Since Wendt et al (*24*). have shown that this trend is followed over the complete concentration range, it is possible to generate a fairly accurate curve for the OH chemical shift versus concentration from 5 to 10 measurements - minimizing the number of solutions and NMR spectra needed to produce a reasonable set of data. If the relationship between chemical shift and mole fraction were rigorously linear, the solution would be ideal and $\Lambda_{ij} = \Lambda_{ji} = 1$ giving $\gamma_i = 1$. A fit of the data in the dilute methanol region using equation (14) indicated $\Lambda_{ji} = 1.05$. Assuming Λ_{ij} can be approximated as $1/\Lambda_{ji}$ for this system, Λ_{ij} is 0.952. This produces a value for $\gamma_i = 0.999$, and the differences in interaction energies determined using equation (12) is 1.5 kJ mol^{-1}.

Conclusions

Modern high field NMR spectrometers can obtain the ^1H spectrum of very dilute solutions and vapor molecules using common NMR sample tubes, thus making it a valuable instrument for use in the physical chemistry laboratory. Coupling the NMR spectrum with DFT calculations further expands the experiments that can be done, and exposes the students to the power (and limitations) of these techniques. The exercises described earlier integrate NMR spectroscopy and DFT-B3LYP calculations and show the synergy between these techniques. While each of these exercises has been presented using methanol, other alcohols were used for each as a part of two undergraduate projects. It should be noted that alcohols may not be the best choice for some of these exercises. For example, acetic acid is probably a better choice for determining the thermodynamics of dimerization since the vapor contains a larger fraction of dimer and a smaller fraction of larger clusters. It is likely that each of these exercises could be expanded to make measurements on different systems, at different temperatures, or in ways we haven't considered. One final word of caution, while students responded well to these experiments and their use of state of the art instrumentation to make research quality measurements, our students rebelled when they thought we were doing too many NMR experiments during the semester. While all NMR all the time is not recommended, including a meaningful NMR experiment or two into the physical chemistry lab is highly recommended.

Acknowledgments

We gratefully acknowledge the Research Corporation Departmental Development Grant #7957, the NSF-REU-CHE-1062629, the DOD-ASSURE/ NSF-REU- DMR-0851367 and the JMU Department of Chemistry and Biochemistry for supporting this research.

References

1. See *Modern NMR Spectroscopy in Education*; Rovnyak, D., Stockland, R., Jr., Eds.; ACS Symposium Series 969; American Chemical Society: Washington, DC, 2007.
2. Drahus, C.; Gallaher, T. N.; DeVore, T. C. In *Modern NMR Spectroscopy in Education*; Rovnyak, D., Stockland, R., Jr., Eds.; ACS Symposium Series 969; American Chemical Society: Washington, DC, 2007; pp 143−154.
3. Grushow, A.; Zielinski, T. J. *J. Chem. Educ.* **2002**, *79*, 707–714.
4. Grushow, A.; Sheats, J. E. In *Modern NMR Spectroscopy in Education*; Rovnyak, D., Stockland, R., Jr., Eds.; ACS Symposium Series 969; American Chemical Society: Washington, DC, 2007; pp 128−142.
5. Gaede, H. C. In *Modern NMR Spectroscopy in Education*; Rovnyak, D., Stockland, R., Jr., Eds.; ACS Symposium Series 969; American Chemical Society: Washington, DC, 2007; pp 176−189.
6. Evans, D. F. *J. Chem. Soc.* **1959**, 2003–2005.
7. Deutsch, J. L.; Poling, S. M. *J. Chem. Educ.* **1969**, *46*, 167–168.
8. Crawford, T. H; Swanson, J. *J. Chem. Educ.* **1971**, *48*, 382–386.
9. Garland, C. W.; Nibler, J. W.; Shoemaker, D. P. *Experiments in Physical Chemistry*, 8th ed.; McGraw-Hill: Boston, 2009; pp 371−379.
10. Gottlieb, H. E; Kotlyar, V.; Nudelman, A. *J. Org. Chem.* **1997**, *62*, 7512–7515.
11. Levine, I. R. *Physical Chemistry*, 6th ed.; McGraw Hill: Boston, 2009; pp 711−717.
12. Atkins, P. W.; DePaula, J. *Physical Chemistry*, 9th ed.; Freeman: New York, 2010; p 404.
13. Engel, T.; Reid, P. J. *Physical Chemistry*; Pearson: San Francisco, 2006; pp 614−615.

14. Cheeseman, J. R.; Trucks, G. W.; Keith, T. A.; Frisch, M. J. *J. Chem. Phys.* **1996**, *104*, 5497–5509.
15. Schreckenbach, G.; Ziegler, T. *Theor. Chem. Acc.* **1998**, *99*, 71–82.
16. Krykunov, M.; Ziegler, T.; Van Lenthe, E. *J. Phys. Chem.* **2009**, *A113*, 11495–11500.
17. Loibl, S.; Manby, F. R.; Schutz, M. *Molec. Phys.* **2010**, *108*, 477–485.
18. Burris, A.; Hause, C. D. *J. Chem. Phys.* **1943**, *11*, 442–445.
19. Weltner, W., Jr.; Pitzer, K. S. *J. Am. Chem. Soc.* **1951**, *73*, 2606–2609.
20. Kretschmer, C. B.; Wiebe, R. *J. Am. Chem. Soc.* **1954**, *76*, 2579–2583.
21. Kell, G. S.; McLaurin, G. E. *J. Chem. Phys.* **1969**, *51*, 4345–4352.
22. Renner, T. A.; Kucera, G. H.; Blander, M. *J. Chem. Phys.* **1977**, *66*, 177–184.
23. Goodman, R. D. *J. Phys. Chem. Ref. Data* **1987**, *16*, 799–892.
24. Ludwig, R. *Chem. Phys. Phys. Chem.* **2005**, *6*, 1369–1375.
25. Wendt, M. A.; Meiler, J.; Weinhold, F.; Farrar, T. C. *Molec. Phys.* **1998**, *93*, 145–151.
26. Deng, J. O.; Lipson, R. H. *Can. J. Chem.* **2006**, *84*, 886–892.
27. Kostko, O.; Belau, L.; Wilson, K. R.; Ahmed, M. *J. Phys. Chem. A* **2008**, *112*, 9555–9562.
28. Ammann, C.; Meier, P; Merbach, A. E. *J. Magn. Reson.* **1982**, *46*, 319–321.
29. webbook.nist.gov (accessed July 2012).
30. Provencal, R. A.; Casaes, R. N.; Roth, K.; Paul, J. B.; Chapo, C. N.; Tschumper, G. S.; Schaefer, H. F., III *J. Phys. Chem A* **2000**, *104*, 1423–1429.
31. Kristiansson, O. *J. Molec. Struct.* **1999**, *477*, 105–111.
32. Xu, Y.; Li, H.; Wang, C.; Ma, L.; Han, S. *Ind. Eng. Res.* **2005**, *44*, 408–415.
33. Deng, D.; Li, H.; Yao, J.; Han, S. *Chem. Phys. Lett.* **2003**, *376*, 125–126.

Chapter 16

^{1}H NMR MAS Investigations of Phase Behavior in Lipid Membranes

Holly C. Gaede*

Department of Chemistry, Texas A&M University, College Station, Texas 77843
***E-mail: hgaede@chem.tamu.edu**

Solid-state nuclear magnetic resonance experiments that characterize the phase behavior of phospholipids are described. In particular ^{1}H magic angle spinning NMR spectroscopy experiments on mixtures of 1,2-dimyristoyl-*sn*-glycero-3-phosphocholine (DMPC) and cholesterol can be used to determine the phases present at different temperatures and compositions.

Introduction

Phospholipids are the most abundant component of biomembranes. As amphipathic molecules with both polar headgroups and non-polar hydrocarbon tails, these phospholipids spontaneously form bilayer sheets 3 - 4 nm thick, with the polar headgroups facing outward toward the water and the hydrocarbon chains forming a hydrophobic core. Another important membrane lipid in mammalian cell membranes is cholesterol. Insertion of the rigid cholesterol molecule into a bilayer can decrease the mobility of the first few hydrocarbon groups along the phospholipid chain, decreasing membrane fluidity. Conversely, cholesterol can interfere with the close packing of these acyl chains, inhibiting its transition to a crystalline state.

Other components of cell membranes include integral and peripheral membrane proteins and carbohydrates. All of the components of the membrane together make up what was termed the "fluid mosaic model." In the late 1980s, the presence of submicron-size domains in cell membranes that are enriched in cholesterol was proposed in areas now called "lipid rafts." It is now clear that lipids are laterally segregated throughout the cell (*1*). Because certain proteins involved in cellular signal processes have been shown to associate with lipid rafts, it is thought that these rafts may play a role in cell signal transduction (*2*). Rafts have also been suggested to play a role in the entrance of certain pathogens, including HIV (*3*), into the cell. Being able to understand and regulate lipid rafts may open methods for preventing or alleviating a variety of infectious diseases. However, the existence of lipid rafts remains controversial, in part because lipid rafts are difficult to observe in actual membranes (*4*).

Lipid Phases

To understand the phenomenon of lipid rafts, simpler model systems without proteins are often studied so as to develop phase diagrams. In biological membranes, lipids are in the liquid crystalline state, where the molecules are free to rotate about their long axes and laterally diffuse in the plane of the membrane. This state has also been referred to as the *liquid disordered state* (l_d), which is characterized by rapid gauche/trans isomerizations of the lipid hydrocarbon chain. If the temperature is lowered below the phase transition temperature, T_m, the lipid enters the gel state, also known as the *solid ordered state* (s_o). The hydrocarbon chain is more rigid and ordered in this state, with strong van der Waals interactions among the chains, and the gauche/trans isomerization is mostly suppressed. The rotational diffusion about the bilayer normal is slow, and the lateral diffusion in the membrane plane is slowed by orders of magnitude compared to the l_d state. In systems containing saturated lipids and cholesterol, a third phase has been observed, called the *liquid-ordered state* (l_o). The exact nature of this state is still under debate, but early evidence shows that cholesterol hinders gauche rotamer formation at certain carbons close to the carbonyl because of the cholesterol's preferred location near the headgroup region. However, rates of rotational and lateral diffusion remain largely unchanged from the l_d state (*5*). Biological rafts are thought to be small accumulations of liquid ordered regions surrounded by pools of liquid disordered region, though these regions are dynamic and the composition may change. The size and lifetime of the rafts in real biological membranes is still unknown.

Several different physical techniques have been used to study the phase behavior of lipids. Nuclear magnetic resonance (NMR) spectroscopy is an attractive method to use since these molecules contain many nuclei that are accessible to NMR spectroscopic investigation, including hydrogen, carbon and phosphorous, and therefore no phase-perturbing labels are required. Though conventional solution-state NMR spectra may be readily obtained for a lipid dissolved in a solvent like chloroform or methanol, these spectra do not give information about the lipids in a membrane environment. Because these bilayer samples do not tumble isotropically like small molecules in solution, conventional solution-state NMR spectra will show broad featureless lines because of the anisotropic chemical shift and dipole-dipole interactions. However, solid-state NMR techniques, such as magic angle spinning, can be used to modulate these interactions, discussed in detail below, to gain structural and dynamic information.

This chapter will outline a series of NMR experiments used in a physical chemistry laboratory, but appropriate for other upper division laboratories, that investigates the phase behavior of the lipids 1,2-dimyristoyl-*sn*-glycero-3-phosphocholine (DMPC) and cholesterol using magic angle spinning NMR spectroscopy. These experiments not only introduce students to the important biophysical system of lipids, but also expose them to the increasingly important technique of magic angle spinning NMR.

Anisotropic Magnetic Interactions

This section will introduce the important anisotropic –angularly dependent – interactions that, while averaged in the rapid isotropic tumbling of molecules in solution, generally cause additional broadening in solid samples.

Dipole–Dipole Interaction

The *dipolar interaction* results from interaction of one nuclear spin (I_1) with the magnetic field generated by another nuclear spin (I_2), and vice versa. In general, the energy of interaction between two magnetic dipoles $\overrightarrow{\mu_1}$ and $\overrightarrow{\mu_2}$ separated by a distance r is given by

$$E = \frac{\mu_0}{4\pi}[(\overrightarrow{\mu_1}\cdot\overrightarrow{\mu_2})r^{-3} - 3(\overrightarrow{\mu_1}\cdot\vec{r})(\overrightarrow{\mu_2}\cdot\vec{r})r^{-5}], \quad (1)$$

where μ_0 is the permeability constant. The Hamiltonian is obtained by replacing the vectors $\overrightarrow{\mu_1}$ and $\overrightarrow{\mu_2}$ by their corresponding operators $\gamma_1\hbar\vec{\hat{I}}_1$ and $\gamma_2\hbar\vec{\hat{I}}_2$, where γ_1 and γ_2 are the gyromagnetic ratios of nuclei 1 and 2.

Specifically, the interaction between dipolar coupled spins is governed by the dipolar Hamiltonian.

$$H_{DD} = \frac{\hbar^2\mu_0\gamma_1\gamma_2}{4\pi r_{12}^3}\left[\left(\vec{\hat{I}}_1\cdot\vec{\hat{I}}_2\right) - 3\frac{\left(\vec{\hat{I}}_1\cdot\vec{r}_{12}\right)\left(\vec{\hat{I}}_2\cdot\vec{r}_{12}\right)}{r_{12}^2}\right]. \quad (2)$$

Equation 2 can be expanded into several terms, each containing a spin part and a spatial part that can be considered separately. For homonuclear spins (e.g., both protons) in high field, it simplifies to

$$H_{DD} = \frac{\hbar^2\mu_0\gamma_1\gamma_2}{4\pi r_{12}^3}\frac{1}{2}\left(3\cos^2\theta - 1\right)\left[3\hat{I}_{1z}\hat{I}_{2z} - \vec{\hat{I}}_1\cdot\vec{\hat{I}}_2\right] \quad (3)$$

The group of constants is called *dipolar coupling constant* and is defined as

$$R_{DD} = \frac{\hbar^2 \mu_0 \gamma_1 \gamma_2}{4\pi r_{12}^{\ 3}} = \frac{\hbar \mu_0 \gamma_1 \gamma_2}{8\pi^2 r_{12}^{\ 3}} (in \quad Hz) \tag{4}$$

In solution molecules reorient quickly and the nuclear spins feel a time average of the spatial part of the dipolar interaction over all orientations, which is zero. Therefore, the dipolar effects only manifest themselves through relaxation. These relaxation effects are exploited in important two-dimensional NMR spectroscopic techniques that include NOESY. However, the dipolar interaction can cause severe broadening in solids, where every spin is coupled to every other spin. The degree of broadening depends on the nuclei involved and the distance between them (Table 1).

Table 1. Dipolar Coupling Constants, R_{DD}

Nuclei	*Distance (Å)*	*Coupling (kHz)*
^{1}H - ^{1}H	1	120
^{1}H - ^{13}C	1	30
^{1}H - ^{13}C	2	3.8

The coupling between two protons is typically 120 kHz, equivalent to 240 ppm in a spectrometer operating with a proton frequency of 500 MHz! This broadening obscures the much smaller chemical shift and *J*-coupling interactions which are exploited in solution for structural determination. However, this dipolar coupling can also be a rich source of information.

Chemical Shift Anisotropy

Another important source of broadening in solids arises from *chemical shift anisotropy*. Recall from solution NMR that the nucleus is shielded from the external magnetic field by the electrons. This phenomenon is why an aldehyde proton has a different characteristic frequency than a methyl proton. The chemical shift is critically important in structural elucidation. In solids, both the local electronic environment *and* the orientation of the molecule relative to the magnetic field play a role in the amount of shielding experienced by a nucleus. The orientation is important because the electronic distribution around a nucleus within a molecule is rarely spherically symmetrical.

In general, a nuclear environment will have its shielding characterized by three distinct values. These three values are referred to as the *principal components* and occur for orientations specified by the *principal axes*. The three components are summarized by the *shielding tensor*, σ.

$$\sigma = \begin{pmatrix} \sigma_{11} & 0 & 0 \\ 0 & \sigma_{22} & 0 \\ 0 & 0 & \sigma_{33} \end{pmatrix}. \qquad (5)$$

The principal components are assigned such that $\sigma_{11} \leq \sigma_{22} \leq \sigma_{33}$, and the observed shielding σ_{obs} is denoted

$$\sigma_{obs} = \sum_{j=1}^{3} \sigma_{jj} \cos^2 \theta_j , \qquad (6)$$

where the angles, θ_j, are those between the principal components and the magnetic field. In solution, the molecule typically tumbles rapidly and only the average chemical shift is observed. The *isotropic shielding*, σ_{iso}, which is observed in solution NMR, is given by

$$\sigma_{iso} = \tfrac{1}{3}\left(\sigma_{11} + \sigma_{22} + \sigma_{33}\right) \qquad (7)$$

Equation 6 and equation 7 can be combined to give

$$\sigma_{obs} = \sigma_{iso} + \tfrac{1}{3} \sum_{j=1}^{3} \sigma_{jj} \left(3\cos^2 \theta_j - 1\right) \qquad (8)$$

As a result a randomly oriented solid would result in a powder spectrum with signal intensities corresponding to the probability of a particular orientation. However, neglecting the dipolar interaction discussed above, it is possible to observe narrow lines if all the nuclei have the same orientation relative to the magnetic field, as in the case of single crystals or otherwise oriented samples. Only two extrema in the powder patterns will be observed for a molecule with axial symmetry, as is case for ^{31}P NMR of lipid headgroup resonances. A molecule lacking this symmetry will have three extrema.

Quadrupolar Interactions

A third interaction can cause broadening for certain nuclei in solids. The most commonly investigated nuclei in lipids (^{1}H, ^{13}C, and ^{31}P) are all spin 1/2 nuclei, and have a spherical charge distribution. As a result, spin 1/2 nuclei do not interact with *electric field gradients* (spatial variations in the electric field) found in the sample. However, **quadrupolar nuclei** have a spin > 1/2, and a non-spherical electric charge distribution. This anisotropy can cause extra splitting and broadening, but is not a factor in the following experiments.

Magic-Angle Spinning

The anisotropic interactions described above have a ($3\cos^2\theta$-1) term called the second-order Legendre Polynomial, $P_2(\theta)$ that is averaged by rapid isotropic tumbling in solution. The same effect can be achieved in the solid state if the sample is rotated rapidly at an angle relative to the magnetic field such that $\cos^2\theta = \frac{1}{3}$. This angle, θ=54.7°, is called the *magic angle* and as a result the technique is called *magic angle spinning (MAS)*. In order for MAS to be effective, the spinning frequency must be greater than the magnitude of the anisotropic interactions (i.e., if the broadening due to dipolar coupling is 15 kHz, then the spinning frequency must be greater than 15 kHz). If the anisotropic interactions are greater than the spinning frequency, the spectrum will be reduced to the isotropic resonances (called *centerbands*) and *spinning sidebands*, spaced at intervals of the spinning frequencies. These spinning sidebands map out the powder pattern that would be observed in the absence of spinning. These spinning sidebands can be reduced by spinning at a frequency greater than the strength of the anisotropic interactions. MAS and other solid state NMR techniques have been reviewed by Laws (*6*).

For lipid samples in the l_d phase at spinning frequencies of 10 kHz, 99% of the proton signal intensities lie in the centerbands. The centerband integral resonance intensities can therefore be related to the number of contributing protons. At 10 kHz, the methylene resonance at 1.3 ppm has a linewidth of ~50 Hz in the l_d phase. This linewidth increases dramatically for the l_o and s_o phases, as does the intensity of the sideband resonances. Therefore, the intensities of the lipid resonances, as well as the ratio of sideband to centerband intensity, can be used to determine the phase of the lipid system. MAS is useful for determining the presence of l_o and s_o, but it can be difficult to distinguish between those two phases (*5*). One advantage of MAS versus other spectral techniques is that MAS requires no special sample preparation or isotopic labeling.

Note that the achievable spinning frequencies are still well below the magnitude of typical ^1H-^1H dipolar interactions, and so ^1H detection in a MAS experiment is rare. However, dipolar interactions in liquid crystalline membrane samples are partially averaged by molecular motion, so ^1H MAS of these samples allows for high resolution spectra.

Equipment

Special NMR sample containers and probes are required for MAS experiments. These probes are no longer limited to specialty research laboratories, and because of the increasing application of MAS to biological samples they are often referred to as High Resolution Magic Angle Spinning (HR MAS) probes. The sample is packed into a special sample container called a *rotor*, typically a ceramic cylinder with a ceramic or plastic cap with fluted edges. The rotor is spun by passing air or N_2 over the cap. Since the membrane samples are liquid crystalline, the samples need to be loaded with a special insert inside the MAS rotor to prevent sample leakage. The sample volume in the inserts depends on the rotor diameter, which in turn is dictated by the NMR probe. For example, Bruker (Billerica, MA) sells inserts in 12 μL or 50 μL volumes for 4 mm diameter rotors.

The rotor sits inside a housing called a *stator* that is inside the probe and whose angle relative to the magnetic field can be precisely adjusted. The stator contains the NMR coil and delivers the spinning gas. Typically, there are two separate gas lines to achieve rapid spinning of the rotor, called *bearing* and *drive*. The pressure in each of these lines can be controlled with the controller unit in the console. Modern MAS probes monitor the spinning frequency by optical detection of a mark on the rotor. Spinning frequencies of 1-35 kHz can be achieved routinely, depending on the diameter of the rotor.

Experimental

Preparation of DMPC: Cholesterol Samples

1,2-Dimyristoyl-*sn*-glycero-3-phosphocholine (DMPC) and cholesterol (Avanti Polar Lipids; Alabaster, AL) were dispersed from stock solution of chloroform or methanol into a microcentrifuge tube to give a total lipid mass of 5.0 mg. Note that not all microcentrifuge tubes are compatible with chloroform. The solvent was removed with a gentle stream of nitrogen gas and placed under a vacuum for 30 minutes to remove residual solvent. Deuterated water (Aldrich) (5 μL) was added to the dried lipid to make the sample 50 wt% water. The sample was homogenized by using a vortex mixer and quantitatively transferred to the rotor (4 mm Bruker, B3799) using a centrifuge. Samples were prepared in various DMPC:cholesterol mole ratios, ranging from 1:0 to 1:1.

^{1}H MAS NMR Experiments

Instrument

NMR measurements were performed on a Bruker DMX400 widebore spectrometer (^{1}H operating frequency of 400.1 MHz) running TOPSPIN1.2. Sample spinning at 2-10 kHz was accomplished with a Bruker double-gas-bearing MAS probe for 4 mm rotors. The temperature was controlled with the built-in temperature control unit with house nitrogen gas flowing to the stator through a heat exchanger containing liquid nitrogen.

Spinning-Speed Dependence

^{1}H NMR spectra were acquired at 25°C at spinning frequencies of 2–10 kHz on a 1:1 DMPC:cholesterol sample. 16k data points were acquired in a one-pulse experiment, with a 90° pulse time of 2.5 μs and a spectral width of 60 ppm. 8 scans were acquired with a 3 s delay between scans.

Temperature Dependence

^1H NMR spectra were acquired at several temperatures from 15–32°C at spinning frequencies of 5 kHz on a number of DMPC:cholesterol samples, with mole ratios ranging from 1:0 to 1:1. 16k data points were acquired in a one-pulse experiment, with a 90° pulse time of 2.5 μs and a spectral width of 11 ppm. 8 scans were acquired with a 3 s delay between scans. The spectra were acquired in automation, with the temperature being systematically incremented with a 900 s equilibration time before signal acquisition.

Results and Discussion

DMPC Spectral Assignment

The ^1H NMR MAS centerband spectrum of 1,2-dimyristoyl-*sn*-glycero-3-phosphocholine (DMPC) along with the chemical shift assignment is given in Figure 1. Though not as high resolution as a corresponding solution spectrum, different functional groups are resolved clearly. In this case, the chemical shift assignment was made on the basis of comparison to literature assignments of other liquid crystalline lipid spectra.

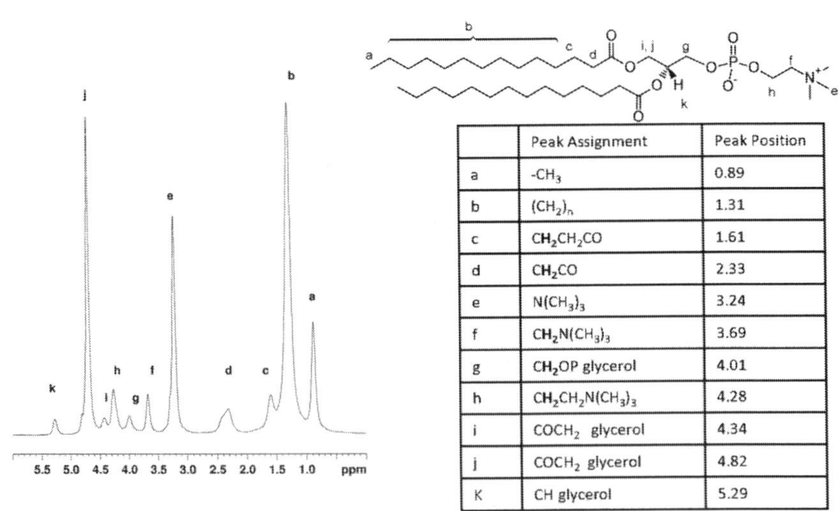

	Peak Assignment	Peak Position
a	$-CH_3$	0.89
b	$(CH_2)_n$	1.31
c	CH_2CH_2CO	1.61
d	CH_2CO	2.33
e	$N(CH_3)_3$	3.24
f	$CH_2N(CH_3)_3$	3.69
g	CH_2OP glycerol	4.01
h	$CH_2CH_2N(CH_3)_3$	4.28
i	$COCH_2$ glycerol	4.34
j	$COCH_2$ glycerol	4.82
K	CH glycerol	5.29

Figure 1. Chemical shift assignment for the 400.1 MHz ^1H MAS NMR spectrum of multilamellar liposomes of 1,2-dimyristoyl-sn-glycero-3-phosphocholine (DMPC) in deuterated water at a rotor spinning frequency of 5 kHz. The structure of DMPC is shown.

Though the length and degree of unsaturation of the hydrocarbon chains can vary in phospholipids, the chemical shifts do not vary much from lipid to lipid. Myristoyl is a saturated 14-carbon chain (14:0), which is convenient to study because of amenable physical properties, including relative stability because of the saturated acyl chains and a near-room temperature phase transition. The solution spectrum has very similar chemical shifts, or if desired, students can perform a series of 2D NMR spectroscopy experiments in the solid-state to fully assign the spectrum instead of relying on a comparison to literature assignments (*7*). It is interesting to note that the methylene in the glycerol group closest to the choline has inequivalent hydrogen atoms. This unusual feature can be difficult to detect since one of the hydrogen atoms (j) is obscured by the resonance signal of HOD.

Effect of Rotor Spinning Frequency

A stacked plot of ^{1}H MAS NMR spectra of a 1:1 DMPC:cholesterol mixture taken at rotor spinning frequencies of 2.5, 5, and 10 kHz at 25°C is shown in Figure 2. The centerband spectrum is surrounded by spinning sidebands at intervals of the rotor spinning frequency. The intensity ratio of the centerband:spinning sidebands increases with an increased spinning speed, and is shown in the inset to Figure 2. This spinning speed dependence can be used to determine the magnitude of the interactions of the anisotropic interactions. Clearly, a rotor spinning frequency of 2.5 kHz is not sufficient to remove the anisotropic interactions of this sample. Note that a cholesterol-containing sample was chosen for this study because the lipid alone has weaker anisotropic interactions, and the spinning sidebands are harder to observe.

Determination of Phase Transition Temperature

The spectrum acquired at 10 kHz is the most desirable as far as having the most spectral intensity in the centerband. However, at these spinning speeds, considerable frictional heating can make the temperature of the cooling gas quite different from the actual temperature of the sample. Accurate measurements of temperature require careful temperature calibration (*7*). In this system, preliminary observation of phase transition behavior of lipids at MAS frequencies of 5 kHz indicated that the thermocouple readings at 5 kHz were accurate, but the actual and recorded temperature varied at MAS spinning frequencies of 10 kHz. Therefore, the study of phase transition was carried out at spinning frequencies of 5 kHz.

Figure 3 shows the temperature dependence of the ^{1}H NMR MAS spectra of samples of (A) DMPC and (B) DMPC with 50% cholesterol.

The sample of DMPC alone shows a dramatic decrease in signal intensity between 296 and 294 K. In samples below 296 K, the only remaining resolved peaks are from HOD and the headgroup methyl groups, which still retain mobility in the solid-ordered phase. In the sample containing cholesterol, the changes are more subtle. The spectral lines are broader from the overlap of cholesterol resonances and the increased anisotropic interactions. The signal intensity, particularly in the acyl chain region, is also reduced in going from high temperature to low temperature, but never disappears.

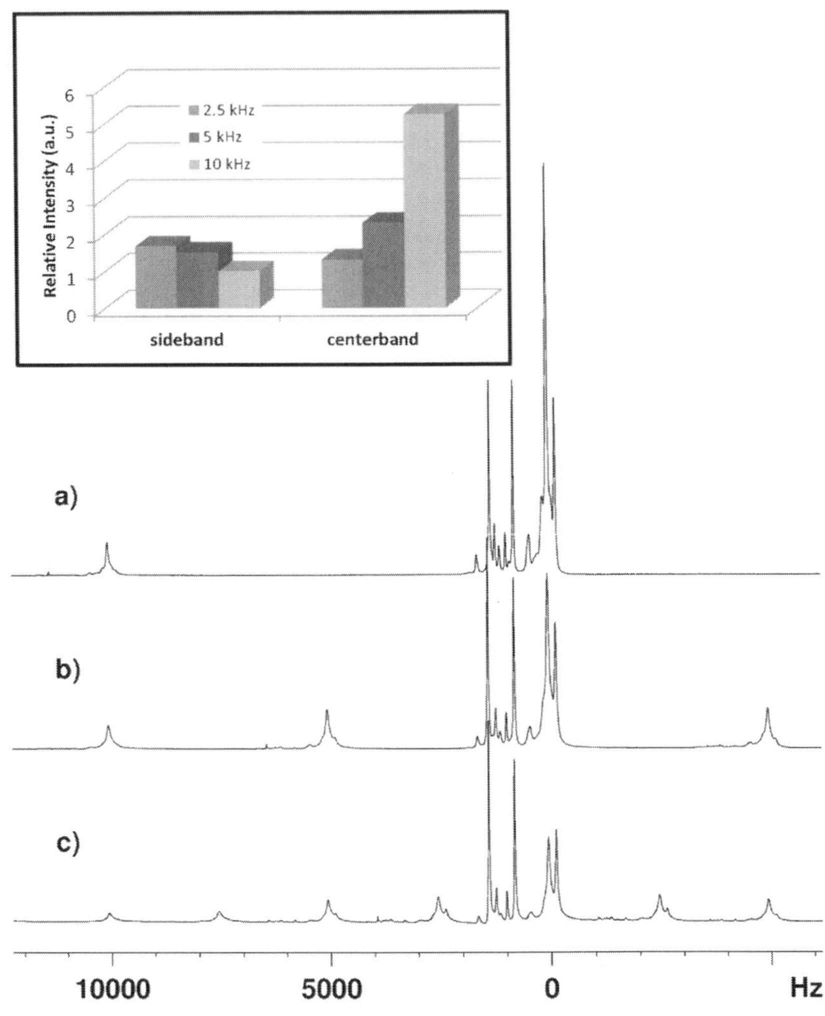

Figure 2. ^{1}H NMR MAS spectra of a 1:1 DMPC:cholesterol mixture taken at 25 °C at rotor spinning frequencies of a) 10, b) 5, and c) 2.5 kHz. The spectra are plotted in Hertz instead of ppm to emphasize the spinning frequency dependence of the spinning sidebands. Inset: Total sideband and centerband intensity as function of rotor spinning speed.

The signal intensity of the $(CH_2)_n$ can be plotted to determine the phase transition temperature, T_m. Figure 4 shows the normalized $(CH_2)_n$ intensity for both the DMPC and the DMPC:cholesterol sample. Again, it is clear that the phase transition between a liquid disordered (l_d) phase and a solid ordered (s_o) phase for DMPC occurs at 295±1 K.

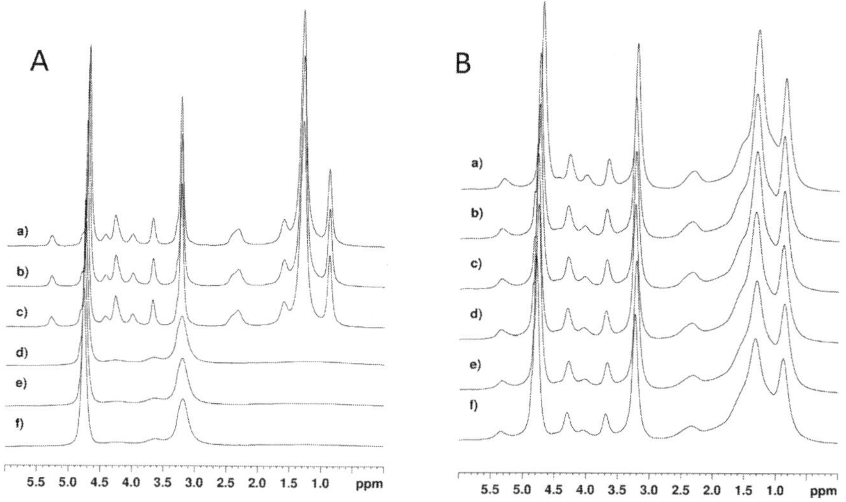

Figure 3. Temperature dependence of ^{1}H NMR MAS spectra of A) DMPC at a) 300 K, b) 298 K, c) 296 K, d) 294 K, e) 292 K, f) 290 K, and B) DMPC:cholesterol at a) 305 K, b) 300 K, c) 297 K, d) 294 K, e) 291 K, f) 288 K.

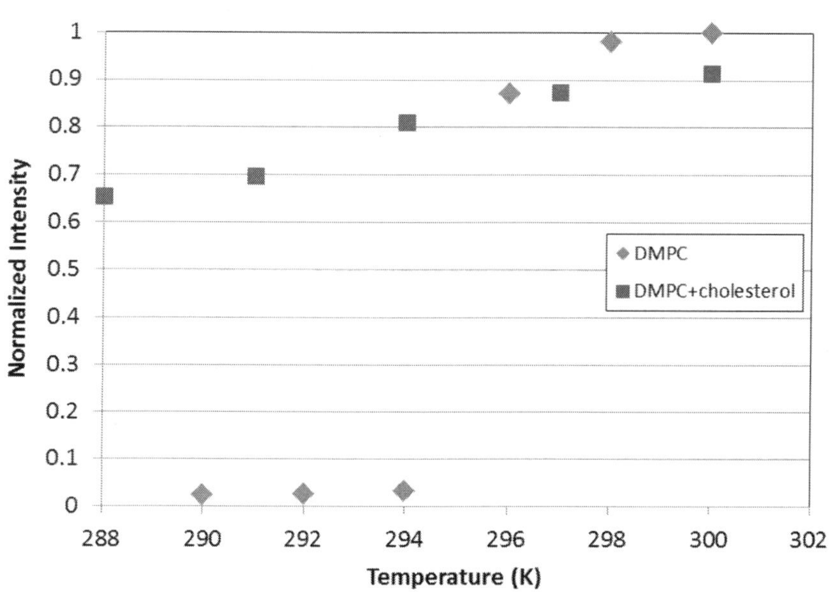

Figure 4. Normalized $(CH_2)_n$ intensity for DMPC and DMPC:cholesterol samples as function of temperature.

This value compares favorably to the literature value of 297 K (*8*). Conceivably, a first-derivative plot could be used to determine the T_m value more precisely; however, there are too few points acquired to benefit from this approach in this example except perhaps as a pedagogical exercise. Acquisition of more temperature points would allow a more precise determination of T_m. Note that temperature equilibration times of longer than 900 s would be desirable to ensure that the sample temperature readings are accurate, but limited laboratory time forced upon us this compromise. Since there are automation programs to systematically increment the temperature, it would be possible to set up the experiments to run overnight, allowing for more data points with a longer equilibration time for each point.

The intensity changes for the cholesterol-containing sample are noticeably different and suggest that there is no phase transition occurring over the temperature range studied. This observation is consistent with the published phase diagram for DMPC:cholesterol (*9*), which shows that at this ratio of cholesterol, the sample is in the liquid ordered phase (l_o) at all temperatures.

Further Studies

This experiment can be expanded in several different ways. As mentioned previously, these solid state experiments can be coupled with solution state experiments aimed at spectral assignment of the phospholipid. Alternatively, additional lipids can be investigated with MAS NMR spectroscopy. The temperature at which the phase transition takes place depends upon the length and the degree of unsaturation of the hydrocarbon chain. Shorter chains have less intense van der Waals interactions and undergo the transition at lower temperature. Likewise, the kinks that result from double bonds disrupt the packing and result in lower phase transition temperatures than saturated chains of comparable length. Lipids with a variety of chain lengths and degrees of unsaturation are available commercially.

Further temperature and cholesterol compositions could be investigated to map out a binary composition diagram (*10*). For saturated lipid cholesterol mixtures, at low temperatures and cholesterol concentrations, the solid ordered (s_o) phase is observed. Raising the temperature above the phase transition temperature (T_m) will result in a transition to the liquid disordered (l_d) state. At high cholesterol concentrations, the phase transition disappears, and only liquid ordered (l_o) phase is observed, regardless of the temperature. At intermediate cholesterol concentrations, a mixture of s_o and l_o phase is observed below T_m, and a mixture of l_o and l_d phases is observed above T_m. Distinguishing between s_o and l_o phase can be difficult with MAS, so other complementary techniques may be used. For example, ^2H wideline NMR spectroscopy has been used to map out lipid-cholesterol phase diagrams (*11*). Other nuclei that can be investigated are ^{31}P, either with or without MAS, and ^{13}C (*12–14*).

Acknowledgments

The author would like to thank the Texas A&M University students in the Physical Chemistry laboratory courses from 2007 – 2012 for their feedback on these experiments. The author gratefully acknowledges helpful discussions with Dr. Sean Bowen, Dr. Christian Hilty and Dr. Ivan Polozov in the development of these experiments.

References

1. Lingwood, D.; Simons, K. *Science* **2010**, *327*, 46–50.
2. Simons, K.; Toomre, D. *Nat. Rev. Mol. Cell Biol.* **2000**, *1*, 31–39.
3. Manes, S.; del Real, G.; Martinez-A, C. *Nat. Rev. Immunol.* **2003**, *3*, 557–568.
4. McMullen, T. P. W.; Lewis, R.; McElhaney, R. N. *Curr. Opin. Colloid Interface Sci.* **2004**, *8*, 459–468.
5. Polozov, I. V.; Gawrisch, K. *Biophys. J.* **2006**, *90*, 2051–2061.
6. Laws, D. D.; Bitter, H. M. L.; Jerschow, A. *Angew. Chem., Int. Ed.* **2002**, *41*, 3096–3129.
7. Gawrisch, K.; Eldho, N. V.; Polozov, I. V. *Chem. Phys. Lipids* **2002**, *116*, 135–151.
8. Mabrey, S.; Sturtevant, J. M. *Proc. Natl. Acad. Sci. U.S.A.* **1976**, *73*, 3862–3866.
9. Almeida, P. F. F.; Vaz, W. L. C.; Thompson, T. E. *Biochemistry* **1992**, *31*, 6739–6747.
10. Ipsen, J. H.; Karlstrom, G.; Mouritsen, O. G.; Wennerstrom, H.; Zuckermann, M. J. *Biochim. Biophys. Acta* **1987**, *905*, 162–172.
11. Hsueh, Y. W.; Zuckermann, M.; Thewalt, J. *Conc. Magn. Reson. A* **2005**, *26A*, 35–46.
12. Gaede, H. C.; Stark, R. E. *J. Chem. Educ.* **2001**, *78*, 1248–1250.
13. Urbina, J. A.; Pekerar, S.; Le, H. B.; Patterson, J.; Montez, B.; Oldfield, E. *Biochim. Biophys. Acta* **1995**, *1238*, 163–176.
14. Zee, B.; Howard, K. Probing the Phase Behavior of Membrane Bilayers Using ^{31}P NMR Spectroscopy. In *Modern NMR Spectroscopy in Education*; American Chemical Society: Washington, DC, 2007; Vol. 969, pp 234–244.

NMR Spectroscopy Across the Undergraduate Curriculum

Chapter 17

Vertical Integration of NMR in the Chemistry Curriculum: A Collaborative Advanced Laboratory Experiment Examining the Structure-Reactivity Relationships in Carbonyl Reduction

Sheila R. Smith* and Simona Marincean

Department of Natural Sciences, University of Michigan- Dearborn, 4901 Evergreen Road, Dearborn, Michigan 48128
***E-mail: sheilars@umd.umich.edu**

With the long-term goal of integrating NMR vertically across the ACS approved chemistry curriculum, we have developed a senior level collaborative experiment that examines the kinetics of the reduction of carbonyls and the structure-activity relationships in that reaction. Through data-sharing in a collaborative research-like experiment, students are able to form a complete picture of the factors that influence the rate of this common organic reaction, while honing their research and communication skills.

Introduction

NMR spectroscopy is a fundamental technique in chemistry. The only explicit instrumentation requirement of ACS for approval of a Bachelor's degree program in chemistry is that of a working NMR instrument (*1*). Though the importance of this technique is unquestioned, the undergraduate curriculum at many institutions tends to ignore much of the facility of the technique beyond organic structural characterization. At these institutions, the only way for a student to learn of the usefulness of NMR spectroscopy for inorganic characterization, kinetics, and even quantitative analysis is through participation in undergraduate research. More importantly, at many institutions that included the University of Michigan-Dearborn until recently, NMR spectroscopy is first introduced at the sophomore organic level and then not used again until advanced synthetic courses or undergraduate research taught at the senior level. At this point, when the need for this technique is at its highest, students' skills are rusty and their knowledge of NMR spectroscopy for any use other than basic structural characterization is lacking. When asked to apply NMR spectroscopy to confirm their products or to solve other chemical problems, a worthy pedagogical goal, students are often flummoxed. The need for retraining seriously limits our ability to accomplish the learning objectives that we set out for the advanced courses. Thus we have begun to explore the idea of applying a vertical integration to the technique of NMR spectroscopy.

Vertical Integration

> "Vertical integration can be defined as coordination and sequencing of content areas longitudinally across a curriculum. Often, vertical integration efforts focus on areas that do not fall along traditional disciplinary lines (*2*)."

There are many examples in the literature arguing the effectiveness of vertical integration in disciplines as distinct as medicine (*2*) and chemical engineering (*3*). The approach is considered particularly effective in combination with problem based learning (PBL) techniques and active learning, and therefore should work admirably in laboratory situations. At the same time, there seem to be as many definitions of curricular integration as there are examples. Abbas and Romagnoli write that

> "Integration is defined as the process that requires units of study to work together in a fashion that enables the student to see the application of the theory and/or practice in one unit of study to the theory or practice in another unit of study allowing for construction of knowledge and skills which are applicable in real life situations (*3*)."

By its very nature, chemistry is very vertical; we build on the general chemistry knowledge base in organic, analytical, physical, inorganic and biochemistry, and generally speaking, we have a set order in which we expect students to take their coursework. Vertical integration was the focus of a session at the recent 2012 Biennial Conference on Chemical Education. The description of this session defined vertical integration as

> "the continuous development of specific topics longitudinally through multiple courses. Each subsequent student exposure increases in the level of detail and complexity (*4*)."

This is the definition of vertical integration that we have adopted. Thus our pedagogical objective is to develop a series of NMR spectroscopy experiences from the vantage point of the various disciplines of chemistry that continuously reinforce and expand the students' working knowledge of NMR spectroscopy.

Our recent acquisition of a new Bruker Avance 400 MHz NMR instrument in December 2009 has made it possible for us to engage in a more concerted effort to vertically integrate NMR spectroscopy across the chemistry curriculum. The basics were already in place with the traditional coverage of structural characterization at the sophomore level in organic lecture and laboratory. Our learning objectives for NMR spectroscopy at this level are fairly standard; at the conclusion of these courses students should be able to predict the splitting patterns of simple to more complex organic compounds, along with the chemical shifts associated with specific organic moieties; students should also be able to prepare an NMR sample of their own compound and run a spectrum in order to confirm their product.

The first step of our expansion of NMR spectroscopy in the curriculum was the inclusion of more hands-on NMR spectroscopy experience at the sophomore level; in this aspect, we were actually playing catch-up to many of our peer institutions. The challenges of using our 20 year old 300 MHz NMR spectrometer, purchased in September 1992 with funds from NSF (USE-9152958) had left us with no choice but to limit student access to the senior chemistry majors. Prior to the acquisition of the new instrument, the students in the Organic Chemistry Laboratory prepared samples, but their spectra were run for them by senior student technicians. Presently, students at the sophomore level in the NMR laboratory acquire their own spectra.

Now our goal is to develop laboratory and classroom experiences that can build and expand upon the basic knowledge that students acquire at the sophomore level in a continuous and consistent fashion. Our first effort in this arena has been the development of an experiment aimed at senior level chemistry majors that uses NMR spectroscopy to determine the kinetics of a simple reduction reaction in our coupled Advanced Inorganic and Organic Synthesis and Characterization laboratories. These separate, co-requisite, one credit hour laboratory courses have served as a capstone experience for our ACS approved chemistry curriculum. Each of the two courses occupies an assigned four hour block in our advanced laboratory space. In practice, each student is assigned a hood that is theirs for the entire semester, along with a drawer containing basic equipment and glassware. Additionally, they may check out glassware and equipment from the stockroom. Instrumentation laboratories are available to the students anytime the stockroom staff is available; thus, even though the assigned hours for the two laboratory courses total only eight hours a week, students enrolled in the courses have much greater access to laboratory time and materials. Admittedly, this is a luxury that we can afford because of the manageable size of our graduating class of chemistry majors, typically 12-14 students per year.

Experiments in these courses have emphasized the application of methods (including NMR spectroscopy) learned in previous courses for the characterization of organic and inorganic products. In addition, students learn new approaches to both synthesis (e.g., air-free methodology, one-pot multi-step synthesis) and characterization (e.g., magnetic studies). The courses also strengthen communication and research skills through record maintenance (laboratory notebooks) and oral and written reports. Concomitantly, we hope to use these capstone courses to give students a taste of real world research problems and collaboration. This is a lofty set of pedagogical goals for any course, especially for two single credit courses in the final senior semester, so the success of these courses rely upon providing the students with relevant experiences that capture the students' interest and drive them to invest their time and efforts in the experiments.

Fortunately, the development of a new experiment focused on integration of NMR paralleled nicely with these goals. We were able to design an experiment that:

- applied NMR spectroscopy (and other techniques) to the characterization of chemical compounds (review);
- applied NMR spectroscopy to the new task of determining kinetics in a chemical reaction (new application of NMR spectroscopy);
- required students to develop their own procedure for a microscale reduction, including workup and isolation (research skills);
- required collaborative skills in sharing of results between student groups;
- required communication via both written formal reports and oral presentations.

The result was a nominal two-week collaborative project applying NMR spectroscopy to the kinetics of a reaction to which students had been previously exposed, the reduction of carbonyls by borohydride reagents (*5*).

Carbonyl Reduction Using Borohydride Reagents

> "Research-supportive curricula also build in experiences that provide scaffolding for undergraduate research, allowing students to acquire and practice transferable skills that can be later applied to independent or faculty-student research. A research-supportive curriculum will expose all students to the importance of research and result in students gaining an appreciation for research methodology in their area of study, even if they do not participate in undergraduate research (*6*)."

Carbonyl reduction is an important transformation in organic chemistry since it may lead to a wide range of products (*7*). Since the discovery of $NaBH_4$ in 1953 (*8*), reductions using borohydride compounds (*9*) have evolved into a mature methodology that is applied in academia and industry due to its robustness (*10*). Typically, the reaction consists of treatment of the carbonyl substrate with a borohydride followed by work-up to yield the corresponding alcohol, Scheme 1. It has been reported, based on both experimental and computational endeavors, that the reaction is first order in carbonyl and that the rate determining step is the attack of the first hydride (from borohydride). Additionally, the reaction rate has been found to be sensitive to several aspects: the structure of the carbonyl substrate, borohydride's counterion, and solvent. If the carbonyl compound possesses moieties capable of dihydrogen bonding (*11*) with the reducing borohydride, the activation barrier is lowered by roughly 6 kcal/mol (*12*). Acceleration is observed when the reducing agent is $NaBH_4$, as opposed to tetrabutylammonium borohydride (NBu_4BH_4), due to complexation between the sodium cation and the oxygen of the carbonyl or other heteroatom present in the carbonyl's structure (*13*). Protic solvents are also included in the reaction rate law (*14*). The reaction rates have been measured for several categories of carbonyl compounds over a wide pH range (*14–17*).

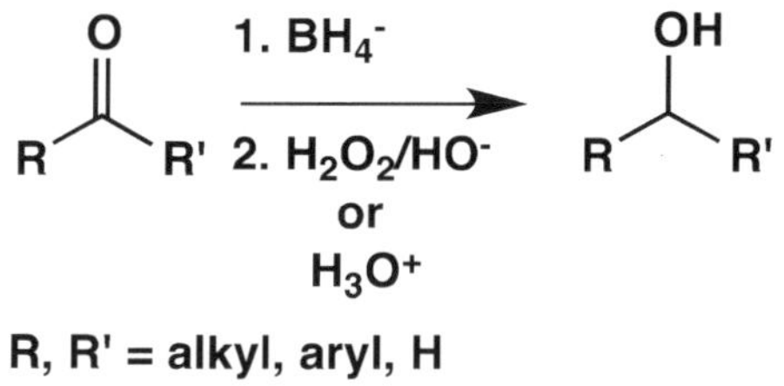

$$RC(=O)R' \xrightarrow[\text{2. } H_2O_2/HO^- \text{ or } H_3O^+]{\text{1. } BH_4^-} RCH(OH)R'$$

R, R' = alkyl, aryl, H

Scheme 1. Reduction of carbonyl compounds by borohydride reagents

Based upon the mechanistic and kinetic information available in the primary literature, we seized the opportunity to use this reaction to develop an experiment in which the structure-reactivity relationship could be examined through *in situ* NMR spectroscopy measurements. The novelty of the experiment is twofold. In contrast with previous experiments regarding reduction of carbonyl compounds by borohydride reagents developed for undergraduate curricula, the current experiment focuses on physical organic chemistry concepts rather than synthesis or green chemistry (*18–20*). This experiment also allows for the use of NMR spectroscopy for an application other than structure elucidation, in this case kinetic determinations. At the same time we recognized that in order to generate a relevant set of data that would allow for trends to be examined, we needed to make a judicious selection of the reducing reagents, solvents, and carbonyl substrates combinations. With respect to borohydride, we considered two choices, $NaBH_4$ and NBu_4BH_4, as the effect of complexation between the counterions could be evaluated. In the solvent case, the choice was apparently straightforward, protic or aprotic (i.e., the ability to interact via hydrogen or dihydrogen bonding with either carbonyl substrate or borohydride, respectively). We chose methanol and dichloromethane as they are available in deuterated versions at accessible prices. However, we were faced with an immediate challenge; a direct comparison between the two hydrides could not be made since $NaBH_4$ is insoluble in CH_2Cl_2/CD_2Cl_2 in any significant amount. Additionally, the reduction of a substrate like acetone, which does not contain activating functional groups, was too fast for effective monitoring when $NaBH_4$ in methanol was used. We considered using a different alcohol solvent such as ethanol or iso-propanol, but the combination of the complexity of the NMR spectra of the reaction mixture and costs led to the decision to focus on NBu_4BH_4 as the reducing borohydride with two solvents, CH_3OH and CH_2Cl_2 and their respective fully deuterated versions. Our reference system was acetone/ NBu_4BH_4/CH_2Cl_2 in which no intermolecular interactions such as hydrogen/dihydrogen bonds are possible between the two components. Substitution of CH_2Cl_2 with CH_3OH does allow for hydrogen and dihydrogen bonds between carbonyl's oxygen and solvent and borohydride and solvent, respectively. Selection of the carbonyl substrates involved several criteria: nature of the carbonyl functionality (aldehyde or ketone), unsaturation or aromaticity in the carbon backbone, and presence of moieties that could participate in intermolecular interactions such as hydrogen or dihydrogen bonds (see Table 1). In order to isolate the effect of the structure of the carbonyl on the reaction rate, the other two components, NBu_4BH_4 and CH_2Cl_2, were kept constant. To assess the aldehyde versus ketone effect, the compared substrates were benzaldehyde and acetophenone, benzaldehyde and benzophenone, and 4-hydroxybenzaldehyde and 4-hydroxyacetophenone. The students examined the role played by presence of C=C bonds in the carbonyl's backbone by comparisons between acetone and benzaldehyde on one side, and methyl vinyl ketone and *trans*-cinnamaldehyde on the other side. The methyl vinyl ketone and *trans*-cinnamaldehyde are able to undergo competing Michael addition reactions, but we did not examine the extent of this reaction as the students based their calculations on the decay of the carbonyl. The effect of aromaticity was evaluated through substrates such as acetophenone and benzophenone. Finally, the students compared substrates that contained functional groups that were able to participate only in hydrogen but not dihydrogen bonds: chloroacetone or dihydrogen bonds: hydroxyacetone to acetone.

Table 1. Examined Trends

Entry	*Trend*	*Compared Pair*
1	Structure of carbonyl: Ketone/aldehyde	Acetophenone/Benzaldehyde
2	Structure of carbonyl: Electron withdrawing group	Chloroacetone/Acetone
3	Structure of carbonyl: Dihydrogen bonding	Hydroxyacetone/Acetone
4	Structure of carbonyl: Alkyl/Aryl	Acetone/Acetophenone
5	Structure of carbonyl: Alkenyl/Alkyl	Acetone/Methyl vinyl ketone
6	Structure of carbonyl: Aryl/Alkenyl	Benzaldehyde/trans-Cinnamaldehyde
7	Solvent	CD_2Cl_2/CD_3OD

Clearly, no single student can collect enough data to analyze all of these effects in the two week period allotted for the experiment. Instead, analysis of the data required collaboration between students, Table 1. The student pairs for comparison were set such that each student discussed their data in reference to those obtained by a colleague and had their data serve as the control system for a different colleague so that no overlaps occurred. In order to add a layer of complexity and have the students refresh their skills at structural determination by NMR, the carbonyl containing compound was handed out as an unknown.

Several learning objectives were set for the experiment:

- Students were expected to apply NMR and IR spectroscopy to the elucidation of the structure of both their unknown substrate and to their product.
- Students were expected to independently develop an experimental procedure based on available primary literature when provided with a list of reagents.
- Students were expected to determine experimentally the kinetic characteristics (e.g., reaction order with respect to carbonyl and reaction rate constant) using NMR spectroscopy.
- Students were expected to analyze and determine structure-reactivity trends for the reaction in question through collaboration with their peers;
- Students were required to present their results through both an oral peer-reviewed presentation and written formal report.

Developing an Experimental Procedure

The students were given a sample with a carbonyl unknown, along with its percent composition and asked to determine its structure based on NMR and IR spectra, and melting or boiling point. Once the carbonyl substrate was identified, sets of conditions (borohydride reagent and solvent) were assigned and the students were charged to develop experimental procedures for the reaction, complete with isolation and characterization of the expected alcohol product. Among the borohydrides, NBu_4BH_4 is generally used less frequently as compared to the $NaBH_4$ or $LiBH_4$, and thus the primary literature was the only source for reports of reduction using NBu_4BH_4. This allowed for a setting in which the students had to consult the literature and adapt the experimental procedures found there to their specific system. Upon approval from the instructor, the students ran a microscale experiment and characterized the isolated product.

Subsequently, the students performed kinetic studies using 0.25 M concentrations of both carbonyl and borohydride reagent in the deuterated version of the solvent that they used in the microscale procedure. The students were responsible for both the selection of the signal that was to be monitored throughout the NMR determination and the time intervals at which the spectra were collected. However, instructors verified that the chosen signal that was to be followed throughout the reaction was relevant to the kinetic study. If this was not the case, then the student was led, via discussion, towards a more appropriate signal. With regards to reaction time, the students were informed that it could be between 15 minutes and 24 hours at room temperature and that they should collect spectra at time intervals such that relevant change was observed. The experimental procedure, information regarding recommended time intervals for data collection, and monitored signals for different substrates, has been reported elsewhere (*5*). Finally, the reaction rate constants were shared among the class through a Google Doc and on the online course management website associated with the course. Each student was assigned a colleague that had a system in which one of the two components, carbonyl, and solvent was identical, and asked to interpret the difference in rate constants in both a peer-reviewed oral presentation and formal written report.

This experiment is a superb one for our capstone laboratory courses, as it requires students to apply knowledge of kinetics that they have acquired in their general, biochemistry and physical chemistry courses. In fact, kinetics is a topic which is already vertically integrated into the curriculum. In General Chemistry, we teach the three-plot method for determining the order of a reaction with respect to a certain reactant, and students in the General Chemistry program at UM-Dearborn put this theory into practice with the classic iodine starch clock reaction. In Biochemistry lecture, they learn of Michaelis-Menten kinetics, and in Physical Chemistry laboratory they monitor the kinetics of a reaction using the volume of a gas produced. In the current experiment, students must make the connection between NMR peak height and amount of analyte in order to measure the kinetics of a classic organic chemistry reaction; they have to start thinking of NMR as an analytical technique.

Results and Discussion

This experiment was performed as a common project in our advanced laboratories over two semesters, and the reaction rate constants for the successful runs are presented in Table 2. The results are based on average of three student runs for each substrate and solvent pair.

The structure elucidation of the carbonyl unknown posed little challenge to the students, as the ^1H NMR spectra and the percent composition were sufficient for this task. This stage of the experiment served as a preliminary step in which the students' practical skills regarding NMR sample preparation, as well as spectral acquisition, processing, and interpretation were developed. More than one student was asked to perform additional spectra acquisition due to poor quality spectra. We addressed these issues through group discussion during which possible causes such as improper sample preparation and poor shimming were tackled.

Development and execution of the experimental microscale procedure based on primary literature served both to develop their independent investigative skills and to offer a research-like experience. The students located without difficulty relevant papers regarding use of NBu_4BH_4 as a reducing agent for carbonyl substrates (typically, references (*21–23*) at the time of the experiment, and very recently 5 and 12). The procedures that were outlined by the students contained the required steps including the work-up but in several cases did not have the level of detail required (e.g., moles and amounts of reagents adjusted for the specific substrate). Upon running their individual reaction, the students isolated and characterized the alcohol product. We emphasized the importance of identifying all the signals present in the ^{1}H NMR spectrum of the alcohol, as they would give information about not only the alcohol but also the extent of the reaction as well as the success of the work-up and isolation steps. We conducted individual discussions with the students to ensure that their spectral analysis was correct and detailed since the next step involved the selection of the signal to be monitored for the kinetic study. At this stage we considered the students ready to tackle the experimental core of this project, the kinetic studies of the effects of the structure-reactivity on the reaction rate constants. Upon selection of time intervals and a signal to be monitored, the students collected a set of spectra, Figure 1.

Table 2. Rate constants for reduction reactions

Entry	*Substrate*	*Solvent*	*Rate Constant x 10^6, s^{-1}*
1	Acetone	CD_2Cl_2	10.0±0.2
2	Hydroxyacetone	CD_2Cl_2	2500±110
3	Chloroacetone	CD_2Cl_2	9.4±0.7
4	Acetone	CD_3OD	4700±200
5	Benzaldehyde	CD_2Cl_2	64.0±3.8
6	Acetophenone	CD_2Cl_2	7.3±0.4
7	*trans*-Cinnamaldehyde	CD_2Cl_2	32±2

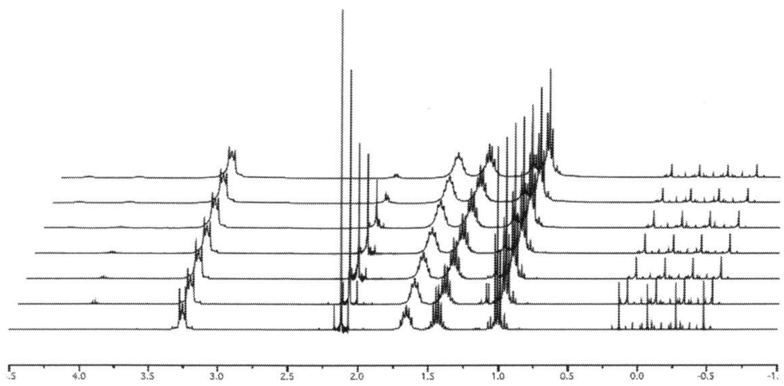

Figure 1. Selected ^{1}H NMR spectra for the reduction of acetone by NBu_4BH_4 in CD_2Cl_2.

Among the systems initially selected, 4-hydroxybenzaldehyde proved to be challenging due to its very low solubility in the solvents of choice. Furthermore, the reduction of 4-hydroxyacetophenone could not be fitted to familiar kinetics plots even after several trials, presumably due to the presence of the phenolic group on the aromatic ring that may be acidic enough to react with the NBu_4BH_4.

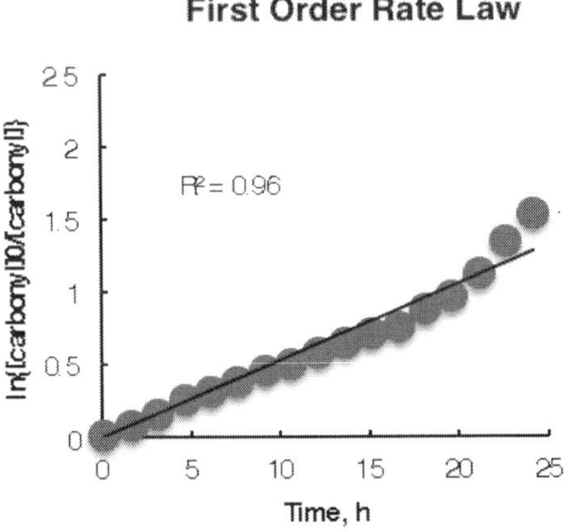

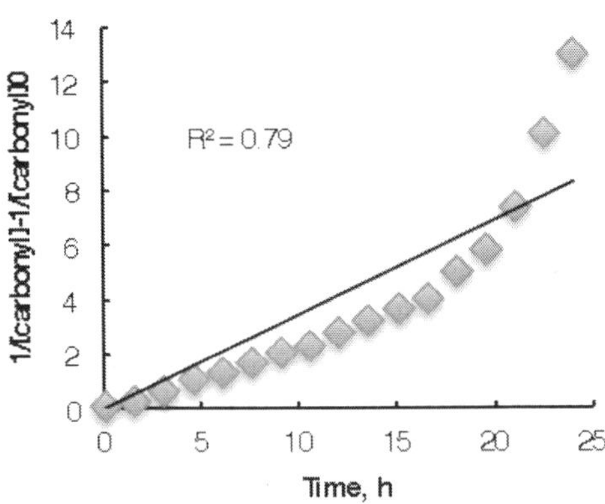

Figure 2. Reaction orders for reduction of acetone by NBu_4BH_4 in CD_2Cl_2.

The first step in the kinetic data analysis was determination of the reaction order with respect to carbonyl. The students were expected to know that the reaction order of 1 or 2, with respect to a particular reagent (the carbonyl in this case), can be determined graphically through plots of $\ln\{[carbonyl]_0/[carbonyl]\}$ versus t, or $\{1/[carbonyl]-1/[carbonyl]_0\}$ versus t. By examining the corresponding representations for reduction of acetone, Figure 2, it can be seen that only the $\ln\{[\text{carbonyl}]_0/[\text{carbonyl}]\}$ versus t can be fitted to a line, suggesting that the reaction is first order with respect to carbonyl. The fast reactions, entries 2 and 4 in Table 2, required additional care because they were completed in less than 30 minutes and fewer spectra could be collected.

Once the rate constants were determined, each student compared his or her data to those of a colleague assigned by the instructor. The structural differences among the carbonyl substrates were examined from the perspective of intrinsic thermodynamic characteristics or electronic effects at the reaction site (i.e., the carbonyl functionality). In the absence of any activating functional groups, an aldehyde such as benzaldehyde was found to be reduced faster by one order of magnitude than a ketone, acetophenone (Table 2, entries 5 and 6, respectively), and the students connected this experimental result to the higher thermodynamic stability of a ketone compared to an aldehyde, a concept routinely taught in sophomore organic chemistry.

The students found that the presence of electron-withdrawing groups, such as chlorine and hydroxyl in close proximity to the carbonyl, have quite different effects on the reduction rate. When compared to a control substrate, acetone, no significant change was noted for the chloroacetone, while a large acceleration for hydroxyacetone was observed. These contrasting results allowed for a group discussion with the goal of reconciling these experimental data. Based on literature reports of rate enhancement by electron-withdrawing groups (methoxy and hydroxyl), when the reducing species is $NaBH_4$ in a protic solvent (*24*), the discussion included concepts such as coordination between the sodium cation and carbonyl's oxygen as a reason for the larger rate (*13*), an interaction that is absent when the borohydride counterion is tetrabutylammonium. This set the stage for the introduction of a novel interaction, dihydrogen bonding, in which conventional proton donors such as hydroxyl or amino moieties interact with hydrides, such as BH_4^- where the H atom is more electronegative than the heteroatom. While the ubiquitous traditional hydrogen bonds are an important topic in the chemistry curriculum, dihydrogen interactions belong to the research realm, where in the last two decades their effects in supramolecular chemistry, catalysis, stereochemistry and selectivity have been widely reported (*11*). The students were able to identify the dihydrogen bonding interaction between the hydroxyl moiety of hydroxyacetone and borohydride and tie it to the rate acceleration compared to the chloroacetone case where such an interaction is not possible. Additionally, the rate acceleration due to a protic solvent versus an aprotic one, entries 4 and 1 Table 2, respectively, was attributed to dihydrogen bonding between the methanol and BH_4^-. The students were able to analyze these trends **only** because each individual contributed her/his data to the group and thus, in the relatively short amount of time of only two weeks, were able to generate a relevant set of data.

The students presented both their experimental data and analysis in an oral presentation, during which they answered questions from their peers and were evaluated by them. The pertinence of the audience questions and comments on the evaluations illustrated students' familiarity with the topic, presumably from the required literature survey and the group discussions.

Performance Evaluation

The students were graded based on achieving the learning objectives enumerated above, with an emphasis on the discussion component of the trends obtained. As mentioned earlier, the first two learning objectives of structure elucidation and experimental procedure development were completed successfully by all students. In regards to the kinetic determinations, a significant portion of the grade was attached to the choice of adequate time intervals between subsequent spectral acquisition and the quality of data that allowed for a good linear fit. The ability to analyze and discuss a specific trend were considered an important aspect of this experiment and graded accordingly in both their oral presentation and formal report. The students were evaluated on their ability to justify their proposed hypothesis using their experimental results and primary literature reports. Finally, the students had the opportunity to evaluate their peers on both the quality of their presentations as well as the ability to answer questions from the audience. While the elocution, length of presentation, and level of detail of background information were seldom criticized, their performance during question/answer session generally received the lowest marks from the audience.

In summary, this experiment exposed students to an application of NMR spectroscopy other than structure elucidation. Additionally, the students had a research-like experience as they developed and/or adapted an experimental procedure based on literature survey, and learned of the importance of both collaboration and interaction with colleagues for successful trend determination.

Conclusions

> "There have been quite a large number of theoretical proposals on curriculum integration but very few deal with the methods to be adopted in the implementation of the integrated curriculum. This is because the implementation represents quite a challenge. It is also difficult to persuade academics to change the way they have been teaching their courses adding the personal dimension to the problem (*3*)."

We feel confident that we have developed an excellent capstone NMR spectroscopy experience for our undergraduates that reinforce and expand upon the knowledge base laid in sophomore organic chemistry. This is only one early step in the integration of NMR throughout the curriculum at the University of Michigan-Dearborn. We tackled these courses first simply because they were the ones in which we had the most control. A continuing goal is to develop more experiments like this one that can challenge seniors to use NMR spectroscopy beyond simple structure determination in research-like environments. We envision several possible modifications to this experiment. The hydroxyacetone reduction was very fast at room temperature compared to acetone for instance, so this substrate may be suitable for a future study of the temperature dependence of the reaction since our instrument is capable of variable temperature experiments.

The effect of the borohydride's cation on the reaction rate could also be explored by comparisons of $NaBH_4$ and NBu_4BH_4 in a polar aprotic solvent (e.g., tetrahydrofuran) in which they are both soluble.

This still leaves the issue of the intervening years between this sophomore experience and the senior capstone experience. Vertical integration requires that the topic be taught at several points throughout the curriculum and that each experience build on the previous ones in complexity. We have been able to affect a change in the sophomore level organic courses and in the senior level advanced labs because these are the courses that we teach. The more pressing question is can we get enough buy-in from our colleagues to make this a truly iterative approach? For us, and this may be a testimony to how fortunate we are in our colleagues, the answer has been yes. In the year after we first introduced this advanced lab experiment, our physical chemistry colleague was convinced to run an experiment where students used NMR spectroscopy and the Evans method to determine the magnetic susceptibility of a range of compounds in the junior level physical chemistry laboratory. This is not a new experiment, but it was new to us at UM Dearborn and especially so to our physical chemistry colleague who has no expertise in NMR spectroscopy. Still, he was persuaded of the importance of including the technique at the junior level and with our assistance plans to include more NMR spectroscopy-based experiments in the physical chemistry laboratory in the future. In fact, the kinetics experiment may shift into this course, with the added temperature dependence component that we described above. One might also imagine experiments where the equilibrium constant of a tautomerization reaction or a ligand substitution reaction could be determined by NMR spectroscopy.

In the classroom, we have begun to include more coverage of NMR spectroscopy in the two inorganic lecture courses, especially that of nuclei other than carbon and hydrogen. Again, we suspect that we were behind many of our colleagues in this respect, but by increasing the exposure to NMR spectroscopy that our students have in their junior level coursework, we can move on to more challenging experiments in the junior and senior level laboratories, and eventually build a curriculum that truly features vertical integration of NMR spectroscopy.

In summary, we have laid a good foundation for continuing to vertically integrate NMR spectroscopy into our curriculum at UM Dearborn and we hope to continue to develop experiments and experiences that benefit our students' knowledge of this important technique.

Acknowledgments

We would like to express our gratitude to the anonymous alumnus whose generous donation allowed for the purchase of our new NMR spectrometer, and to Professor Daniel Lawson for incorporation of new NMR spectroscopy-based experiments in the Physical Chemistry Laboratory. We would also like to acknowledge the efforts of the students who enrolled in our Advanced Inorganic and Organic Synthesis and Characterization courses during the development of the experiment.

References

1. *ACS Guidelines and Evaluation Procedures for Bachelor's Degree Programs*; Committee on Professional Training, American Chemical Society: Washington, DC, 2008.
2. Baum, K. D.; Axtell, S. *Keio J. Med.* **2005**, *54*, 22–28.
3. Abbas, A.; Romagnoli, J. *Trans IChemE, Part D* **2007**, *2*, 46–55.
4. *2012 Biennial Conference on Chemical Education*; http://www.2012bcce.com (accessed December 2012).
5. Marincean, S.; Smith, S. R.; Fritz, M.; Lee, B.J. E; Rizk, Z. *J. Chem. Educ.* **2012**, *89*, 1591–1594.
6. *Characteristics of Excellence in Undergraduate Research*; Hensel, N., Ed.; The Council on Undergraduate Research: Washington, DC, 2012.
7. *Reductions in Organic Synthesis. Recent Advances and Practical Applications*; Abdel-Magid, A. F., Ed.; ACS Symposium Series 641; American Chemical Society: Washington, DC, 1996.
8. Schlesinger, H. I.; Brown, H. C.; Finholt, A. E. *J. Am. Chem. Soc.* **1953**, *75*, 205.
9. Brown, H. C.; Ramachandran, P. V. Sixty Years of Hydride Reductions; In *Reductions in Organic Synthesis*; Abdel-Magid, A. F., Ed.; ACS Symposium Series 641; American Chemical Society: Washington, DC, 1996; pp 1−30.
10. Burkhardt, E. R.; Matos, K. *Chem. Rev.* **2006**, *106*, 2617–2650.
11. Custelcean, R.; Jackson, J. E. *Chem. Rev.* **2001**, *101*, 1963–1980.
12. Marincean, S.; Fritz, M.; Scamp, R.; Jackson, J. E. *J. Phys. Org.* **2012** DOI:10.1002/poc.2986.
13. Suzuki, Y.; Kaeneno, D.; Tomoda, S. *J. Phys. Chem. A* **2009**, *113*, 2578–2583.
14. Wigfield, D. C.; Gowland, F. W. *J. Org. Chem.* **1977**, *42*, 1108–1109.
15. Rickborn, B.; Wuesthoff, M. T. *J. Am. Chem. Soc.* **1970**, *92*, 6894–6904.
16. Geribaldi, S.; Decouzon, M.; Boyer, B.; Moreau, C. *J. Chem. Soc., Perkin Trans. II* **1986**, 1327–1330.
17. Cho, B. T.; Kang, S. K.; Kim, M. S.; Ryn, S. R.; An, D. K. *Tetrahedron* **2006**, *62*, 8164–8168.
18. Cunningham, A. D.; Ham, E. Y.; Vosburg, D. A. *J. Chem. Educ.* **2011**, *88*, 322–324.
19. Peeters, C. M.; Deliever, R.; De Vos, D. *J. Chem. Educ.* **2009**, *86*, 87–90.
20. Pohl, N.; Clague, A.; Schwarz, K. *J. Chem. Educ.* **2002**, *79*, 727–728.
21. Raber, D. J.; Guida, W. C.; Shoenberger, D. C. *Tetrahedron Lett.* **1981**, *51*, 5107–5110.
22. Raber, D. J.; Guida, W. C. *J. Org. Chem.* **1976**, *41*, 690–696.
23. Gatling, S. C.; Jackson, J. E. *J. Am. Chem. Soc.* **1999**, *121*, 8655–8656.
24. Krishnan, K.; Chandrasekaran, J. *Indian J. Chem.* **1982**, *21B*, 595–597.

Chapter 18

NMR Spectroscopy in the Undergraduate Curriculum at the University of Notre Dame

Steven M. Wietstock,* Kathleen A. Peterson, DeeAnne M. Goodenough Lashua, Douglas A. Miller, and James F. Johnson

Department of Chemistry and Biochemistry, University of Notre Dame, Notre Dame, Indiana 46556
***E-mail: swietsto@nd.edu**

The Department of Chemistry and Biochemistry at the University of Notre Dame acquired a Bruker 400 MHz Avance III NMR spectrometer with a 120-position BACS sample changer in the fall of 2006 exclusively for use in the undergraduate teaching laboratories. Since that time, the use of this instrument has been incorporated throughout the curriculum. It is routinely used in the majors' laboratory sequence beginning with General Chemistry and continuing through to its use in more advanced laboratories. In addition, the NMR spectrometer has been used with the non-majors organic chemistry courses. This chapter describes how the instrument is used in these courses, the types of experiments that are performed, how the students are trained in the instrument's use, and how they are introduced to using Topspin, the Bruker NMR spectrometer acquisition and processing program, to perform data analysis on the acquired spectra. The challenges and successes of having this instrumentation in our teaching facility are also described.

The Notre Dame Facilities

In 2006, the University of Notre Dame opened the 202,000 square foot Jordan Hall of Science. This seventy million dollar facility with 40 laboratories, an observatory, and a 136-seat Digital Visualization Theater (with a 50-foot dome screen) is dedicated to undergraduate laboratory teaching for the Departments of Chemistry and Biochemistry, Physics, and Biological Sciences.

One of the items at the top of our instrumentation budget for the new facility was a nuclear magnetic resonance (NMR) spectrometer for the exclusive use in the undergraduate teaching laboratories. The instrument purchased was a 400 MHz Bruker Avance III instrument outfitted with a BBEO broadband probe, 120-position BACS sample-changer robot, and a variable temperature accessory (Figures 1 and 2).

Figure 1. The Avance III 400 MHz NMR console.

The department also owns a 100-seat floating license of Topspin 2.1, Bruker's acquisition and processing software allowing up to 100 concurrent users to access the program at one time. The software is available on all the departmental computers (3 carts each containing 24 computers in the data analysis rooms and 50 desk-top computers located in instrumentation rooms) and can be loaded on departmental majors' personal computers for after-hours use while they are connected to the Notre Dame network. Non-majors use the departmental computers for reprocessing data during laboratory hours and laboratory office hours.

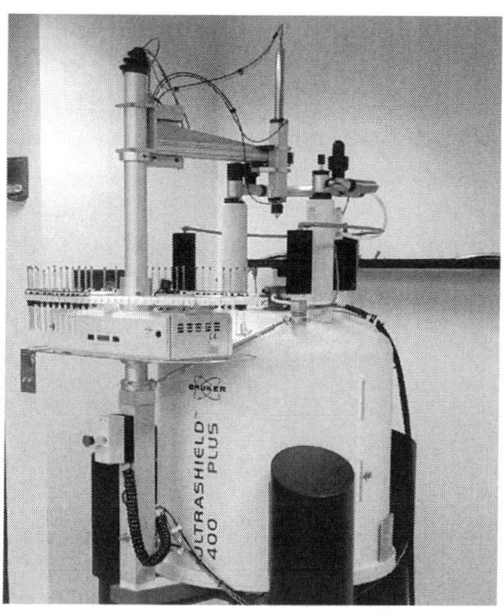

Figure 2. The superconducting magnet and BACS sample changer for the Bruker Avance III NMR.

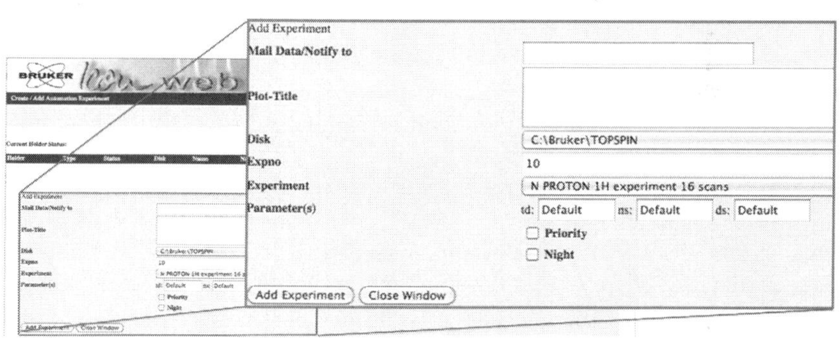

Figure 3. Screen shot of IconNMR sample entry screen.

Each course that uses the NMR spectrometer has a separate account in the system. This allows for the selection of a subset of the experiments and experimental parameters that are relevant for each course. Samples are logged into the master console computer using IconNMR, Bruker's web-based software for setting up and submitting experiments to the spectrometer. There are four pieces of information that a student must enter into the sample screen (Figure 3): their email address, a title/sample description, the experiment, and the sample solvent. This software can be configured to run the samples in numerical order in the sample changer or in the order submitted. Certain samples can be set to run at night or at high priority if needed. At the completion of an experiment, the IconNMR software autoprocesses the data and sends three messages to each student: (1) a status message (experiment completed with either the condition of "done" or "error"), (2) a message with the PDF of the processed data, and (3) a message with the FID in a JDX compressed file. Students open the JDX file with the Topspin software to reprocess the data as needed for their assignment.

Notre Dame Chemistry Courses

In the 2007-2008 academic year, the department moved all the courses for the first two years into a 1:2:1 format. This sequence begins with a semester of general chemistry, two semesters of organic chemistry, and finishes with a semester of intermediate inorganic chemistry (including kinetics and electrochemistry). This followed the same change made to the Chemistry and Biochemistry majors sequence to the 1:2:1 format in the 2004-2005 academic year.

The first course in this sequence, Introduction to Chemical Principles, enrolls about 1000 students each fall. This represents 50% of the entering class of all Notre Dame freshmen. From this number, approximately 100 students start in the majors' track. This "atoms first" curricular approach includes a discussion of molecular spectroscopy that integrates infrared (IR) and NMR spectroscopy and mass spectrometry (MS).

The second course in the sequence, typically taken in the spring of the freshman year, is Organic Structure and Reactivity. This course is representative of a normal organic chemistry course that covers alkanes, alkenes, alkynes, stereochemistry, substitution/elimination reactions, and equilibria. There is little to no discussion of spectroscopy in this course, with an introduction to practical spectroscopy being integrated into the accompanying laboratory course.

In the fall of their sophomore year students take Organic Reactions and Applications. As the title of this course suggests, the majority of the semester is spent covering organic reactions and their applications. A brief discussion of theoretical MS, IR and NMR specotroscopy is revisited early in the semester at a more advanced level than was previously seen in the Introduction to Chemical Principles course. This discussion covers these topics as presented in most standard sophomore level organic texts.

The last course taken by students in the introductory chemistry sequence is Chemistry Across the Periodic Table. This intermediate inorganic chemistry course provides a look at the elements grouped by their block designation – s, p, d, and f blocks. In addition, topics such as electrochemistry and kinetics, that are omitted in the first semester general chemistry course, are covered in this course most often in the context of inorganic chemical reactions. Once students complete this course they are finished with the core courses and there is more flexibility in the sequencing of upper-level chemistry courses and electives.

NMR Spectrometer Use in Majors Courses

Introduction to Chemical Principles Laboratory (Semester 1)

Students are first exposed to principles of NMR, IR spectroscopy and mass spectrometry during the lecture portion of this course near the mid-point of the semester. These concepts are then reinforced in laboratory experiments that occur near the end of the semester when students are given one of ten unknown solids. Those unknowns include adipic acid, 3-(dimethylamino)benzoic acid, 4-fluoroacetanilide, 9-hydroxyfluorene, 5-methylsalicylic acid, naphthoic acid, naproxen, 4-nitrophenylacetic acid, salicylic acid and toluenesulfonamide. In the first period of this two-week experiment the students are asked to obtain a melting point, an IR spectrum, and a ^{1}H NMR spectrum of their compound. During the second week of the laboratory experiment, students are walked through the use of Topspin to process their NMR data. The goal for each student in using the NMR spectrometer in this experiment is to learn how to (i) prepare the sample, (ii) load samples into the NMR robot, (iii) enter experiments into IconNMR, and (iv) process the data in Topspin. The students quickly realize that a melting point will not be of much use since all of the compounds melt between 150 °C and 160 °C. Therefore, students will need to rely on their IR and NMR spectra to determine which other student(s) in their laboratory section have the same compound. This gives instructors the ability to talk about functional groups (to which they have been introduced to in the lecture part of the course) and some very basic peak assignments. This is perennially one of the students' favorite experiments of the semester as they are eager to work with the advanced instrumentation in the laboratory.

Organic Structure and Reactivity Laboratory (Semester 2)

Most organic texts do not delve into a discussion of NMR spectroscopy until the mid-point of the book. Therefore these topics are often not covered until the second semester of organic chemistry. However, students in the majors level Organic Structure and Reactivity Labortory course are introduced to practical aspects of NMR spectroscopy through a staged approach where different aspects of NMR spectroscopy are introduced in sequential laboratory experiments.

The first experiment using NMR spectroscopy in this laboratory course is a two-week experiment that involves the separation of a two-component mixture of fluorene and fluorenone by column chromatography (*1*). Once the students recover the two products from their silica gel column and remove the hexane and ethyl acetate solvents by rotoary evaporation, NMR samples are prepared and ^{1}H and ^{13}C NMR spectra are collected. During the second period of this laboratory, students process the data in Topspin and the entire class works through the peak assignment analysis of the spectra together in a lecture setting. The focus of the discussion is identifying the residual NMR solvent signals, peaks due to solvents used in the experiment, and a generalized discussion of functional group chemical shifts.

The second experiment is a traditional acetylation of ferrocene experiment that uses a microwave as the heating source (*2*). In this two-week project, the students isolate the monoacetylated ferrocene product using column chromatography using conditions that they determined from thin layer chromatography analysis, and obtain the ^{1}H NMR spectrum of their product. The students now have enough experience to do their own peak assignments of their spectrum with the aid of an experimental handout.

In the third experiment students explore how ^{1}H NMR spectroscopy can be used to quantitate the amount of two products using the reaction of iodoethane with the ambident sodium saccharin (*3*). This reaction gives two different acylation products, *O*-ethylsaccharin, and *N*-ethylsaccharin. The students process the NMR data to obtain the integration of the *O*-ethylsaccharin signal at 4.67 ppm and the *N*-ethylsaccharin signal at 3.86 ppm. By comparing the ratios of these two integrations students can quantitate how much of each product was produced in their reaction. Figure 4 gives a sample spectrum from one of the student experiments which show that about 85% of the product is the *N*-ethylsaccharin. This is a one-week experiment, with students processing their NMR data on their own during the following experiment.

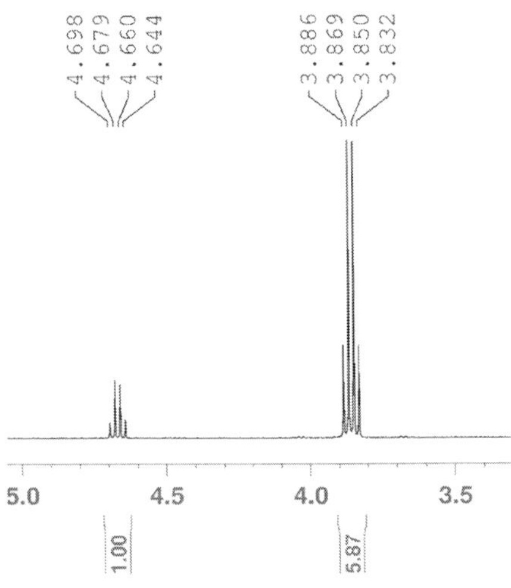

Figure 4. Results of a saccharin alkylation experiment.

The fourth experiment is the investigation of the elimination products obtained from *erythro*-2,3-dibromo-3-phenylpropanoic acid under three different reaction conditions (*4–8*). Students rely on IR, ^{1}H, ^{13}C NMR spectra to determine whether their reaction product is (*Z*)-2-bromo-1-phenylethene, a mixture of (*Z*)- and (*E*)-2-bromo-1-phenylethene, or 3-phenylprop-2-ynoic acid. This experiment is the first time in the laboratory that students are formally introduced to splitting patterns and coupling constants, which allows them to determine the structure of their product(s) including the stereochemistry around the double bond. Students who used reaction conditions of a weak base in a polar protic solvent can also determine the ratio of (*Z*)- and (*E*)-2-bromo-1-phenylethene reaction products by comparing the integration of the peaks due to the isomers (Figure 5).

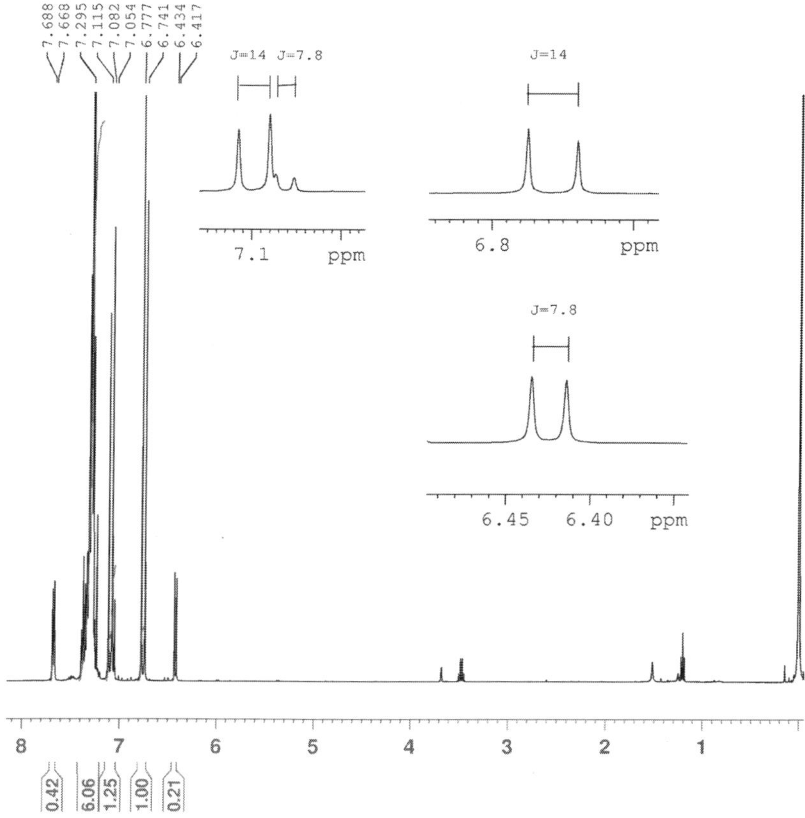

Figure 5. ^{1}H NMR spectrum of a mixture of (Z)- and (E)-2-bromo-1-phenylethene.

The fifth experiment is designed to teach students to work with liquids and perform a microdistillation that involves cracking dicyclopentadiene using a Hickman still and distilling off the resulting cyclopentadiene (*9*). Students aquire a ^{1}H NMR spectrum to determine the success of their purification by looking at the integration of the signal at 2.97 ppm for cyclopentadiene in comparison to the signal at 2.87 ppm for the dicyclopentadiene starting material (shown in Figure 6). Although students are required to draw these structures in advance, many students overlook the fact that cyclopentadiene has an integration value for two protons at 2.97 ppm whereas dicyclopentadiene has an integration value of a single proton at 2.87 ppm. Figure 6 presents the student results showing about 27% of the recovered material was cyclopentadiene.

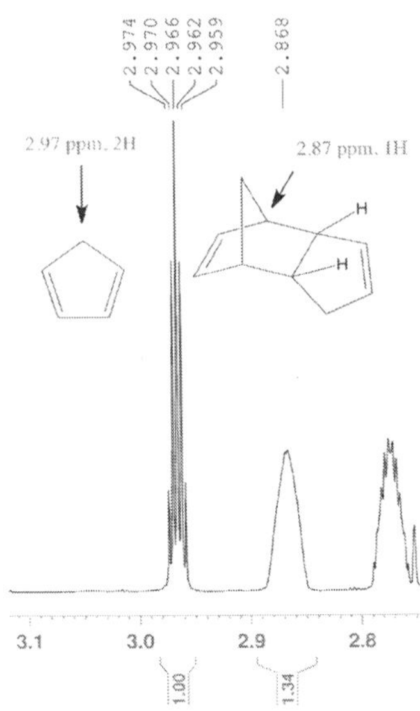

Figure 6. Student results from the dicyclopentadiene cracking cyclopentadiene distillation experiment.

In the final project of the semester, Diels-Alder Puzzles, students carry out either a microwave reaction between maleic anhydride and anthracene in diglyme (*10*), or a standard reflux reaction between maleic anhydride and 3-sulfolene in xylene (*11*). Students work in small groups in these experiments and use ^{1}H and ^{13}C NMR spectroscopy to determine the structures of their products and the mechanism of their assigned reaction. Each group then presents their findings to the rest of the class.

Organic Reactions and Applications Laboratory (Semester 3)

By the time students get to this course, it is expected they are obtaining melting/boiling points, IR spectra, ^{1}H and ^{13}C NMR spectra, and GC-MS data on a routine basis for all the compounds produced in the laboratory. This includes the assignment of all peaks in their spectra. This course starts with a project involving a solid and a liquid unknown. Students use various spectroscopic methods to determine their unknowns. For each of the unknowns, students propose a reaction utilizing their unknown that they then perform in the laboratory with faculty and teaching assistant approval. These proposals include: appropriate reaction schemes (with stoichiometry and theoretical yields), required materials including quantities and CAS numbers, safety information for all materials including the product(s), reaction conditions (i.e., procedures, reaction workup, isolation methods, etc.), reaction metrics, waste disposal issues, and reference(s).

The products of these reactions are characterized and are used to confirm the identity of their initial unknown compounds. Work on this project is performed throughout the semester along side the other assigned projects in an effort to begin developing the students' ability to multi-task in the laboratory.

There are a few isolated cases in the unknown project where students must aquire 2D-NMR spectra (COSY, HMQCGP [Heteronuclear Multiple Quantum Correlation Gated Pulse; this experiment provides correlation between protons and their attached heteronuclei], etc.) to assist in the identification of their products from the proposed reactions. Teaching assistants and the faculty assist these students individually with the analysis of the results. Future plans for this course include the development of a project to introduce all students to 2D NMR spectroscopy experiments.

Chemistry Across the Periodic Table Laboratory (Semester 4)

In the majority of the multiweek experiments of this course students must prepare organic ligands to which metals are intercalated. Students continue to utilize standard NMR experiments to characterize and confirm the structure of the products they prepare.

In the first experiment of the semester, students prepare tin (IV) iodide (*12*). This compound is analyzed for the tin (II) iodide byproduct by performing ^{119}Sn NMR spectroscopy. The saturated SnI_4 in DMSO-d_6 NMR samples contain a capillary with 1 M tetramethyl tin in $CDCl_3$ as a reference standard. The SnI_4 has a chemical shift of -2024 ppm relative to the tetramethyl tin standard (Figure 7), whereas the SnI_2 byproduct has a chemical shift of -779 ppm.

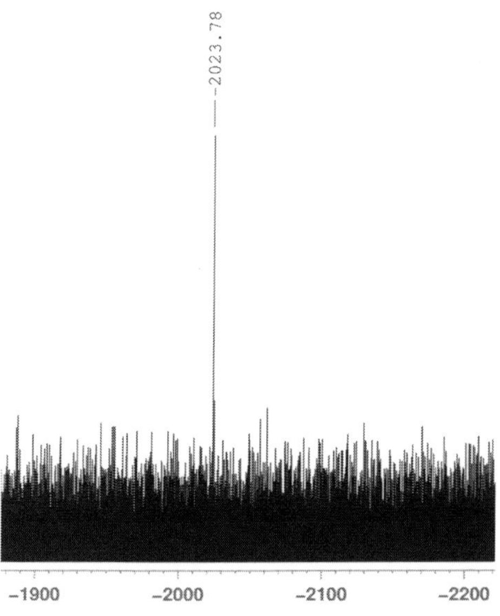

Figure 7. ^{119}Sn NMR spectrum of Tin (IV) Iodide in DMSO-d_6.

The largest project in this course involves the synthesis of sterically hindered square-planar nickel complexes to study the kinetics of substitution of a bromine ligand with thiocyanate to form the *N*-thiocyanate complexes (*13*). The experiment requires students to use ^{31}P NMR spectroscopy to look at the ratios of the complexes containing chlorine vs. bromine in the fourth ligand position and to ensure their final products are pure brominated complexes.

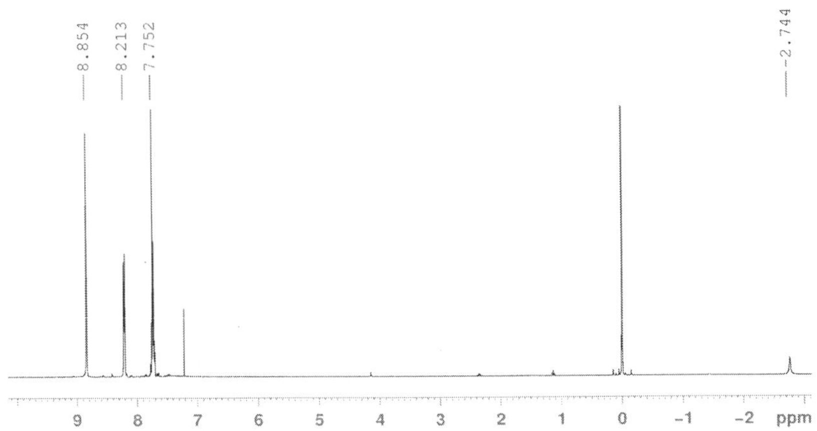

Figure 8. ^{1}H NMR spectrum of H_2-tetraphenylporphyrin prior to metallation in $CDCl_3$. Note the endocyclic proton signal at -2.74 ppm.

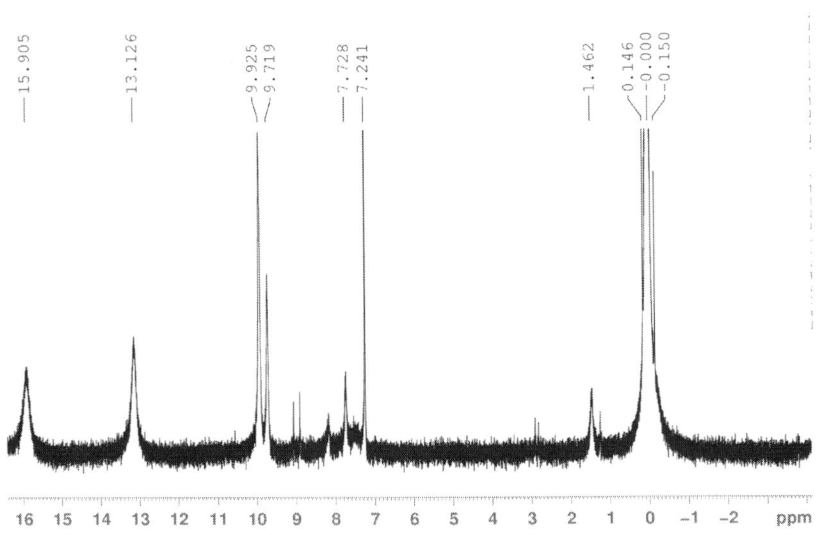

Figure 9. ^{1}H NMR spectrum of the paramagnetic cobalt tetraphenylporphyrin. Note the loss of the endocyclic proton signal due to metallation of the porphyrin and the significant line broadening due to the paramagnetic cobalt center.

In an experiment to produce metallotetraphenylporphyrins synthesized by a microwave procedure developed in-house, ^{1}H NMR spectroscopy is used to study the extent of metallation of the product (*14*, *15*). This is accomplished through comparison of the endocyclic proton signal before and after the metallation of the porphyrin (Figure 8). Students are able to distinguish between diamagnetic and paramagnetic systems (Figure 9) by looking at the NMR spectra of the metallotetraphenylporphyrins.

Students use ^{1}H NMR spectroscopy to determine the magnetic susceptibility of 1,1′-bis(diphenylphosphino)ferrocene metal complexes by the Evans method (*16*, *17*) and compare that to values obtained via a magnetic susceptibility balance. Students also characterized their complexes by use of ^{31}P NMR spectroscopy and cyclic voltammetry in this experiment.

NMR Spectrometer Use in Non-Majors Introductory and Advanced-Level Courses

Organic Reactions and Applications Laboratory (Semester 3)

For students in the non-majors chemistry sequence, the faculty desired that students not be introduced to NMR spectroscopy in the laboratory until it is covered in lecture course. Three experiments use NMR spectroscopy extensively, with the first use of NMR spectroscopy coming in a lab where students are assigned unknowns. In this experiment students only obtain the ^{1}H NMR spectrum for their compound and are provided copies of the ^{13}C NMR spectrum by the faculty. This is due to the fact that with nearly 400 students enrolled in this course and it would take over a week of NMR spectrometer time to run the ^{1}H and ^{13}C NMR samples just for this course. However, this still permits students to prepare and submit their own samples for analysis. In the second week of this experiment, students are taught how to use Topspin to analyze their data and create appropriate spectra for their laboratory reports.

In the second experiment, students study the isomerization of menthone using ^{13}C NMR spectroscopy to monitor the shift in the carbonyl signal (*18*). Samples are run in the absence of solvent, allowing for a detectable ^{13}C NMR signal with as little as a single scan. An inverse gated-decoupling experiment is used to allow for the integration of the ^{13}C NMR signals.

In the final multi-week project that utilizes NMR spectroscopy, students are asked to propose an experiment based on the reduction of an aldehyde or ketone using literature sources. After approval by their TA, the students carry out the reaction in the laboratory. The products of these reactions are characterized by ^{1}H and ^{13}C NMR spectroscopy. Students then process the data on their own using the Bruker Topspin program.

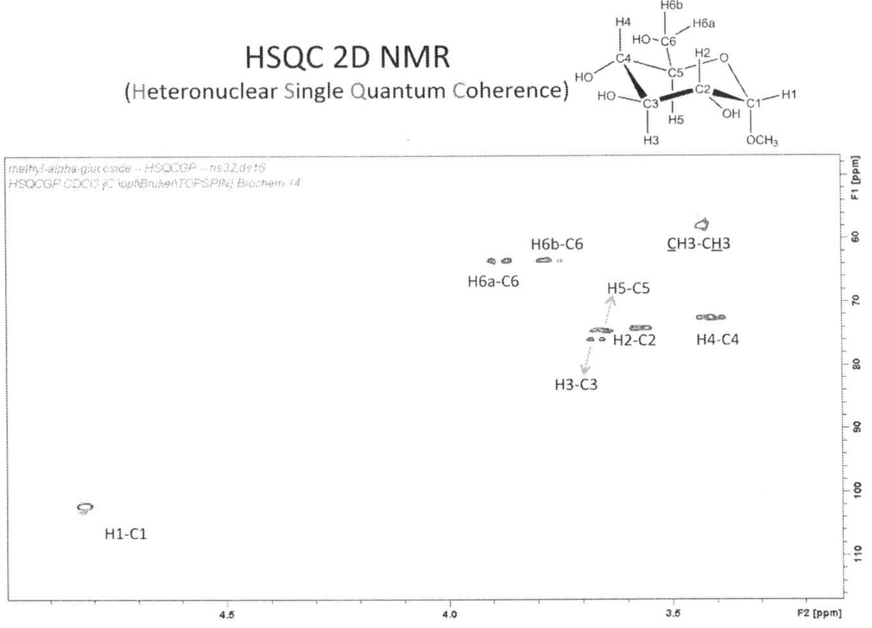

Figure 10. HSQC NMR spectrum of methyl-α-glucose.

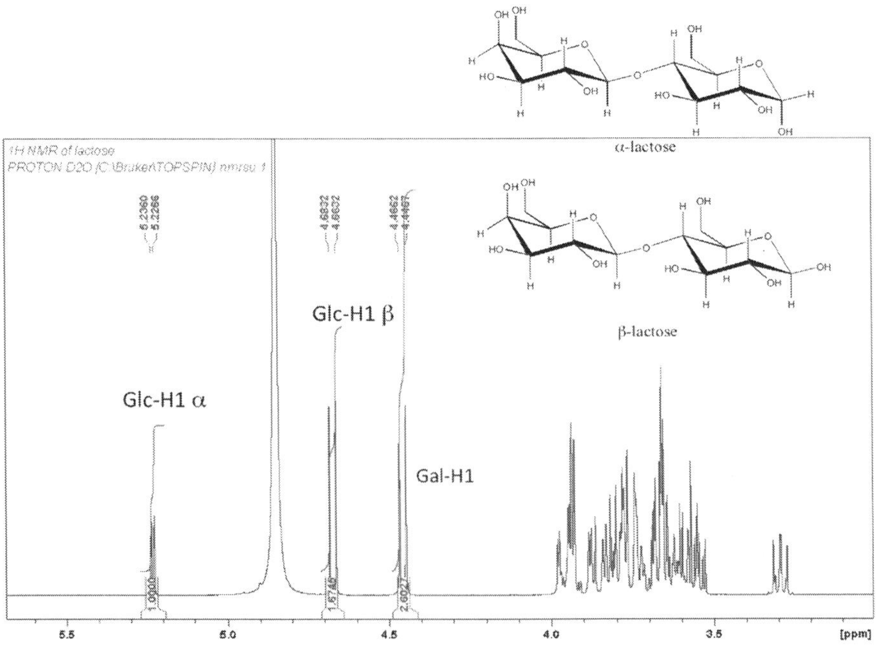

Figure 11. ^{1}H NMR of lactose showing the anomeric proton signals of lactose.

Biochemistry Laboratory

Students in the majors-track biochemistry laboratory course use the NMR spectrometer to obtain ^{1}H and ^{13}C NMR spectra of unknown disaccharides (*19*). During the prelab, methyl-α-glucoside is used as an example to demonstrate how the structure of complex biological molecules can be elucidated using 2D NMR experiments, COSY and HSQC (Figure 10). Students are provided with ^{1}H and ^{13}C NMR spectra of all of the possible monosaccharides that might be in their unknown sample. The students determine the constituent monomers of their disaccharides by HPLC analysis of the polysaccharide hydrolysis products. ^{1}H and ^{13}C NMR spectra of the unknowns are acquired by students and two pieces of information are obtained from these spectra (Figure 11). The ^{1}H NMR spectrum provides chemical shifts and coupling constants of the anomeric protons necessary to determine whether the linkage is an α or β glycosidic bond using the Karplus relationship. In general, the β anomers are found at lower chemical shifts with a larger coupling constant than the corresponding α anomer. From the ^{13}C NMR spectrum, the position of the glycosidic linkages in the monosaccharides can be determined.

Conclusions

Since the installation of the Bruker NMR spectrometer for the teaching laboratories in the late fall of 2006, the system has run 2000-3500 samples per semester (many of these samples had multiple experiments run upon them). In fact there are many weeks the sample changer is filled to capacity throughout the week. Given the heavy use of the instrument throughout the academic year there have been few technical issues with the instrumentation, with only two samples breaking in the sample changing robot due to student error.

Students have been very responsible with the instrumentation and most issues encountered deal with incorrect email addresses, selecting the wrong sample position in the sample changer, selecting the wrong solvent in IconNMR, or using too little solvent in sample preparation.

The teaching faculty are responsible for the day-to-day operation and upkeep of the instrument that includes liquid nitrogen and helium fills, software updates, fixing basic instrument errors, etc. The NMR facility staff is also available to assist with the issues the teaching faculty does not have the expertise to handle.

The biggest challenge we have faced is in scheduling the instrument. There are times when the NMR spectrometer is in use 24 hours a day for several weeks in a row. This means there are times when faculty and preparatory staff are constantly removing and loading samples in the robot for their students (especially in the non-majors course). Since so much time is taken up with relatively straightforward experiments in can be difficult to schedule and perform some of the more advanced NMR experiments that require longer data collection times.

Currently, we only have anecdotal evidence that student performance has improved since the installation of this instrumentation. Prior to 2006, students submitted their samples for several select experiments to their TAs who aquired the NMR spectra for the students. Students then received printouts of their spectra and had no opportunity to manipulate the data on their own. For other experiments, faculty provided copies of NMR spectra for students to analyze. Since the installation of a dedicated NMR spectrometer for the teaching labs, students are now responsible for preparing their samples, running their own NMR experiments, and processing their data using the software. We have observed an increase in their critical thinking skills as it relates to determining which materials are present in their samples (e.g., starting material, products, and solvents from the reaction and isolation of products), trouble-shooting sample preparation issues, and critically looking at their data to decide when a reaction needs to be repeated or if an NMR sample needs to be rerun. Research faculty, especially those who rely on NMR characterization in their projects, have noted that students are better prepared to begin working in the research laboratories. This has allowed students to move beyond learning the "nuts and bolts" of spectroscopy to being able to spend more time learning the chemistry required for research. We have found that our students appreciate the high degree of access they have to this instrumentation. Finally, curricular innovation continues among the teaching faculty as they continually develop new laboratory experiences to further utilize and teach the capability of this powerful instrument.

References

1. Wietstock, S. M.; Goodenough Lashua, D. University of Notre Dame. Unpublished work.
2. Mohrig, J. R.; Hammond, C. N.; Schatz, P. F.; Morrill, T. C. *Modern Projects and Experiments in Organic Chemistry: Miniscale and Standard Taper Microscale*, 2nd ed.; W. H. Freeman and Company: New York, 2003; pp 171−176.
3. Greenberg, F. H. *J. Chem. Educ.* **1990**, *67*, 611.
4. Grovenstein, E., Jr.; Lee, D. E. *J. Am. Chem. Soc.* **1953**, *75*, 2639.
5. Cristol, S. J.; Norris, W. P. *J. Am. Chem. Soc.* **1953**, *75*, 2645.
6. Corvari, L.; McKee, J. R.; Zanger, M. *J. Chem. Educ.* **1991**, *68*, 161.
7. Mestdagh, H.; Puechberty, A. *J. Chem. Educ.* **1991**, *68*, 515.
8. Reimer, M. *J. Am. Chem. Soc.* **1942**, *64*, 2510.
9. Wietstock, S. M.; Goodenough Lashua, D. University of Notre Dame. Unpublished work.
10. Bose, A. K.; Manhas, M. S.; Ghosh, M.; Shah, M.; Raju, V. S.; Bari, S. S.; Newaz, S. N.; Banik, B. K.; Chaudhary, A. G.; Barakat, K. J. *J. Org. Chem.* **1991**, *56*, 6968.
11. Sample, T. E.; Hatch, L. F. *J. Chem. Educ.* **1968**, *45*, 55.
12. Schaeffer, R. W.; Chan, B.; Molinaro, M.; Morissey, S.; Yoder, C. H.; Yoder, C. S.; Shenk, S. *J. Chem. Educ.* **1997**, *74*, 575.
13. Martinez, M.; Muller, G.; Rocamora, M.; Rodriguez, C. *J. Chem. Educ.* **2007**, *84*, 485.
14. Wietstock, S. M. University of Notre Dame. Unpublished work.
15. Nascimento, B. F. O.; Pineiro, M.; Rocha Gonsalves, A. M. d'A.; Silva, M. R.; Beja, A. M.; Paixao, J. A. *J. Porphyrins Phthalocyanines* **2007**, *11*, 77.

16. Corain, B.; Longato, B.; Favero, G.; Ajo, D.; Pilloni, G.; Russo, U.; Kreissl, F. R. *Inorg. Chim. Acta* **1989**, *157*, 259.
17. Nataro, C.; Fosbenner, S. M. *J. Chem. Educ.* **2009**, *86*, 1412.
18. Alonso, D. E.; Logan, A.; Kim, A.; Peterson, K.; Goodenough Lashua, D. University of Notre Dame. Unpublished work.
19. Hu, X.; Seriani, A.; Goodenough Lashua, D. University of Notre Dame. Unpublished work.

NMR Spectroscopy: Collaborative Strategies, Resources, and Grant Writing

Chapter 19

Oregon NMR Consortium: A Collaboratory for NMR Data Acquisition and Processing

R. Carlisle Chambers*

Department of Chemistry, George Fox University, 414 N. Meridian Street, Newberg, Oregon 97132
***E-mail:cchamber@georgefox.edu**

The Oregon NMR consortium was created to provide access to a modern, high-field 400 MHz NMR spectrometer for students and faculty at several two-year and small four-year institutions. Students use both on-site and remote access to conduct their NMR experiments. Remote access involves connecting to the NMR console over the Internet. Features of the consortium are described, including the details of remote access, sample transport, and data processing. This paper also discusses the impact that the access to the NMR spectrometer has had on student achievement of several learning goals in the organic chemistry courses at the partner institutions.

Introduction

Chemical research and chemical laboratory education have become increasingly dependent upon instrumentation. A quick survey of the *Journal of Chemical Education* reveals a large number of laboratory experiments that involve some type of chemical instrumentation. Chemical educators need to keep pace with this development in order to prepare their students for future success in graduate school and chemical industry careers. However, the high cost of some instruments has limited the access of many students and faculty, primarily at small undergraduate institutions. In particular, high-field nuclear magnetic resonance (NMR) spectroscopy is not available at many colleges and universities due to the high cost (> $200K) of a new instrument as well as the ongoing cost of cryogens for the maintenance of the superconducting magnet. Even used or reconditioned NMR spectrometers are beyond the budget of many small schools.

The lack of access to an NMR spectrometer is unfortunate as NMR spectroscopy has become one of the most important methods for structural analysis and determination of compounds. The central importance of NMR spectroscopy to the undergraduate chemistry curriculum is underscored by the fact that an NMR spectrometer is the only instrument required for program certification by the American Chemical Society Committee on Professional Training (*1*). Many computer and Internet-based simulations and databases have been developed that provide users access to virtual NMR spectroscopy data (*2*). These simulations no doubt help students develop the skills of spectroscopic analysis and data interpretation. However, they do not provide students with the opportunity to execute experiments or to analyze samples that they have prepared themselves.

The increased reliance on chemical instrumentation has been matched by an increase in educational computer-based networks (*3*). Most college campuses have local area networks as well as access to the World Wide Web. As almost all modern chemical instrumentation involves a computer interface and digital data acquisition and storage, instruments have increasingly been placed on networked systems. "Collaboratories" or shared instrumentation facilities have been designed to meet the needs of regional (*4–8*) and international users (*9, 10*). Indeed, a number of collaboratories have been developed over the last decade, such as the Integrated Laboratory Network (ILN) at Western Washington University (*11*). Collaboratories have lowered the barrier to access for many instruments, and students and faculty can now perform experiments in real time through the remote control of instruments at a distant location.

While access to an instrument through a collaboratory does provide some opportunities, there are still several barriers that limit the impact of these networked instruments on chemical education. Remote access is possible and has been achieved in a number of places, but delivery of samples from a remote site is an issue. The distance from a remote site to the host facility may limit the opportunity to perform experiments on samples that students have prepared themselves within a short time frame. In addition, many faculty members may lack the training to operate the equipment, the expertise to guide students through the data interpretation steps, or the confidence to update their chemistry curriculum with new experiments.

To address these and other issues the Oregon NMR Consortium was created with four objectives: 1) increase understanding of chemistry through hands-on access to modern NMR technology; 2) increase competency in using modern chemical instrumentation in teaching laboratories; 3) develop technical expertise among chemistry faculty; and 4) enhance chemical instruction at all consortium institutions by sharing curriculum ideas, teaching pedagogies, and assessment information.

The Oregon NMR Consortium was not designed to serve the NMR spectrometer needs of all chemical education institutions in the region, but rather the membership of the Oregon NMR Consortium was intentionally drawn from institutions that serve primarily undergraduate student populations and that did not have an NMR spectrometer or convenient access to an instrument. The schools in the Oregon NMR consortium are small, private four-year institutions and a two-year public community college. In 2010, private four-year and public two-year institutions enrolled 3.9 million and 7.2 million students, respectively, which represented 63 percent of the total students enrolled in undergraduate programs (*12*). While many of the students at private four-year and public two-year institutions do have access to NMR spectroscopy, it is safe to assume that some percentage of these students have no access to NMR spectroscopy instrumentation at all.

Table 1. Overview of Oregon NMR Consortium member institutions

	GFU	*PCC*	*CU*	*WPC*
Location	Newberg	Portland	Salem	Portland
Type	Private, 4-yr	Public, 2-yr	Private, 4-yr	Private, 4-yr
Total undergraduate student population	1500	27,000	750	500
Offers BS/BA in chemistry	Yes	No	No	No
Highest CHEM offering	Physical	Organic	Organic	Organic
Typical Organic Chem. enrollment	45	35	12	5, alt yrs
Distance from GFU, Newberg		17 miles	38 miles	30 miles
Distance from GFU, branch		2 miles	6 miles	10 miles

Through the support of the Course Curriculum and Laboratory Improvement (CCLI) program of the National Science Foundation (NSF), the Chemistry Department at George Fox University in Newberg, Ore. purchased a new NMR spectrometer. George Fox is responsible for the maintenance and upkeep of the spectrometer. The first section of the proposal to the NSF-CCLI program dealt with curricular changes that took place at George Fox. As a result of the CCLI support a number of laboratory experiences were upgraded or introduced to take advantage of the new instrument. The second section of the CCLI proposal, which is the focus of this report, concerns the development of the Oregon NMR Consortium which has had a significant impact on the regional chemistry educational community.

Program Design

Overview

The Oregon NMR Consortium initially involved three institutions: George Fox University (GFU), Warner Pacific College (WPC) and Corban University (CU), with Portland Community College (PCC) joining shortly after the consortium was launched. The four institutions of the consortium represent a mix of large and small, public and private, urban and rural institutions with diverse chemistry curricula and student populations (Table 1). Of the consortium members, GFU, PCC and Corban University have made significant use of the instrument during the project. Access to the NMR spectrometer is free for all consortium members, with each user group being responsible for purchasing their own sample tubes and deuterated solvents, as well as preparing and labeling their samples.

Equipment

The equipment selected and purchased for this project was intentionally chosen to meet the program objectives. The NMR spectrometer is a JEOL ECS-400 which has the standard features of most modern NMR spectrometers, including a 9.39 T superconducting magnet, a two channel broadband probe, gradient shimming, an autotune unit and a low-temperature/variable-temperature accessory. While these items are certainly important with respect to project success, other features of the JEOL instrument proved just as crucial. The instrument also was equipped with a 24-sample autosample carousel, JEOL data processing software (Delta) that can be loaded and used separately from the acquisition software, unlimited acquisition and processing software licenses for Mac, Windows or Linux operating systems, and a hardware/software interface that uses an Ethernet connection to the NMR console from any workstation equipped with the JEOL software. Any workstation can be used to access and control the JEOL NMR spectrometer so long as it is connected to the Internet and hosts the JEOL Delta acquisition software (Figure 1). In fact, all users connect to the NMR spectrometer through the network, including those who initiate experiments and collect data locally at the NMR facility. In this sense, everyone is a remote or client user of the instrument. The appearance and features of the JEOL acquisition software are the same for all users at all locations. Consequently, students who operate the instrument from both the local workstation and from a remote location will experience the same software interface. To maintain network security, the George Fox institutional technology department has configured the institutional firewall to permit access to the NMR spectrometer by known clients who attempt to connect to the instrument from a remote site.

The NMR spectrometer console can be owned only by one workstation at a time. Moreover, experiments can be initiated only when the spectrometer is owned by a workstation. At the conclusion of an NMR experiment, the final package of spectroscopic data is transmitted to whichever workstation initiated the experiment. This approach to data storage eliminates the need for any subsequent data transfer or retrieval from a primary server. We have restricted the acquisition software to one workstation at each consortium site to avoid too many users trying to initiate experiments at the same time. However, the processing software is available to anyone in the consortium who wishes to have it, and the unlimited number of software processing licenses means that all faculty and students in the consortium are able to load the JEOL software on their personal computers. Spectroscopic data can also be sent by email from the workstation that initiated the experiment, which allows students to view and process spectra without waiting for access to a single, central workstation.

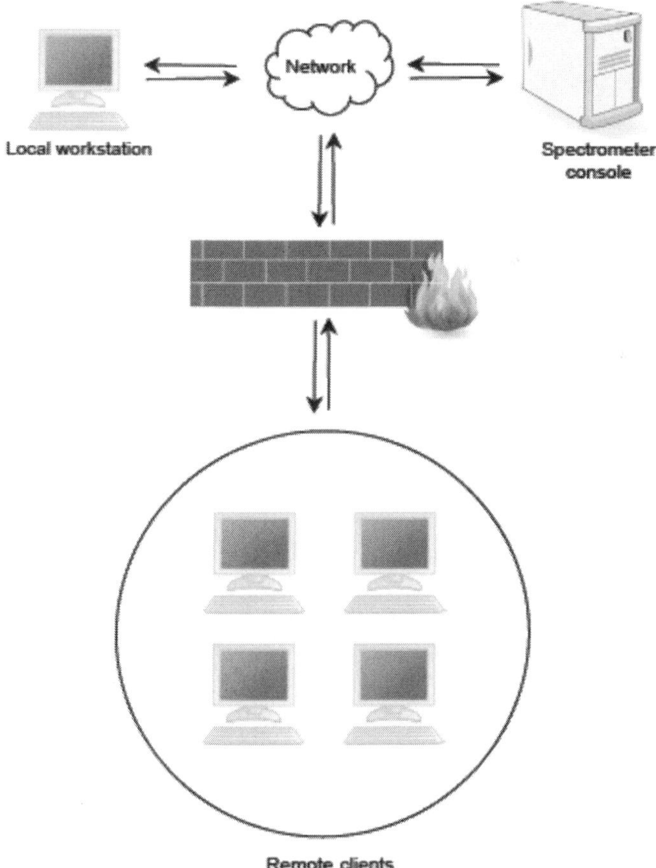

Figure 1. Consortium network configuration for users.

Activities

A planning session was held at George Fox University before the NMR spectrometer was installed. At that meeting the consortium members discussed the features of the instrument, the logistics of on-site and remote access, the types of experiments that could be shared among the member institutions, and plans for common assessment instruments. After the NMR spectrometer had been delivered and installed, an initial training session was held at which representatives from each of the consortium members were trained on basic procedures, such as loading and unloading samples, locking and shimming samples, and selecting and executing a simple ^{1}H NMR experiment while operating the instrument on-site. On the basis of the initial training session, several stand-alone modules were developed by GFU faculty to provide more detailed training in sample preparation, remote access of the instrument, obtaining an FID, generating a processed spectrum and obtaining integration and peak information from a ^{1}H spectrum. The training modules were initially developed for faculty training on the NMR spectrometer and have since been modified for introducing students to the basic principles of NMR data acquisition and interpretation. All of the training materials and tutorials are available on the consortium website that is hosted by George Fox University.

In subsequent meetings, the consortium has discussed specific experiments that can be added to the chemistry curriculum, along with the development of assessment instruments to evaluate the NMR project. Both curricular changes and assessment data will be discussed below. Member institutions have continued to meet at the conclusion of each academic year, with evaluation of assessment data, discussion of best teaching practices in the undergraduate laboratory, and faculty development in NMR spectral data interpretation being topics at these annual consortium workshops.

Curriculum

The theory and practice of NMR spectroscopy has been integrated into the chemistry curricula at the various consortium institutions. The faculty at each college determines the specific curricular changes, such as experiment choice and scheduling. In general, each institution introduces compound identification using a variety of techniques, including NMR spectroscopy, during the first semester or term of the organic chemistry sequence. Specific experiments or laboratory activities are then incorporated into the curricula throughout the remainder of the academic year according to the schedules prepared by the individual instructors. There are several common experiments that are performed by all members of the consortium (Table 2). One experiment (Table 2, Experiment 1) that has been shared, is an active learning experiment developed at GFU for general chemistry students that relates the chemical shifts of a homologous set of small organic molecules to electronegativity trends (*13*). In addition, a set of NMR samples, containing simple organic molecules that are used to introduce the principles of ^{1}H and ^{13}C NMR spectral interpretation, were prepared and are available for use by the various consortium members (Table 2, Experiment 2).

Table 2. Common NMR Spectroscopy Experiments

Experiment	*Course*
The effect of atom structure on electron clouds	General
Introduction to organic structure determination using ^{1}H and ^{13}C NMR data	Organic
^{13}C NMR spectroscopy identification of alcohols	Organic
^{1}H and ^{13}C NMR analysis in banana oil synthesis	Organic

Sample Transport and Analysis at Remote Locations

The transport of samples to the NMR spectrometer facility at George Fox from the other consortium members is one of the key issues in the success of the consortium. Typically, students and faculty from the consortium partners travel to the GFU-Newberg campus for their first NMR experiment in the Organic Chemistry course. They have the opportunity to see the instrument, directly experience inserting samples into the magnet bore, execute an NMR experiment, and process the spectral data. The Chemistry Department at GFU has a computer workroom with several workstations that contain the JEOL Delta processing software. However, while there is value in hands-on use of the NMR spectrometer, the lack of a credible sample transport system could lessen some of the benefits of real-time access and control of the instrument from a remote location since the consortium members are too far from the GFU-Newberg campus to travel to the NMR facility for sample analysis on a regular basis. Other collaboratory arrangements have used mail delivery to transport samples to the host NMR facility (*5*, *9*). However, in addition to the cost associated with mail delivery, this system would likely involve several days between sample preparation and arrival at the NMR facility.

The Oregon NMR Consortium has developed a system in which a one-day turnaround time is possible for samples sent to the NMR facility at George Fox University. In addition to the main campus in Newberg, George Fox University also has branch sites at other locations throughout Oregon, including Portland and Salem. These branch locations are very close to the home institutions of the consortium partners (Table 1). George Fox University employs a courier system through its mail services department for transporting internal mail, supplies and other items to the branch locations. The GFU courier makes daily trips to the branch campuses from Newberg, and we have used this system to transport samples between the remote partners and the NMR facility.

The U.S. Department of Transportation allows for the transport of small amounts (< 30 mL) of hazardous materials by courier under a limited quantity regulation (*14*). Under this arrangement, NMR samples are placed inside a metal canister that is lined with adsorbent material. The canister is then placed inside a sturdy five-gallon bucket that is packed with additional adsorbent material. When a consortium member is ready to send samples to the NMR facility they contact the project director who sends the empty transport container through the GFU courier system to the appropriate branch location. The samples are then placed in the container and returned to the Newberg campus via the GFU courier system. When the samples are received in Newberg they are loaded into the 24-sample carousel for the remote partners to access. The faculty and students at the remote location then connect to the NMR and conduct the experiments on their samples using the Delta data acquisition software running on a remote client workstation. For large laboratory sections, the gradient shimming routine and software macros available with the JEOL Delta system streamline the acquisition process. Again, the spectroscopy data is owned by the workstation that initiates the experiment and can be subsequently sent to a student email account for later processing. Upon completion of the NMR analysis the GFU courier system is used to return the samples from the GFU-Newberg campus to the appropriate branch location for pick up by a representative of the cohort institution.

Project Assessment

Overview

The assessment plan involved an evaluation of active consortium members over the length of the project. Three general areas were evaluated in the project: (i) student success in achieving the NMR learning goals; (ii) student learning through remote access to the NMR spectrometer; (iii) student confidence in using modern chemical technology. The assessment plan involved administering questionnaires to students before and after installation of the NMR spectrometer. Baseline responses (Year 0) were collected from George Fox students before the installation of the NMR spectrometer. The populations evaluated in this project were the student cohorts in the organic chemistry courses at all of the participating consortium institutions. Evaluation of the students in organic chemistry occurred at the end of the last semester or term of the course, and the baseline results were likewise collected at an analogous point in the spring semester of Year 0. The assessment plan was not designed as a rigorous evaluation of an academic experiment, but rather as an examination of the extent to which the project goals were achieved.

Assessment of NMR Learning Goals

Several NMR spectroscopy learning goals were surveyed in the various consortium Organic Chemistry courses (Table 3). The assessment instrument used a 5-pt Lickert scale (1-strongly agree, 5-strongly disagree) for evaluating the extent to which students achieved the learning goals.

Table 3. Assessment scores of NMR learning goals by various student cohorts in Organic Chemistry[a]

		GFU		PCC		CU	
Learning Goal	*Year 0*	*Year 1*	*Year 2*	*Year 1*	*Year 2*	*Year 1*	*Year 2*
I have a good understanding of the meaning of chemical shift in an NMR spectrum.	1.7	1.8	1.4	1.6	1.5	2.5	2.2
I have a good understanding of the meaning of peak integration in an NMR spectrum.	1.7	2.0	1.5	1.6	1.4	2.3	2.0
I am able to interpret correctly a one-dimensional ^{1}H NMR spectrum using chemical shift and peak integration information.	1.7	1.5	1.5	1.5	1.4	3.1	1.4
I am able to interpret correctly a one-dimensional ^{13}C NMR spectrum using chemical shift information.	3.1	1.6	1.5	2.3	2.2	3.1	2.4

[a] Students were asked to respond to the Learning Goal statements with the following scale: 1-strongly agree; 2-agree; 3-neither agree nor disagree; 4-disagree; 5-strongly disagree.

The learning goals (1 – 3) that involve the analysis of ^{1}H NMR spectroscopy data are essentially the same for the study groups as for the baseline cohort (Year 0). As the analysis of ^{1}H NMR data was part of the curricula before the installation of the high field spectrometer it is not surprising that the scores are similar. A significant increase in the ability to analyze ^{13}C NMR data (Goal 4) was observed in all years of the project. The theory of ^{13}C NMR spectroscopy was discussed in lecture settings prior to the installation of the spectrometer. Students in Year 0 and preceding years received instruction in the analysis of ^{13}C NMR data by using spectra from textbooks or databases and online simulations. However, the practice of ^{13}C NMR spectroscopy was not covered at any of the consortium institutions before the new NMR spectrometer was installed due to the lack of an instrument that was capable of this type of analysis. The gains in learning related to ^{13}C NMR spectroscopy have been realized over several years. Moreover, the first cohort of students at Corban University (CU, Year 1) did not take advantage of the ^{13}C capabilities of the new instrument. However, in Year 2, experiments that involved ^{13}C NMR data analysis were added to the Organic Chemistry curriculum at CU and the score for this learning goal showed a corresponding increase. The results in Table 3 indicate that the actual practice of NMR analysis is important in achieving learning goals associated with NMR data interpretation.

Assessment of Student Attitudes on Using Remote Access To Operate the NMR Spectrometer

We were also interested in the attitudes of consortium students who primarily used remote control from their home institution to operate the NMR spectrometer (Table 4). The assessment instrument for evaluating this project goal again used a 5-pt Lickert scale (1-strongly agree, 5-strongly disagree.) Thus far, only the students at PCC have made extensive use of the remote access capabilities of the NMR spectrometer. These students typically come to the GFU-Newberg campus for their first experience with the NMR technology, and in this initial meeting they have the opportunity to load samples, shim the instrument, execute experiments, and process data. All of their subsequent experiments are conducted by remote operation of the instrument from their home campus. It's worth noting that the experiments conducted by remote control involve samples the PCC students have prepared themselves and sent to GFU by courier. At the end of the organic chemistry course, and after they had conducted several other NMR experiments, they were asked if the remote access had hampered their achieving the learning objectives (Table 4). According to the student responses the remote access is a viable method for achieving the desired NMR outcomes. They reported that remote execution of experiments was very similar to direct operation of the NMR spectrometer, and that the remote access did not hamper their understanding of NMR technology or their interpretation of NMR spectroscopic data.

Table 4. Assessment scores of student attitudes on using remote access by student cohorts in Organic Chemistry at PCC[a]

Question	*Year 1*	*Year 2*
Operating the NMR by remote control from PCC was as easy as operating the instrument by direct control at GFU.	1.7	2.0
Using the instrument by remote control did not hamper my understanding of the NMR technology.	1.7	2.0
Using the instrument by remote control did not hamper my understanding of the NMR data.	1.6	1.6

[a] Students were asked to respond to the statements with the following scale: 1-strongly agree; 2-agree; 3-neither agree nor disagree; 4-disagree; 5-strongly disagree.

Assessment of Confidence in Using Modern Chemical Technology

The extent to which students gained confidence in using modern chemical technology as a consequence of their hands-on experience with the 400 MHz NMR was also assessed. (Table 5) The assessment instrument once again used a 5-pt Lickert scale (1-very confident, 5-not confident) for evaluating this project goal. Prior to the installation of the high-field spectrometer at George Fox, all NMR experiments were performed by the course instructor or laboratory teaching assistant rather than the students in organic chemistry. The baseline score of 4.0 expressed by students in Year 0 for confidence in their ability to correctly operate the NMR indicates the low confidence that students had in this skill. We also asked the students to express their confidence in using several other instruments, FT-IR, fluorescence spectroscopy, and electrochemistry. The students in Organic Chemistry at George Fox routinely use infrared spectroscopy to analyze laboratory products and the Year 0 baseline score indicates a high level of confidence with this instrument. Experiments that use fluorescence spectroscopy or electrochemistry techniques are not part of the GFU organic chemistry curriculum, nor are these techniques used in the organic chemistry courses at the other consortium institutions. Not surprisingly, the students expressed a low confidence in their ability to correctly operate these instruments in Year 0.

The increased hands-on experience in using the 400 MHz NMR spectrometer corresponded to a higher level of expressed confidence by students at all of the consortium schools. Prior to their participation in the Oregon NMR consortium, the students at PCC and Corban did not have any direct experience in operating a modern NMR spectrometer. Their scores in both years 1 and 2 closely track the values reported by students at GFU.

Table 5. Assessment scores of student cohorts in organic chemistry when asked "For the following instrumental techniques I am confident in my ability to operate the instrument correctly."[a]

	GFU			*PCC*		*CU*	
	Year 0	*Year 1*	*Year 2*	*Year 1*	*Year 2*	*Year 1*	*Year 2*
NMR	4.0	1.3	1.4	1.9	1.7	1.4	1.3
FT-IR	1.5	1.4	1.6	1.4	1.5	4.5	- -
Fluorescence	4.4	3.1	3.2	5.0	4.0	4.5	5.0
Electrochemistry	4.4	3.0	3.2	5.0	4.8	4.0	5.0

[a] Students were asked to respond to the confidence statement with the following scale: 1-very confident; 2-confident; 3-somewhat confident; 4-low confidence; 5-not confident.

An interesting result from this assessment data involves the changes in student confidence in operating other chemical instruments. As noted earlier, none of the students at the consortium schools used fluorescence or electrochemistry techniques in the organic chemistry curriculum prior to the beginning of the NMR project. The NMR project did not change the Organic Chemistry curriculum with respect to fluorescence and electrochemistry techniques, however, GFU students in Organic Chemistry Year 1 and 2 cohorts expressed a higher level of confidence in their ability to conduct experiments using these techniques. We don't know if these differences are statistically significant, but the experience of using chemical instrumentation in the form of a modern NMR spectrometer may have had a positive impact on students' confidence in using unfamiliar instrumentation. A possible explanation is that student experience with the NMR spectrometer has made modern chemical instrumentation more accessible and less of a black box.

Conclusions

The Oregon NMR Consortium has provided access to a modern, high-field NMR spectrometer for a large number of students in our region who previously could not easily use this type of instrumentation. Many of the barriers that limit the impact of remote access to instrumentation have been overcome. We have developed a convenient strategy for transporting samples between institutions. The instrumentation and software used in this project permit real-time access for students at remote institutions. The combination of sample transport and computer access has made it possible for the students at remote sites to analyze samples they have prepared with one-day wait times. The assessment information strongly indicates that the consortium has had a positive impact on the laboratory experiences of the students at the partner institutions, with students having an increased understanding of NMR interpretation and, we hope, the underlying, fundamental chemical principles. We have also found that actual experience of operating the NMR spectrometer appears to have led to an increased student confidence in using any modern chemical instrument. Anecdotal information from interviews with faculty members at consortium institutions confirms that real-time access to a modern NMR spectrometer has also positively impacted faculty development. In ways that parallel the student responses, faculty note a better understanding of chemical and NMR concepts, as well as a higher level of confidence in using modern chemical instrumentation. The NMR consortium has helped develop and strengthen relationships between the participating institutions, and we are continuing to explore other ways in which these relationships could be used to enhance chemical education in our region.

Acknowledgments

Funding for this project was provided by the NSF-CCLI program, (DUE CCLI 0633346). The author acknowledges the contributions of faculty and staff at the consortium institutions: GFU – Paul Chamberlain, Jeffrey Vargason, Michael Everest, Sherrie Frost; PCC – Patty Maazouz; CU – James Dyer; WPC – David Terrell. The author also acknowledges the technical assistance provided by representatives of JEOL USA, Inc., Ashok Krishnaswami and Chip Detmer.

References

1. *Undergraduate Professional Education in Chemistry: ACS Guidelines and Evaluation Procedures for Bachelor's Degree Programs*; American Chemical Society: Washington, DC, 2008; p 6.
2. A number of web-based NMR simulators, tutorials and databases are available including: the Virtual NMR Spectrometer (http://www.vsnmr.org); Organic Spectroscopy Online (http://ospecweb.boisestate.edu); East-NMR (http://www.east-nmr.eu/databases-and-links.html); Spectral Database for Organic Compounds SDBS (http://riodb01.ibase.aist.go.jp/sdbs/cgi-bin/direct_frame_top.cgi) (accessed November 2012).
3. Harris, C. M. *Anal. Chem.* **2002**, *74*, 535–538A.
4. Mills, N. S.; Shanklin, M. *J. Chem. Educ.* **2011**, *88*, 835–839.
5. Benefiel, C.; Newton, R.; Crouch, G. J. *J. Chem. Educ.* **2003**, *80*, 1494–1496.
6. Alonso, D.; Mutch, G. W.; Wong, P.; Warren, S.; Barot, B.; Kosinski, J.; Sinton, M. *J. Chem. Educ.* **2005**, *82*, 1342–1344.
7. Baran, J.; Currie, R.; Kennepohl, D. *J. Chem. Educ.* **2004**, *81*, 1814–1816.
8. Kennepohl, D.; Baran, J.; Connors, M.; Quigley, K.; Currie, R. *Int. Rev. Res. Open Distance Learning* **2005**, *6* (online journal, ISSN: 1492-3831).
9. Fitch, A.; Mavura, W.; Kishimba, M.; Muriithi, A. *Anal. Bioanal. Chem.* **2006**, *384*, 7–10.
10. Albon, S. P.; Cancilla, D. A.; Hubball, H. *Am. J. Pharm. Ed.* **2006**, *70*, 1–8.
11. Western Washington University's Integrated Laboratory Network. http://www.wwu.edu/iln/ (accessed November 9, 2012).
12. Aud, S.; Hussar, W.; Johnson, F.; Kena, G.; Roth, E.; Manning, E.; Wang, X.; Zhang, J. *The Condition of Education 2012 (NCES 2012-045)*; U.S. Department of Education, National Center for Education Statistics: Washington, DC, 2012; p163.
13. Everest, M. A.; Vargason, J. M. *J. Chem. Educ.* manuscript in review.
14. U.S. Department of Transportation, Title 49:173.4.

Chapter 20

ChemSpider: How a Free Community Resource of Data Can Support the Teaching of NMR Spectroscopy

Antony J. Williams,* Valery Tkachenko, and Alexey Pshenichnov

Royal Society of Chemistry, U.S. Office, Department of eScience, 904 Tamaras Circle, Wake Forest, North Carolina 27587
***E-mail: williamsa@rsc.org**

The Royal Society of Chemistry (RSC) hosts a number of resources of direct benefit to teaching chemistry. One of these resources, ChemSpider, is a free chemical database containing data for millions of chemical compounds. The data includes thousands of spectra, primarily contributed by members of the scientific community as ChemSpider depends on crowdsourcing for certain types of data. The NMR spectral data hosted on ChemSpider are integrated to various other RSC systems including an educational wiki and a micropublishing platform for reaction syntheses. The data are also available programmatically to the community and have been used as the basis of a game to teach NMR spectroscopy.

Introduction

The Royal Society of Chemistry (RSC) is the largest organization in Europe with the specific mission of advancing the chemical sciences. Supported by a worldwide network of 47,000 members and an international publishing business, RSC activities span education, conferences, science policy and the promotion of chemistry to the public. In keeping with the mission to provide access to resources of value to the chemistry community RSC has delivered a number of informatics resources, primarily in the form of websites, and these have proven to be valuable to the community as evidenced by the high traffic and positive feedback received from chemists around the world.

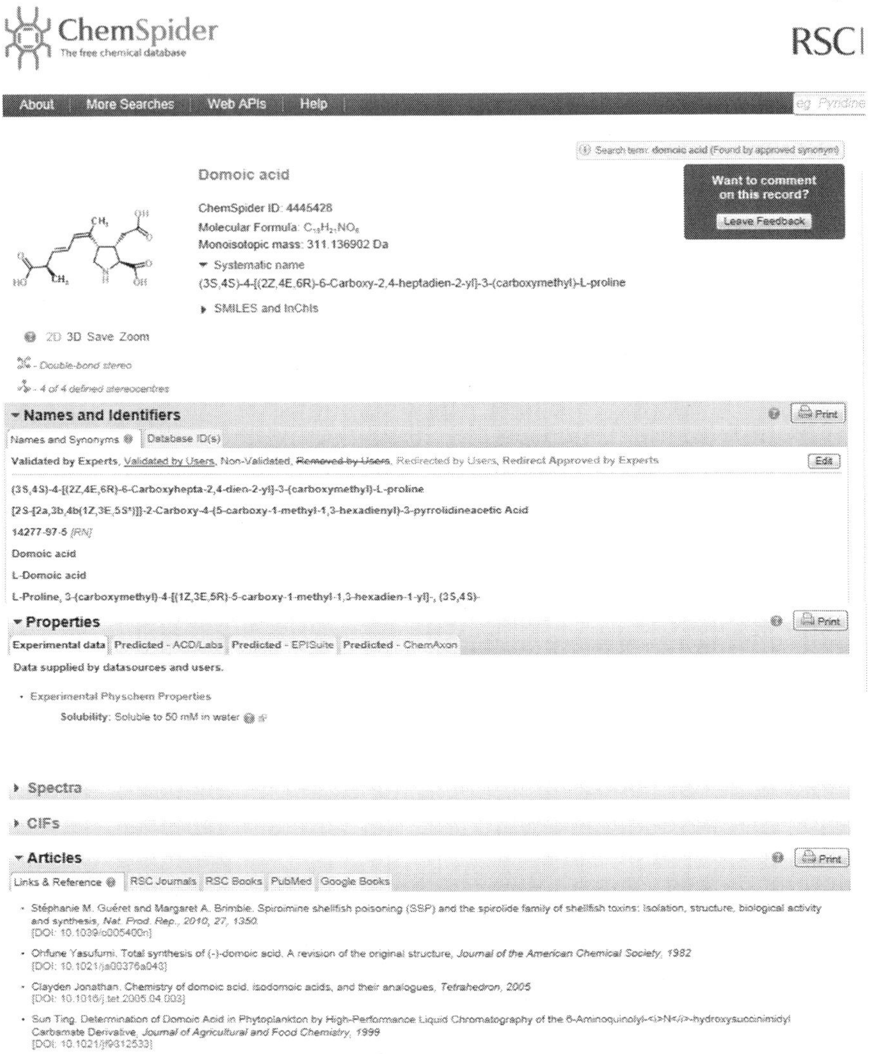

Figure 1. A partial screen capture of the ChemSpider record following a search for "Domoic Acid" (http://www.chemspider.com/4445428). The chemical record shows the structure, multiple identifiers, an experimental property (solubility), links to multiple articles and additional infoboxes. Courtesy of Royal Society of Chemistry.

RSC hosts ChemSpider (*1–3*), one of the chemistry community's primary online resources. It is a free resource developed with the primary intention of aggregating and linking chemical structure based information and data across the web. Expanding in content daily, it currently contains over 28 million unique chemical entities and is linked to well over 400 data sources. The platform is unique relative to the majority of other online offerings in that it allows users to expand, annotate and curate the data contained within the database. ChemSpider is also the foundation of a series of other related projects for the management of community deposited chemical syntheses, serving data to an educational platform for students and for hosting spectral data serving a spectroscopy teaching resource and game.

The majority of data contained within the database were originally aggregated from various contributors including chemical vendors, commercial database vendors, government databases, publishers and a number of individual scientists. The database can be searched using alphanumeric text searching of both intrinsic properties (such as molecular formula and molecular weight), as well as predicted molecular properties and structure/substructure searching. The diversity of searches has expanded since inception to support various types of users such as mass spectrometrists and medicinal chemists. The search system is flexible, fast, and provides access to a lot of data integrated to a particular chemical from across the various depositors. A screenshot of the interface and *partial* results obtained from a search for domoic acid is shown in Figure 1.

Problems with data quality became apparent when aggregating data from various data sources, and so graphical user interface elements were introduced to allow the community to curate the data. The success of this approach, as evidenced by the participation of a number of chemists in validating chemical name-structure relationships in particular, encouraged us to add further capabilities for the annotation and expansion of data. The ability to add analytical data (specifically spectral data and crystallographic information files (CIFs)) to chemical structure records was provided. Users were also given the ability to deposit single chemicals or files containing multiple chemical structures. The database now contains over 8000 spectra, almost half of which were contributed by members of the chemical community. Additional data are being added regularly. The spectral data types include infrared, Raman, mass spectrometric and NMR spectra. The majority of community contributed data are ^{1}H and ^{13}C NMR spectra.

While much of the data contained within ChemSpider has been sourced from other online databases and web resources, an increasing amount of data is now being provided by members of the community. This increase in crowd sourced data indicates the willingness of a growing component of the ChemSpider user base to contribute to the growth of the database. Funding agencies are also starting to require that publicly funded research data are stored in institutional repositories and, where possible, public repositories. While this has been a common situation for biological data, especially for genes, proteins and enzymes, such a situation is far less common for chemistry data as there are no true repositories for chemistry data. We believe that ChemSpider will become one of these public repositories and that the community will continue to contribute as funding agencies encourage the deposition of data.

Spectral data can be submitted as images or, to allow for interactivity such as zooming, addition of an integral curve, etc., can also be submitted in JCAMP format (*4*) and displayed in an open source interactive applet, JSpecView (*5*). The process of depositing data onto the site is very simple (*6*), and at deposition a submitter has the ability to add comments and declare the data as "Open Data" allowing for download and reuse. An example spectrum is shown in Figure 2.

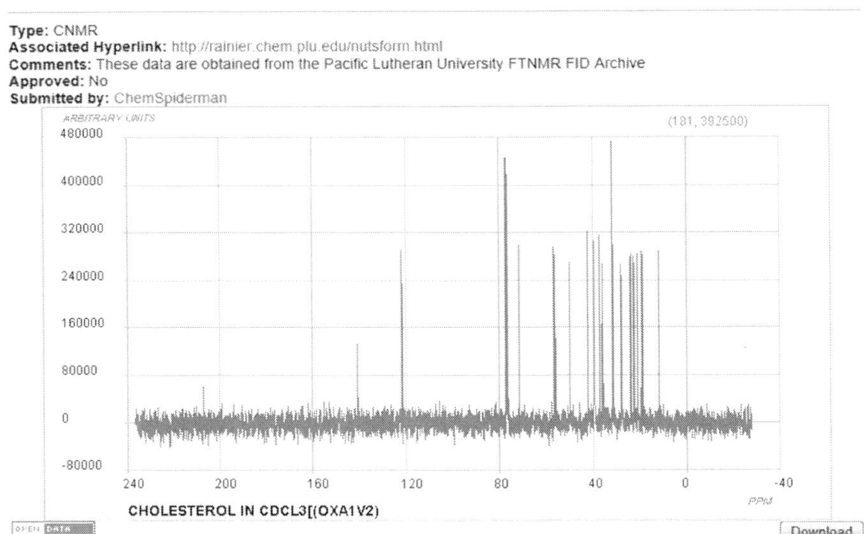

Figure 2. The ^{13}C NMR spectrum of cholesterol. Notice the Open Data icon and the download button to download the spectral data in JCAMP format to the desktop for inclusion in reports, lesson plans, etc. Courtesy of Royal Society of Chemistry.

In regards to the focus of this chapter, there are a number of ways that ChemSpider can support the teaching of NMR spectroscopy. As outlined below the data are made available via a programmatical interface for integration to a number of platforms including a wiki for teaching, to a micro-publishing environment for chemical syntheses, and to a game for teaching NMR interpretation skills. However, it should be noted that the data are also available for downloading and distribution so that they can be used as example teaching and test sets in classroom exercises. When coupled together with other related spectral data they are ideal training data. The sections below discuss the various applications that have been made available.

Integrating NMR Data to Other RSC Platforms

Learn Chemistry Wiki

The RSC's objective is to advance the chemical sciences, not only at a research level but to also provide tools to train the next generation of chemists. The RSC's Learn Chemistry (*7*) platform has been developed to provide a central access point and search facility to make it easier to access the various different chemistry resources that it provides. ChemSpider contains a significant amount of useful information for students learning chemistry, but there is also a lot of information that is not relevant to their studies and might be distracting. As a result, the RSC has developed a teaching resource which restricts the compounds and the properties, spectra, and links displayed for each, to those relevant to their studies. However, students do not just need compound information in isolation since it is most useful when linked to and from study handouts and laboratory exercises. In addition, this resource is not just intended to be read and browsed, but to be interactive – allowing students to answer a variety of quiz questions, and allowing chemical educators to contribute to the content.

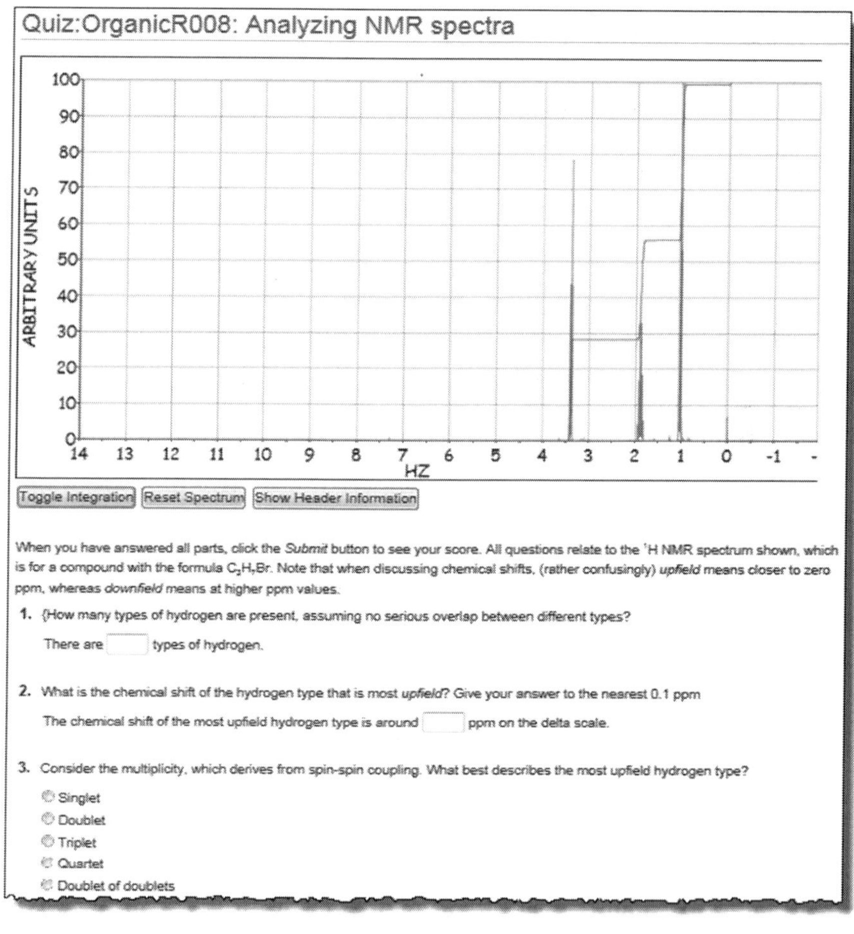

Figure 3. The Analyzing NMR Spectra quiz on the Learn Chemistry wiki. Courtesy of Royal Society of Chemistry.

The Learn Chemistry wiki (see Figure 3) is a mediawiki (*8*) environment integrated to ChemSpider, providing access to data and information that is delivered at a level most appropriate to students in their last years of school, and first years of university (ages 16-19). The wiki restricts the list of compounds shown, the properties listed, and the spectra and links displayed to those most relevant to studies for this age group. The resource is an interactive environment and allows students to answer a variety of quiz questions, and chemical educators to contribute to the content.

The wiki offers a number of tools to help students learn NMR spectroscopy including an introduction to NMR spectroscopy (*9*) and several quizzes (*10*, *11*). The wiki also provides links from chemical compound pages to external sites containing spectral data.

SpectraSchool - Integrated Data for Teaching Spectral Interpretation

At the time of writing there are more than 3000 NMR spectra already deposited to ChemSpider. The majority of these are ^{1}H and ^{13}C NMR spectra, but there are also infrared, near infrared, UV-Visible and mass spectra. Recently the NIST/EPA IR spectral data collection (*12*) was added to the database incorporating over 5200 spectra. Certain ChemSpider records offer rich integrated NMR datasets including ^{1}H, ^{13}C, COSY, HSQC and HMBC spectra (*13*), and such datasets can be very valuable for students to work through spectral assignment problems.

For structure elucidation the teaching of spectroscopy eventually encourages the usage of multiple techniques in order to bring an aggregate of data to bear on the identification of a chemical compound. Students are commonly taught to use a collection of data including NMR spectra (^{1}H and ^{13}C), a mass spectrum, an IR spectrum and a UV-vis spectrum. 2D NMR spectra commonly show up later in a student's training. Such aggregated data are valuable in helping a student learn what spectral features are associated with a chemical compound and, delivered in the appropriate manner, such data can form the basis of a quiz-mode.

The SpectraSchool (*14*) site provides access to a few tens of aggregated data sets. This site allows for browsing and review of chemicals and the associated NMR, MS, IR (see Figure 4) and UV-Vis spectra. An "Identify" mode provides a quiz-based mode where the user is shown a number of spectra of various types from which the user has to determine what the chemical compound is. Recent developments include the display of the assignments associated with both the ^{1}H and ^{13}C NMR spectra, where hovering over a nucleus highlights the associated peak(s) in the spectrum. This capability will be extended to include both IR band assignment and MS fragment assignment in the future.

ChemSpider SyntheticPages

ChemSpider is primarily a chemical compound database integrating data and information associated with chemicals. In order to extend the coverage to syntheses, a new database was established known as ChemSpider SyntheticPages (CSSP) (*15*). The CSSP website hosts community contributed data regarding synthetic procedures and is a form of "micro-publishing platform". The system can host multimedia content, spectral data and links to the ChemSpider database, and contributed NMR spectral data. Figure 5 shows an example reaction from CSSP. This reaction is the final stage of the synthesis of the compound known as Olympicene (*16*). The spectral data are stored in the ChemSpider database and retrieved for display.

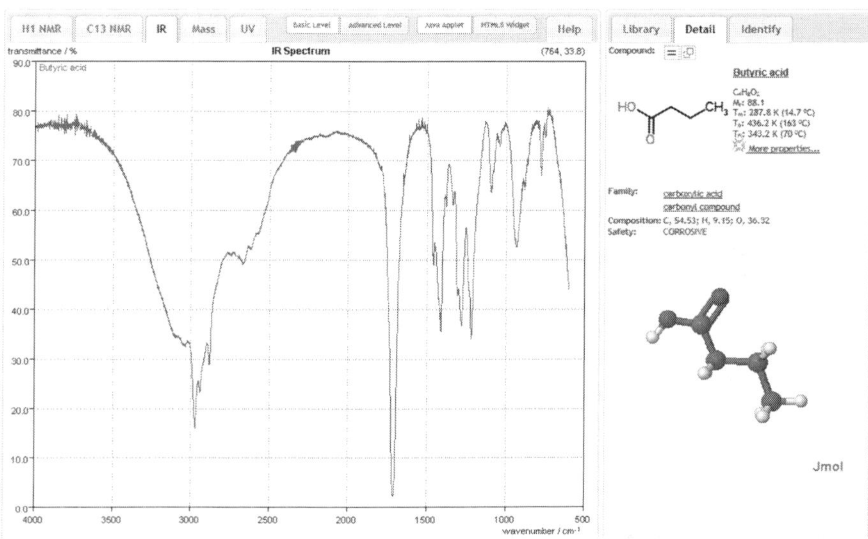

Figure 4. The SpectraSchool website. The infrared spectrum for butyric acid is displayed. The ^{1}H NMR, ^{13}C NMR, IR, MS and UV spectra for butyric acid can be displayed. Courtesy of Royal Society of Chemistry.

CSSP provides chemistry students with an opportunity to develop an online reputation since each SyntheticPage has a single author, the chemist who performed the synthesis. The submission is reviewed by one or more members of the editorial board and comments are made available to the author. The author then makes edits before the article is published online. This feedback between the editorial board and the author is generally very fast relative to classical review, commonly less than 48 hours, and once published, the article can then be commented on by the community. CSSP hosts hundreds of synthetic procedures and new submissions are made regularly. NMR spectral data are deposited into ChemSpider and are therefore available in the compound database for students to download, practice their interpretation skills, and flag potential errors in the comments section of the SyntheticPages article. While this is one way to practice their interpretation skills a more engaging way is to use a gaming approach, discussed below.

Dehydration of 3,4-dihydro-5H-Benzo[cd]pyren-5-ol; 6H-Benzo[cd]pyrene

SyntheticPage 542
DOI: 10.1039/SP542
Submitted Mar 15, 2012, published May 31, 2012
Anish Mistry (a.mistry@warwick.ac.uk)
A contribution from Fox Group, Warwick University

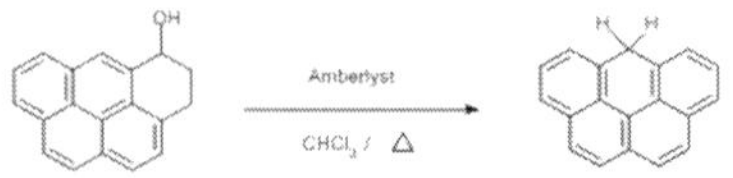

OH — Amberlyst, $CHCl_3$ / Δ → H H

Chemicals Used

3,4-dihydro-5H-Benzo[cd]pyren-5-ol (prepared)
Amberlyst 15 (Sigma-Aldrich)
Chloroform

Procedure

3,4-dihydro-5H-Benzo[cd]pyren-5-ol (0.1 g, 0.39 mmol) was dissolved in chloroform (30 ml) and Aberlyst 15 (0.1 g) added under a dinitrogen atmosphere. The reaction was heated to 30°C and left overnight under the inert atmosphere. The solution was then filtered to seperate the Amberlyst and washed with chloroform. The combined solvents were removed under vacuum using a Rotary evaporator. The crude product was column chromatographed under a dinitrogen atmosphere eluting with 1:1 chloroform:petroleum ether 40-60°C. A white solid was obtained using this method (50 mg, 54%).

Author's Comments

- Degradation of product occurs on the column. The use of a dinitrogen atmosphere slows this down. Firstly the eluent and coloumn were purged using dinitrogen gas, after the crude mixture was loaded on the purged coloumn a continous flow of dinitrogen gas was used for flash chromatography via a nitrogen line.
- The chloroform was washed with sodium bicarbonate before coloumn chromatography.
- The white product on exposure to air became yellow then orange in colour.

Data

δ_H(400MHz, $CDCl_3$) ppm: 5.00 (2H, s, -$\underline{CH_2}$), 7.46 - 7.55 (4H, m, aryl), 7.72 - 7.84 (6H, m, aryl).

Figure 5. A screenshot showing an example reaction from ChemSpider SyntheticPages. This is the final stage of the synthesis of Olympicene. The NMR spectrum of the compound (not shown) is displayed in a JSpecView applet. Courtesy of Royal Society of Chemistry.

Large scale depositions of spectral data available in electronic lab notebooks collected from a number of academic institutions are presently being prepared. Some of these reactions will be deposited as synthetic pages. This increase in spectral data will benefit the academic community by enriching the data available for reuse in teaching NMR.

Programmatic Access to NMR Spectral Data

The Spectral Game

The spectroscopic data contained within ChemSpider are used as the basis for the Spectral Game (*17*) that was created by meshing together the ChemSpider spectral data, a spectrum viewing tool, and workflows for delivering them in a gaming fashion. At the beginning of the game, two structures, one correct and one incorrect, are shown below the spectrum. The user has to select the structure that best matches the spectrum displayed. If the user selects the correct structure then they proceed to the next set of spectrum and structures and the process is repeated. While the player could operate only with guesswork in order to learn, the player needs to review the spectrum to compare various spectral features in order to confirm or reject each of the structures. These include the NMR chemical shifts, multiplicities, peak intensities, functional groups, etc. Users should be able to quickly distinguish aromatic protons from alkyl protons, aldehydic resonances from exchangeable carboxylic acid protons, and methoxy singlets from methylene groups within a chain.

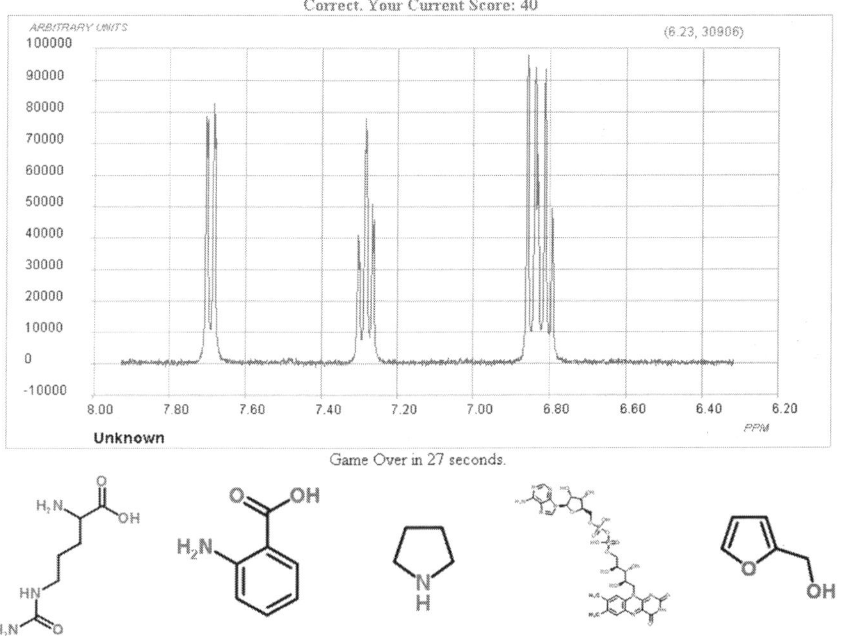

Figure 6. The Spectral Game. The final level of the game offers five structures, one of which matches the spectrum. Courtesy of Royal Society of Chemistry.

The complexity changes as the player progresses, and the game becomes more difficult as the number of associated structures increases to a maximum of five per spectrum (as shown in Figure 6). As the number of structures increases they also become more structurally similar. When a player reaches a score of forty, rounds also become timed and the player must select an answer before the countdown expires, to a minimum of ten seconds. Complexity also increases for the ^{13}C NMR spectra where the number of carbon atoms in all structures is made equivalent to the number of carbon atoms present in the correct structure. This can be confusing until the player takes account of issues other than simply carbon count and chemical shifts: symmetry, peak intensity related to nature of carbon nucleus and so on. The game proceeds until the player gets an answer wrong at which point the player is given their performance relative to the list of both recent and top players.

As reported by Bradley et al. (*17*) the Spectral Game was evaluated in one of the author's (JCB) undergraduate organic chemistry classes. Workshops where the instructor led the class discussion were useful for a larger number of students, especially when students had just started learning to analyze spectra. A class discussion would evolve about the key differences between the expected spectra of the molecules on display, and then the instructor could zoom into relevant regions to explore those details. In this manner all of the key concepts in the course relating to NMR spectral interpretation were repeatedly reviewed. When the opportunities arose, simple coupling patterns, peak shifts, symmetry and diastereotopic groups were highlighted.

An advantage of the Spectral Game, when compared to textbook problems, is that real-world spectra were made available. Large solvent peaks (such as HOD at 4.8 ppm), peak distortions, overlapping peaks in complex coupling patterns, and impurities were pointed out, and the students were shown how to address these. Since the spectra are random, on occasion the instructor may not be able to solve the problems based on simple heuristics taught in class. This provides an opportunity to discuss other techniques that a chemist might use and such discussions are essential to help train students for real world research.

The spectral data identifiers and the properties and identifiers of the chemical structures are obtained from ChemSpider using the freely available web services. The details of the software development required in order to create the spectral game from the ChemSpider content and the available JSpecView open source java applet have been discussed elsewhere (*17*). Since the original reported work, Bradley et al. have integrated a second spectral viewing tool, the iChemLabs ChemDoodle Web Components (*18*) which allows the spectral game to be played on mobile devices such as iPads and Android devices.

A proof-of-concept 2D version of the spectral game was developed involving automated versus human validation of 1D and 2D NMR data sets based on the simultaneous analysis of a 1D ^{1}H NMR and 2D ^{1}H-^{13}C single-bond correlation HSQC spectrum.

The spectral game has been accessed by many thousands of students in over 100 countries. This gaming approach has also helped to identify errors in the spectra, offering an opportunity to curate the data through gaming. From this example it should be clear how important access to experimental data can be for stimulating rapid re-mixing for educational examples. As more data become available, the usefulness of the Spectral Game and similar initiatives will become even greater.

Future Work

We believe that ChemSpider and other related RSC resources will continue to develop in both functionality and content and have a greater impact in terms of serving a pedagogical role in the teaching of spectroscopy. We foresee that in the future we will be able to integrate NMR prediction algorithms allowing chemists to predict NMR spectra for unknowns as well as for the existing content of over 28 million chemical compounds. Such predicted spectra can be used as the basis of spectral comparison approaches and for the identification of chemical compounds when the appropriate algorithms are integrated.

Our hope is that the community will come to see ChemSpider as a platform for the sharing of their spectral data and will deposit both their chemicals and associated analytical data for others to access. In order to encourage participation in terms of contribution of data we are presently developing a "rewards and recognition" scheme where contributors will be recognized for the provision of data to the community. This will initially be rolled out as part of the ChemSpider SyntheticPages system early in 2013 and later extended across other related platforms. We envisage a series of "medals" indicating the types and number of depositions of data to the various platforms, all connected together via an RSC profile for the scientist. These awards will also be made available so that they can be delivered as part of the growing alternative metrics movement measuring the overall contributions of a scientist.

Conclusions

The ChemSpider database is one of the premier chemistry sites on the internet. It has assumed a special role in hosting data contributed by the community and, as a result, has been able to develop a significant collection of spectral data. These data are a valuable resource for students of chemistry. The spectral data contained within ChemSpider also serve a number of other educational projects including SpectraSchool, the Learn Chemistry wiki, ChemSpider SyntheticPages and the SpectraSchool game. Each of these contributes to the education and training of chemistry students in the manipulation, interrogation and assignment of NMR spectral data. Such environments are becoming increasingly important in supporting education. The increase in freely-accessible data and information and the myriad of resources available today for students will, in theory, be highly beneficial to both the teaching and learning processes associated with the pedagogical aspects of NMR spectroscopy.

Acknowledgments

The projects discussed in this article result from the work of many people, not only those within RSC. ChemSpider and its associated projects are developed by the cheminformatics team led by Valery Tkachenko (Chief Technology Officer) and include Colin Batchelor, Aileen Day, Ken Karapetyan, Alexey Pshenichnov, Dmitry Ivanov, David Sharpe and Jon Steele. In particular, Alexey contributed his skills and passion to the SpectraSchool project and Aileen Day was the lead developer on the Learn Chemistry wiki. All projects are supported by a dedicated team of IT specialists.

The Open Source community and the commercial software vendors are acknowledged for the valuable contributions their software has made to the development of our software. In regards to the spectroscopy aspects of the project discussed in this article, we are indebted to the following companies and individuals for their contributions: Robert Lancashire (University of the West Indies in Jamaica), Jean-Claude Bradley (Drexel University), Andrew Lang (Oral Roberts University), Martin Walker (Potsdam University), Leicester University, iChemLabs, ACD/Labs and Synthonix.

References

1. Pence, H.; Williams, A. J. *J. Chem. Educ.* **2010**, *87*, 1123–1124.
2. *ChemSpider*. http://www.chemspider.com (accessed September 2011).
3. Williams, A. J.; Tkachenko, V.; Batchelor, C.; Day, A.; Kidd, R. Utilizing Open Source Software to Facilitate Communication of Chemistry at the Royal Society of Chemistry. In *Free and Open Source Software in Applied Life Science and Industry*; Harland, L., Forster, M. J., Eds.; Biohealthcare Publishing Limited: Oxford, 2012.
4. *JCAMP-DX Protocols*. http://www.jcamp-dx.org/protocols.html (accessed September 2011).
5. Lancashire, R. J. *Chem. Cent. J.* **2007**, *1*, 31.
6. *Depositing spectral data to ChemSpider*. http://www.chemspider.com/help_uploadspectra.aspx (accessed September 2012).
7. *LearnChemistry*. http://www.rsc.org/learnchemistry (accessed September 2011).
8. *MediaWiki*. http://www.mediawiki.org/wiki/MediaWiki (accessed September 2011).
9. *An Introduction to NMR Spectroscopy*. http://www.rsc.org/learn-chemistry/wiki/Introduction_to_NMR_spectroscopy (accessed September 2012).
10. *Learn Chemistry Wiki, Quiz 2, Determining an Unknown Using NMR Spectroscopy*. http://www.rsc.org/learn-chemistry/wiki/Quiz:NMRQ002:_Determine_an_NMR_spectroscopy_unknown (accessed September 2012).
11. *Learn Chemistry Wiki, Quiz 8, Analyzing NMR Spectra*. http://www.rsc.org/learn-chemistry/wiki/Quiz:OrganicR008:_Analyzing_NMR_spectra (accessed September 2012).
12. *NIST/EPA Gas-Phase Infrared Database JCAMP Format*. http://www.nist.gov/srd/nist35.cfm (accessed September 2012).
13. *Spectral Data on ChemSpider, ID 24528095*. http://www.chemspider.com/24528095 (accessed September 2012).
14. *SpectraSchool*. http://spectraschool.rsc.org/ (accessed September 2012).
15. *ChemSpider Synthetic Pages*. http://cssp.chemspider.com. (accessed September 2012).

16. *The Story of Olympicene from Concept to Completion*. http://www.chemconnector.com/2012/05/27/the-story-of-olympicene-from-concept-to-completion/ (accessed September 2012).
17. Bradley, J. C.; Lancashire, R. J.; Lang, A. S.; Williams, A. J. *J. Cheminf.* **2009**, *1*, 9.
18. *ChemDoodle web components.* http://web.chemdoodle.com/ (accessed September 2011).

Chapter 21

Writing More Competitive Grant Proposals for NMR Spectrometers: Research and Curriculum Programs of the National Science Foundation

Thomas J. Wenzel*

Department of Chemistry, Bates College, Lewiston, Maine 04240
***E-mail: twenzel@bates.edu**

The National Science Foundation has two important grant programs that provide support for NMR spectrometers. The Major Research Instrumentation Program operates under the research directorates and is focused on enabling research activities. The Transforming Undergraduate Education in Science, Technology, Engineering and Mathematics Program operates under the Division of Undergraduate Education and is focused on promoting curricular development that enhances undergraduate student learning. Advice for writing successful grant proposals to both of these programs is provided.

Introduction

The National Science Foundation (NSF) has two programs – Major Research Instrumentation (MRI) and Transforming Undergraduate Education in Science, Technology, Engineering and Mathematics (TUES) – that are possible sources of funding for the purchase of NMR spectrometers. The advice offered herein does not reflect an NSF opinion, as I have never served as an NSF program officer. It reflects my experience reviewing proposals on numerous occasions for NSF and other funding organizations as well as my experience in writing successful proposals to NSF research and education programs, including both MRI and TUES. Proposals to the MRI program focus on research activities whereas those to the TUES program focus on educational activities. Even though the aims of these two programs are quite different, all proposals to NSF follow a general format and have specific evaluation criteria that are described in the Grant Proposal Guide (GPG) (*1*). The general components of all NSF proposals are as follows:

Project Summary
Project Description
Literature References
Biographical Sketch(es)
Budget and Justification
Current and Pending Support
Facilities and Equipment
Data Management Plan

The GPG describes in general terms the features of each of these sections of the proposal. For example, the Project Description section of MRI and TUES proposals can be no longer than 15 pages with specific constraints on the margins and font size used in the document. Additional documents for the MRI (*2*) and TUES (*3*) programs describe further expectations within certain sections of the proposal; most often related to aspects of the Project Description. For example, the Project Description for an MRI proposal must describe a management plan for the instrument. Proposals to the TUES program must describe assessment plans for the project. Anyone intending to submit a proposal should scrupulously read the GPG and specific guidelines for the program and become familiar with all of the expectations before starting a proposal.

While this seems like obvious advice, I cannot emphasize enough how essential it is to follow the guidelines. An observation I have made over many years of panel and individual reviewing is how often people fail to follow some aspect of the guidelines. In some cases (e.g., not including an instrument management plan, or focusing on curricular uses of the instrument in an MRI proposal) it can be responsible for rejection of the proposal regardless of the merits of other facets of the work. In other cases (e.g., providing biographical sketches that do not conform to the NSF style and do not provide all the requested categories of information), the error may be more subtle but may still have a profound impact on one or more of the reviewers. Failure to follow the guidelines in any form creates the possibility for doubt on the part of reviewers about the extent to which the investigator(s) will be careful and thorough in the execution of the work. Since I often find myself on the fence about what final ranking to apply (e.g., excellent or very good), a proposal that has not followed some obvious aspect of the guidelines may create enough doubt that I will give it the lower score. In the competitive world of grant-writing, one score going from an excellent to a very good may be sufficient to make the difference between whether the proposal is funded or not. Also, a failure to follow some aspect of the guidelines may become magnified in the panel discussion if a reviewer raises it as an issue. Each proposal only gets a small amount of time for discussion, and it is best if that time is spent discussing the merits of the work rather than an issue with following the guidelines. Similarly, it is essential to scrupulously proofread the proposal to remove typographical and grammatical errors, especially with the features that exist today for electronic spell- and grammar-checking. If a PI writes a sloppy proposal, does it imply that the person will do similarly sloppy work? With that thought in a reviewer's head, it has a good chance of affecting her or his ranking of the proposal.

The GPG also describes the two general criteria used by NSF in evaluating proposals: (i) intellectual merit and (ii) broader impacts. The nature of the intellectual merits and broader impacts is quite different for an MRI and TUES proposal, and will be described within the following sections on each program. A relatively recent change to the evaluation of intellectual merit is the additional requirement that reviewers evaluate whether the work has the potential to be transformative. The goal of this change is to improve the odds that high-risk, high-gain work gets funded. If a reasonable case can be made that some or all facets of the work have the potential to be transformative, it will help make the argument for funding. Since reviewers must complete a separate evaluation for the intellectual merits and broader impacts, give careful consideration to both when writing the proposal.

NSF Major Research Instrumentation (MRI) Program

The MRI program operates through the research directorates of NSF and is the program within NSF that provides the majority of funding for the purchase of NMR spectrometers. It is essential to recognize that the MRI program primarily focuses on the research activities of the investigators who will use the instrument. Reviewers of MRI proposals focus on the quality of the research activities described in the proposal. Unlike many other NSF programs, the MRI program has historically received a separate allocation of funding for research universities and predominantly undergraduate institutions (PUIs). When I have been on MRI panels, we considered all the research universities as a group and ranked them separately from a consideration and ranking of the PUIs. Research universities must provide 30% of the cost of the instrument as a match. PUIs must request the entire amount of money needed to purchase the equipment; cost sharing is not allowed. There is no upper dollar limit on what can be requested, so it is possible through the MRI program to secure funding for any field strength NMR spectrometer. When requesting an NMR spectrometer, there will be an expectation that it will have multiple users, although a consideration of who to include in the proposal as users will be described in more detail in the section on the Project Description.

Project Summary

The Project Summary must have two distinct paragraphs, one of which discusses the intellectual merits and the other the broader impacts of the work. The intellectual merits of an MRI proposal are focused on the quality of the outcomes that can be expected to arise from completion of the research projects.

The paragraph on intellectual merit should summarize the different areas of research that are developed in the proposal for the primary users of the instrument and describe the significant outcomes that may result from their work.

The paragraph on broader impacts in an MRI proposal can address a number of points. An obvious broader impact is the undergraduate students, graduate students, postdoctoral associates, and other users who will benefit either educationally or in advancing their work by having access to the instrument. Involvement of research students from minority and other groups historically underrepresented in the sciences will be viewed favorably, provided it appears to be a credible goal. It will be important in the text of the Project Description to provide data or a specific plan for involvement of students from underrepresented groups that appears likely to succeed.

Perhaps less obvious is that the research described in the intellectual merits section may also have broader impacts. Publications of papers and review articles or talks at conferences represent a broader impact. If work in the proposal may lead to a patent application, then that is a broader impact. Some work may cross fields and provide important insights to investigators not directly in that area of work. Having colleagues from other institutions who may use or benefit from data collected on the instrument is another broader impact. There may also be educational outcomes of the work (e.g., a discovery in the research lab may be incorporated into an experiment in an undergraduate instructional lab). Finally, an NMR spectrometer obtained at a PUI will almost certainly be used in courses and these uses represent a broader impact, although it is necessary to remember that it is a research proposal.

Project Description

The guidelines for MRI proposals specify the inclusion of additional information that must be addressed in the Project Description and recommend the length of each section. These include a description of the research activities that will be enabled (9 pages), the instrument and needs of the investigators (2 pages), the impact that the instrument will have on research and training infrastructure (2 pages), and a management plan (2 pages). It is important to note that these are recommended lengths. For example, the description of and needs for a standard 400 MHz NMR with a single broadband probe for liquid samples might require much less text than a request for a 600 MHz instrument with a solid and liquid probe and other specialized features. Similarly, a department with a full-time instrument technician who is already maintaining, operating and managing an existing but outdated high-field instrument that is being upgraded to a new model might require a shorter management plan than a department with no instrument technician that is trying to upgrade from a 60 to a 400 MHz instrument.

Research Activities

Reviewers will evaluate the quality of the research activities described in the proposal. Given that there are about nine pages for this section, and given that the work of three to six primary investigators might be included, there is a limited amount of space to describe each individual's research project. The situation is further complicated by the makeup of the review panel. Because of the range of research areas that will be included within a single MRI proposal, review panels must be comprised of people who cover many areas of expertise. Proposals will be reviewed by people who are not an expert in every area of work included in it. How do reviewers make an assessment of the quality of the research and what key information belongs in this section? Because of the differences among reviewers' priorities and the different situations that exist across chemistry departments, there is no single answer to this question. However, there are things that reviewers consider when evaluating a proposal.

It is important to explicitly state the significance of each investigator's area of research in a way that is comprehensible to non-experts in the field. I have read too many MRI proposals, especially from departments at PUIs, where people immediately jump into the experimental details of a project and never describe the general importance of the work. Once I appreciate that the research has potentially significant outcomes, I can then go on to examine the brief experimental plan included in the proposal to assess whether it seems likely to succeed. Obviously each investigator's project will need references to appropriate and current literature that is informing the work.

For faculty from PUIs, it is especially important to recognize that the proposal is being considered by a research directorate so the usual outcomes of a research project (e.g., peer-reviewed publications) are beneficial to the review of a proposal. Because of the limited space available to describe several projects, other facets of a person's record become important in assessing the quality of the research. These include the recent publication record of the investigator as well as the record of pursuit and receipt of other external grant support for the work. Considerable care needs to be exercised in deciding who to include as the primary users of the instrument. If too many do not have a recent track record of research success, it may work against the proposal. Reviewers understand that relatively new faculty may not yet have publications from their independent work. Evidence that new investigators are seeking other forms of external support for their research helps. Inclusion of several senior faculty without a recent publication record may be detrimental to the review of an MRI proposal, especially if the department already has a high-field NMR spectrometer that it is trying to upgrade. If these people have not been able to complete work on an existing 300 MHz spectrometer, it can be difficult to claim that a new 400 MHz instrument will lead to an enhancement in productivity.

Proposals can distinguish between primary users – those with full research descriptions, biographical sketches, data on current and pending support – and secondary users – those with much briefer descriptions. There is no single right number of primary users to include on a proposal. In general, primary users should be the most active people, as measured by the extent to which they will use the equipment. To the extent that it is possible, it is best if primary users have an active record of publications and/or grants or be relatively recent hires. Secondary users should be those who will occasionally use the equipment with students. Someone who may regularly use the equipment but has gone a long time without any publications may be better to include as a secondary user, provided there are enough other primary users within the department. Be careful not to include projects that may not even be possible on the instrument being requested. In an effort to increase their user pool, I have read proposals from PUIs where a biochemist with no experience with NMR spectroscopy proposed to use a 400 MHz instrument for structure determination of large proteins. There are many chemistry departments at PUIs that legitimately need a better NMR spectrometer for research and educational purposes, but need alone is not sufficient to justify an instrument through the MRI program. The goal of the MRI program is to provide instrumentation to people whose need, when fulfilled, will allow them to advance an important research agenda.

Instrument and Needs of the Investigators

It is essential to request an instrument that is justified by the needs of the ongoing research included in the proposal. If the goal is to complete protein structure determinations using NMR spectroscopy, a 400 MHz instrument will not be sufficient. A more common problem, especially at PUIs, is that a department requests a 500 or 600 MHz instrument that is not justified by the research described in the proposal. It is not acceptable to say that the higher field strength or capabilities are justified because it will make it easier to hire new faculty or allow people to move into different research areas in the future. The projects included in the proposal must themselves justify the instrument being requested.

It is especially useful to provide spectra of representative samples that demonstrate what will be achievable with the new instrument that cannot be obtained on an existing instrument. These can be obtained from a colleague at another institution or from the vendor. Comparative spectra should be placed and discussed in the research descriptions, but referred to in the section on instrument needs. Investigators who regularly travel to another site to obtain spectra for their research can use that to show their dedication to their work. Reviewers recognize that such travel arrangements are seldom practical in the long-term.

It is also important to include specific data on instrument usage. Information on the number of faculty and students who will use the equipment and estimates of the number of spectra that will be run (this may be based on a log for an existing but outdated instrument) can help justify the need for the instrument. If a department is seeking a second NMR because of high usage of the existing instrument, it will be necessary to show data on the volume of usage.

Impact on Research and Training Infrastructure

For those at PUIs, it is appropriate to describe the general involvement of students in research in the department. If a stated outcome of the project is the involvement of students from minority and underrepresented groups, this section should contain a thorough description of the plan for meeting this goal. Indicate if the department has an active summer research program, since this shows that the instrument will be used throughout the entire year. It can be tempting to mention that faculty at other institutions will use the equipment in a consortial arrangement as a way of enhancing the impact. Reviewers know that establishing shared instrument arrangements is often difficult. Specific plans should be provided to convince reviewers that the consortium will realize its potential. For PUIs, it can be helpful to describe how the instrument will be used in the curriculum. Since the emphasis is on research, descriptions of curricular uses should be brief. If curricular uses are described, it is important that the experiments be modern ones that take advantage of the various features of the equipment. If curricular uses are included, reviewers will evaluate the quality of the curricular plans in making their recommendation.

Management Plan

Reviewers must be convinced that the department has a plan in place that will ensure the continual operation of the instrument. This plan should enable the integration of the instrument into the department's holdings without undue time demands on the faculty. The ideal situation is to have a trained instrument technician who will help with the routine operation and maintenance of the NMR spectrometer. If such a person does not exist within your department, then it will be essential to show that faculty members have the time and expertise to maintain the instrument. This may be obvious if the department is replacing an existing high-field instrument. It will be less so if the department is upgrading from a 60 MHz to a high-field instrument. Issues about where the instrument will be housed, what arrangements are in place for the purchase of cryogens, and whether or not the institution will maintain a service contract and, if not, what arrangements exist for covering the cost of repairs must be addressed. While the institution cannot provide matching funds toward the purchase of the instrument, it will need to cover the maintenance costs beyond the expiration of the grant. A letter from the Dean or Provost that commits the institution to cover the maintenance costs or salary of an instrument technician should be included as an Appendix. Also, since there will be multiple users and the possibility of the need for different probes depending on the nature of the request, a plan for balancing the competing demands for the instrument must be provided.

NSF Transforming Undergraduate Education in Science, Technology, Engineering, and Mathematics Program (TUES)

The TUES program operates through the NSF's Division of Undergraduate Education (DUE). This program supports curricular enhancements that provide more effective learning for undergraduates. A prior article provides general advice for obtaining instructional equipment through curriculum development grants (*4*). Admittedly, only a few departments have gotten grants to purchase NMR spectrometers in recent cycles of the TUES program. Nevertheless, it is still a program where it is possible to get support for an NMR spectrometer, and may be the best choice for a department that would have an especially difficult time justifying an instrument on the basis of research activities. Given the significance of NMR spectroscopy in several chemistry sub-disciplines, a TUES proposal to purchase an NMR spectrometer should describe its use in multiple places in the curriculum. Reviewers will evaluate the value to students and faculty of the curricular activities in determining whether a proposal deserves to be funded. Especially for faculty members at PUIs, involvement of undergraduates in research is a valuable educational opportunity. It is advisable to include brief descriptions of research projects that will use the spectrometer and indicate the extent to which the instrument will be used for research in the academic year and summer. Year-round and frequent use of the instrument will help to make a more compelling case for funding. However, it is important to remember that this is a curriculum proposal, so research uses will likely represent a minor component of the overall text compared to use in courses.

There are Type 1, 2 and 3 TUES awards. Type 2 and 3 awards are for multi-institutional or national initiatives aimed at developing and disseminating effective educational practices; these do not provide funds to purchase instruments. A Type 1 award is an appropriate place to request funds for an NMR spectrometer. Type 1 awards have a maximum of $200K for a single institution and $250K if the project involves significant involvement of one or more two-year colleges. In order to qualify for the extra $50K, curriculum development activities planned for the lead institution should also be implemented in comparable courses taught at the two-year college. In addition, faculty who teach these courses at the two-year college should collaborate with faculty at the lead institution in developing the new curriculum. A situation in which faculty members and students from the two-year college will merely be provided access to the instrument to run spectra is not sufficient to justify the extra $50K. An issue for many departments with the TUES program is that $250K is insufficient to cover the entire cost of many high-field NMR spectrometers and NSF does not permit the use of and discussion of matching funds in the proposal.

Project Summary

The Project Summary for a TUES proposal must have two distinct paragraphs discussing the intellectual merits and broader impacts of the work. Intellectual merits focus on the new curricular activities and the beneficial effect they are expected to have on student learning. If the curricular changes involve significant alterations in the way faculty members deliver courses or laboratories (e.g., replacement of traditional teaching methods with active-learning pedagogies), activities aimed at faculty development fall within the intellectual merit of the project. Broader impacts will include information on the number of students and faculty who will be affected by the curricular changes. It is important not to overstate the number of students who will be impacted by the project. NSF is especially interested in the dissemination of new curricular developments and activities that communicate the outcomes of the project to a larger audience are important broader impacts.

Project Description

The Project Description should provide a thorough discussion of the curricular changes that will be implemented and the rationale for those changes. How the changes are expected to improve student learning must be emphasized. NSF identifies five components of curriculum development in the description for the TUES program.

Creating learning materials and strategies
Implementing new instructional strategies
Developing faculty expertise
Assessing and evaluating student achievement
Conducting research on undergraduate STEM education

A proposal to obtain an NMR spectrometer will likely have some aspects of items 1-4, although generally Type 1 proposals only focus on one or two of the areas. If the primary goal is to institute a new set of laboratory experiences that are facilitated by acquisition of a better NMR spectrometer, the emphasis will likely be the creation of new learning materials and strategies. If faculty members are changing the way they teach laboratories to emphasize more student independence in the execution of experiments, a plan to develop faculty expertise will be important. Every Type 1 proposal requires an assessment of student learning outcomes. However, the focus of a proposal to purchase an NMR spectrometer will not be on the development of new tools or techniques for the assessment of student achievement.

It is essential to remember that reviewers will evaluate the quality of the learning experience for students affected through the project. It is doubtful, if not impossible, to get funding for a new NMR spectrometer that replaces an existing but outdated piece of equipment and continues the same set of experiments already being done. Presumably a new instrument will have capabilities that open up new types of experiments. It is essential to relate new experiments and teaching pedagogies described in the proposal to previous published work. While it is less common today, it is surprising how many education proposals I have reviewed over the years that had no references. Some facility with the education literature on chemistry will be essential.

A proposal requesting an NMR spectrometer will almost certainly describe its use in several courses. This will enhance the broader impact of the project by affecting more students. The experiments will likely increase in sophistication as students advance in the curriculum. It is important to provide either a complete set of experiments and activities that will be implemented or specific examples of the types of experiments and learning activities that will be developed. Reviewers must be impressed by these examples.

There is recognition in the education community that there are better ways to teach chemistry than a lecture approach and "cookbook" labs. Numerous reports have called for changes in the way we teach science (*5–13*). Various studies show the effectiveness of teaching techniques that engage the students more actively in the learning exercise (*14–51*). While it is not essential, it can be tempting to claim that more active forms of learning (e.g., inquiry-based experiments) will be developed as part of the project. However, sprinkling a phrase like "new inquiry-based experiments will be developed" throughout the proposal without providing some well-constructed examples will ring hollow with reviewers. Reviewers must be convinced that an interesting set of activities has been selected or that the project participants have the ability to develop effective inquiry-based activities. If the development of inquiry-based experiments is a goal, examples should include inquiry-based techniques as their approach. If the project will involve developing new experiments for several courses, it is preferable to provide at least one specific example for each course. If one or more of the examples are too long to include in the proposal, they should be summarized in shorter form and it is possible to provide a link to a website where the complete materials are available. Reviewers may go to the website if the nature of the experiments to be developed is a key question they have. I have been on panels where we went to a website to examine materials during the discussion of a proposal.

A curriculum proposal will need a timeline for completion of each activity. This will include when the equipment will be installed, when curriculum development will occur, and when changes will be incorporated into courses. Also, it is important to provide the names of individuals who are committing to develop experiments for each course.

Assessment Plan

A TUES proposal will need an assessment plan that includes formative and summative components. Formative assessment occurs early in the project and is used to refine curricular activities to make them more effective. Summative assessment occurs at the end of the project. Established processes that may already occur at your institution (e.g., end-of-semester course evaluations) can be incorporated into assessment. Since chemistry faculty members rarely have sufficient expertise in the area of assessment, so it is advisable to involve a consultant for this aspect of the project. This individual should have a proven track record in the area of assessment. An institutional research officer or faculty member at your institution with assessment expertise may fill this role. Many people involve someone from off campus. The individual leading the assessment cannot be intimately involved in the curriculum development activities. Instead they are in communication with the project participants to develop assessment activities as the project progresses. Most important is that this person has demonstrable experience in assessment, and can show that in their biographical sketch that will be included in the proposal.

Dissemination Plan

It is essential to have a dissemination plan that communicates your curricular development activities beyond your institution. This should involve more than putting the materials up on a website. Plans to give conference talks and participate in appropriate discipline-specific networking opportunities should be provided. Specific and realistic plans are better. Peer-reviewed publications are especially valuable, although it is important that any claims about plans to publish appear realistic. If earlier in the proposal you state that you have prior experience at curriculum development activities as a way of giving the reviewers more confidence in your future plans, the absence of publications on your previous work may diminish the veracity of claims that you will publish your future work.

Budget

The budget for a TUES grant permits the inclusion of funds in all categories that appear in an NSF budget. In addition to the cost of the spectrometer, common items include summer salary for faculty and student assistants for curriculum development activities, travel to visit someone to observe effective practices that you want to include in your curriculum or to attend a conference for dissemination, and funds for an assessment consultant. Since the NMR spectrometer may use up all or most of the available $200K, there may not be much money available for these other items. You may have to develop the new curricular materials without summer salary. It is essential that the reviewers feel confident that assessment will occur, so having some funds to support an assessment consultant may be essential to incorporate into the request, even if it is listed in the budget justification as an expense being provided by your institution.

Concluding Remarks

The grant process is highly competitive and NSF has insufficient levels of funding to support all deserving proposals. Many proposals are rejected. Some are rejected because the ideas contained in them are inferior and they may never warrant funding. A more common situation is that the proposal is lacking in some important details that the reviewers want to see developed better. In the latter case, the proposal should be revised and resubmitted. If a proposal is rejected, one piece of advice is to let one or more colleagues read the reviews and share with you their thoughts about a resubmission and what needs to be addressed. It is also helpful to talk to the program officer, but never handle such a conversation in a confrontational manner. Let news of the rejection and reviewers' comments sit for a period of time before contacting the program officer. Since the program officers handle many proposals, it is best to contact the person via email to set up a phone appointment and to indicate that you want to talk about your rejected proposal. This will allow the program officer to review your file and be better prepared for the call.

Organizations like the Council on Undergraduate Research offer a variety of programs aimed at furthering people's proposal writing skills (*52*). American Chemical Society meetings also have sessions on NSF-catalyzed curriculum development activities that can help in the generation of ideas for a proposal. One thing I do know for sure is that you will not get a proposal funded if you do not submit it. Try your best and, if it is not funded on the first submission, do not give up.

Note from the Editors: This chapter raises a number of important issues specifically concerning the TUES and MRI NSF programs. Since the guidelines governing these and other programs continue to evolve we strongly encourage all proposal writers to check the latest version of the official guidelines for the program to which you are applying. Readers are also encouraged to contact the appropriate program officer early in the proposal development process if you have questions related to the topics raised in this chapter.

References

1. *Grant Proposal Guide*; http://www.nsf.gov/publications/pub_summ.jsp?ods_key=gpg (accessed August 2012).
2. *Major Research Instrumentation Program: Instrument Acquisition or Development, NSF 11-503*; http://www.nsf.gov/publications/pub_summ.jsp?WT.z_pims_id=5260&ods_key=nsf11503 (accessed August 2012).
3. *Transforming Undergraduate Education in Science, Technology, Engineering and Mathematics (TUES), NSF 10-544*; http://www.nsf.gov/publications/pub_summ.jsp?WT.z_pims_id=5741&ods_key=nsf10544 (accessed August 2012).
4. Wenzel, T. J. *J. Chem. Educ.* **2010**, *87*, 1128–1130.
5. *Shaping the Future: New Expectations for Undergraduate Education in Science, Mathematics, Engineering, and Technology*; National Science Foundation: Arlington, VA, 1996.
6. *Science Teaching Reconsidered: A Handbook*; National Academies Press: Washington, DC, 1997.
7. *Reinventing Undergrauate Education: A Blueprint for America's Research Universities*; Carnegie Foundation for the Advancement of Teaching: Princeton, NJ, 1998.
8. *Project 2061*; American Association for the Advance of Science, 1985; http://www.project2061.org/publications/sfaa/default.htm (accessed October 2012).
9. *BIO 2010, Transforming Undergraduate Education for Future Research Biologists*; National Academies Press: Washington, DC, 2003.
10. Kuwana, T. *Curricular Developments in the Analytical Sciences*; National Science Foundation: Washington, DC 1998; http://www.asdlib.org/files/curricularDevelopment_report.pdf (accessed October 2012).
11. Pienta, N. J.; Cooper, M. M.; Greenbowe, T. J. *Chemists' Guide to Effective Teaching*; Prentice Hall: Upper Saddle River, NJ, 2005.
12. *Rising Above the Gathering Storm: Energizing and Employing America for a Brighter Economic Future*; National Academy Press: Washington, DC, 2005.
13. *How People Learn: Brain, Mind, Experience and School*; Bransford, J., Brown, A. L.. Cocking, R. R.. Eds.; National Academies Press: Washington, DC, 2000.

14. Gilbert, J. K.; de Jong, O.; Justi, R.; Treagust, D. F.; van Driel, J. H. *Chemical Education: Towards Research-Based Practice*; Gilbert, J. K., de Jong, O., Justi, R., Treagust, D. F., van Driel, J. H., Eds.; Kluwer Academic Publishers: Boston, 2002; pp 391−408.
15. *How People Learn: Brain, Mind, Experience and School*; Bransford, J., Brown, A. L., Cocking, R. R., Eds.; National Academies Press: Washington, DC, 2000.
16. Hohlach, J. M.; Grove, N.; Bretz, S. L. *J. Chem. Educ.* **2007**, *84*, 1530–1534.
17. Getzels, J. W. Creative thinking, problem solving and instruction. In *Theories of Learning and Instruction, The 63rd Yearbook of the National Society for the Study of Education*; Hilgard, E. R., Ed.; University of Chicago Press: Chicago, 1963.
18. Duch B. J.; Groh, S. E.; Allen, D. E. *The Power of Problem-Based Learning*; Stylus Publishing: Sterline, VA, 2001.
19. Johnson, D. W.; Johnson, R. T.; Smith, K. A. *Cooperative Learning: Increasing College Faculty Instructional Productivity*; ASHE-ERIC Higher Education Report No 4; The George Washington University, Graduate School of Education and Human Development: Washington, DC, 1991.
20. McKeachie, W. J. *Teaching Tips: Strategies, Research, and Theory for College and University Teachers*, 11th ed.; Houghton Mifflin: Boston, MA, 2002.
21. Mayer, R. E. *Am. Psychol.* **2004**, *59*, 14–19.
22. Astin, H. S.; Astin. A. W.; Bisconti, A.; Frankel, H. *Higher Education and the Disadvantaged Student*; Human Science Press: Washington, DC, 1972.
23. Wales, C.; Stager, R. *The Guided-Design Approach*; Educational Technology Publications: Englewood Cliffs, NJ, 1978.
24. Astin, A. W. *Achieving Education Excellence*; Jossey-Bass: San Francisco, 1985.
25. Murray, H. G. *Using Research to Improve Teaching*; Donald, J. G., Sullivan, A. M., Eds.; Jossey-Bass: San Francisco, 1985.
26. Treisman, P. U. *A Study of the Mathematics Performance of Black Students at the University of California, Berkeley*, Ph.D. Dissertation, University of California, Berkeley, 1985.
27. Tinto, V. *Leaving College: Rethinking the Causes and Cures of Student Attrition*; University of Chicago Press: Chicago, 1987.
28. Johnson, D. W.; Johnson, R. T. *Cooperation and Competition: Theory and Research*; Interaction Book Company: Edina, MN, 1989.
29. Light, R. J. *The Harvard Assessment Seminars*; Harvard University Press: Cambridge, MA, 1990.
30. Slavin, R. E. Group rewards make groupwork work. *Educ. Leadership* **1991**, *48*, 71–82.
31. Blosser, P. E. *The Science Outlook: Using Cooperative Learning in Science Education*; ERIC-CSMEE: Columbus, OH, 1993; pp 1−9,
32. Allen, D. E.; Duch, B. J.; Groh, S. E. *New Directions in Teaching and Learning*, No. 68; Jossey-Bass Publishers: San Francisco, 1996; pp 43−52.
33. Springer L.; Stanne, M. E.; Donovan, S. *Effects of small-group learning on undergraduates in science, mathematics, engineering, and technology: A*

meta-analysis; National Institute for Science Education: Madison, WI, 1997; http://www.wcer.wisc.edu/nise/CL1/CL/resource/R2.htm (accessed October 2012).

34. Hake, R. R. *Am. J. Phys.* **1998**, *66*, 64–74.
35. Towns, M. H. *J. Chem. Educ.* **1998**, *75*, 67–69.
36. Wenzel, T. J. *Anal. Chem.* **1998**, *70*, 790A–795A.
37. Spencer, J. N. *J. Chem. Educ.* **1999**, *76*, 566–569.
38. Farrell, J. J.; Moog, R. S.; Spencer, J. N. *J. Chem. Educ.* **1999**, *76*, 570–574.
39. Johnson, D. W.; Johnson, R. T.; Stanne, M. E. *Cooperative Learning Methods: A meta-analysis*; Cooperative Learning Center, University of Minnesota: Minneapolis, 2000; http://www.tablelearning.com/uploads/File/EXHIBIT-B.pdf (accessed October 2012).
40. Bowen, C. W. *J. Chem. Educ.* **2000**, *77*, 116–119.
41. Towns, M. H.; Kreke, K.; Fields, A. *J. Chem. Educ.* **2000**, *77*, 111–115.
42. *Peer-led Team Learning: A Guidebook*; Gosser, D. K., Cracolice, M. S., Kampmeier, J. A., Roth, V., Strozak, V. S., Varma-Nelson, P., Eds.; Prentice-Hall: Upper Saddle River, NJ, 2001.
43. Cabrera, A. F.; Crissman, J. L.; Bernal, E. M.; Nora, A.; Terenzini, P. T.; Pascarella, E. T. *J. Coll. Stud. Development* **2002**, *43*, 20–34.
44. Smith, K. A.; Sheppard, S. D.; Johnson, D. W.; Johnson, R. T. *J. Eng. Educ.* **2005**, *94*, 87–101.
45. Springer, L.; Stanne, M. E.; Donovan, S. S. *Rev. Educ. Res.* **1999**, *69*, 21–51.
46. Hofstein, A.; Lunetta, V. N. *Rev. Educ. Res.* **1982**, *52*, 201–217.
47. Reif, F; St. John, M. *Am. J. Phys.* **1979**, *47*, 950–957.
48. Raghubir, K. P. *J. Res. Sci. Teach.* **1979**, *16*, 13–18.
49. Wheatley, J. H. *J. Res. Sci. Teach.* **1975**, *12*, 101–109.
50. Wright, J. C. *J. Chem. Educ.* **1996**, *73*, 827–832.
51. Wright, J. C.; Millar, S. B.; Kosciuk, S. A.; Penberthy, D. L.; Williams, P. H.; Wampold, B. E. *J. Chem. Educ.* **1998**, *75*, 986–992.
52. Council on Undergraduate Research Home Page; http://www.cur.org/ (accessed August 2012).

Editors' Biographies

David Soulsby

David Soulsby was born in 1974 in Newcastle, England. After earning his B.Sc. degree in chemistry from the University of Lancaster, England, he attended the University of Colorado Boulder and obtained a Ph.D. in organic chemistry. His research involved the development of ferrocenyl oxazoline catalysts for use in the Heck reaction with Dr. Tarek Sammakia. He began his academic career in 2001 at the University of Redlands where he is currently department chair. Working alongside undergraduate students, his research focuses on the reactions of intermediates generated from the ozonolysis of vinyl ethers and enamines. NMR spectroscopy plays a critical role in identifying the products of these complex reactions.

Laura J. Anna

Laura J. Anna was born in 1969 in Johnstown, Pennsylvania. After earning her B.S. in Chemistry (magna cum laude) from Indiana University of Pennsylvania, she obtained her Ph.D. in organic chemistry from the University of Michigan where she investigated applications of chiral vinyl sulfoxides in the asymmetric synthesis of natural products with Dr. Joseph P. Marino. She began her academic career at Millersville University where she was a professor of chemistry for 14 years before moving to Montgomery College in 2011 where she is now professor of chemistry and department chair. Her current research efforts continue to focus on the integration of NMR spectroscopy in the organic chemistry curriculum and the initiation of an undergraduate research program at Montgomery College.

Anton S. Wallner

Anton S. Wallner (Tony) was born in 1963 in Milwaukee, Wisconsin. After earning his B.S. degree in chemistry from the University of Wisconsin-Milwaukee, he attended the University of Michigan (M.S. analytical chemistry) and obtained a Ph.D. in physical chemistry from Case Western Reserve University where he studied magnetic resonance imaging of ceramics and polymers with Dr. Bill Ritchey. He has conducted research at the Naval Air Warfare Center on MRI of rocket propellants and explosives as well as a sabbatical in 2009 at Monash University in Melbourne, Australia investigating the degradation of creatine followed by NMR. He began his academic career in 1992 at Missouri Western State University. He moved to Barry University in 2000 as chair and professor of chemistry in the Department of Physical Sciences. In 2008, he became Associate Dean of Undergraduate Programs at Barry University. His current research, with active participation from undergraduates, focuses on the synthesis, characterization and degradation followed by NMR of novel creatine salts.

Indexes

Author Index

Subject Index

G

H

I

K

L

M

N

O

P

S

T

U

V

W

Printed in the USA/Agawam, MA
February 11, 2014

585139.041

Withdrawn
CAIRNS LIBRARY
UNIVERSITY OF OXFORD
BODLEIAN HEALTH CARE LIBRARIES
JOHN RADCLIFFE HOSPITAL
OXFORD OX3 9DU